TRANSACTIONS
OF THE
INTERNATIONAL
ASTRONOMICAL UNION

VOLUME XVIIA - PART 2
REPORTS

INTERNATIONAL COUNCIL OF SCIENTIFIC UNIONS

INTERNATIONAL ASTRONOMICAL UNION

UNION ASTRONOMIQUE INTERNATIONALE

TRANSACTIONS

OF THE

INTERNATIONAL ASTRONOMICAL UNION

VOLUME XVIIA - PART 2

REPORTS
ON
ASTRONOMY

Edited by

EDITH A. MÜLLER

General Secretary of the Union

D. REIDEL PUBLISHING COMPANY

DORDRECHT : HOLLAND / BOSTON : U.S.A. / LONDON : ENGLAND

ISBN-13:978-94-009-9472-0 e-ISBN-13:978-94-009-9470-6
DOI: 10.1007/978-94-009-9470-6

Published on behalf of
the International Astronomical Union
by
D. Reidel Publishing Company, P.O. Box 17, Dordrecht, Holland

Sold and distributed in the U.S.A., Canada, and Mexico
by D. Reidel Publishing Company, Inc.
Lincoln Building, 160 Old Derby Street, Hingham,
Mass. 02043, U.S.A.

PREFACE

Three years ago the Reports on Astronomy were for the first time published in three separate parts whereby related astronomical topics were collected in one volume and each volume was kept to a size of no more than about 250 pages. The aim was to facilitate a wider distribution of individual volumes among astronomers and other scientists interested in specific subjects of astronomy.

The same system was adopted for the present volume A of the IAU Transactions which covers the activities in astronomical research from January 1976 to December 1978. The Commission Presidents were invited to restrict their reports, to present a selection of the most important developments in the field of astronomy covered by their Commission, and to avoid overlaps with other Commission reports. As a result of the restriction in the number of pages some Commissions present a large, concentrated list of references with little discussion, whereas others review some of the most exciting recent research developments in their subject.

The distribution of the Commission Reports into the three parts of volume XVII A is as follows:

VOL. XVII A - PART 1

Commission	4.	Ephemerides
"	5.	Documentation
"	6.	Astronomical Telegrams
"	7.	Celestial Mechanics
"	9.	Instruments and Techniques
"	14.	Fundamental Spectroscopic Data
"	15.	Physical Study of Comets, Minor Planets and Meteorites
"	16.	Physical Study of Planets and Satellites
	-	Working Group on Planetary System Nomenclature
"	17.	The Moon
"	19.	Rotation of the Earth
"	20.	Positions and Motions of Minor Planets, Comets and Satellites
"	21.	Light of the Night-Sky
"	22.	Meteors and Interplanetary Dust
"	31.	Time
"	38.	Exchange of Astronomers
"	41.	History of Astronomy
"	46.	Teaching of Astronomy
"	49.	The Interplanetary Plasma and the Heliosphere
"	50.	Identification and Protection of Existing and Potential Observatory Sites.

VOL. XVII A - PART 2

See the Contents for this volume, p. vii

VOL. XVII A - PART 3

In order to minimize the price of each volume the method of offset printing was used which required the preparation of the typescripts of the reports ready for photographic reproduction.

On behalf of the IAU I thank all Presidents of Commissions and all Colleagues who contributed to the reports for the wealth of interesting and useful information they have collected and presented in this volume.

Thanks are also due to the Secretaries of the Commission Presidents for their careful preparation of the camera-ready typescripts. My personal thanks go to my secretaries Mme R. Bertschi and Mme R. Läubli for their most valuable editorial help.

EDITH A. MÜLLER
General Secretary

CONTENTS

REPORTS OF COMMISSIONS

8. POSITIONAL ASTRONOMY (ASTRONOMIE DE POSITION)

PRESIDENT: R.H. Tucker

VICE PRESIDENT: E Høg

ORGANIZING COMMITTEE: C.A. Anguita, S. Débarbat, W. Fricke, B.L. Klock, E. Marcus, A.A. Nemiro, J.L. Schombert, K.N. Tavastsherna, G. Teleki, G. van Herk, H. Yasuda.

1. Introduction

We commemorate four well-known positional astronomers deceased since the last Report:

B.V. Novopashenny (Odessa), A.K. Korol (Goloseevo), Y.E. Gordon (Nikolaiev) and L.S.T. Symms (Herstmonceux).

The following have notified their retirement from the Commission:

W. Dieckvoss, R.H. Stoy, J.L. Schombert, P. Rybka, A. Bohrmann, and A.N. Adams.

The Commission members have displayed their interest in discussing the role of the Commission, debating whether it should become an actively interventionist body or a medium for exchange of information.

Cooperation with other Commissions and bodies has continued in connection with several activities, notably the IAU/IGGU Symposium on Refraction, the Vienna Colloquium on Astrometry, (which recommended a Working Group on Stellar Position Programmes) and the ESA Astrometry Satellite. Accounts of these will be found at the end of this Report.

2. Reports from Observatories and Commission Members

As in the last Report, these are presented under national headings.

A. AUSTRALIA

At Sydney the zones from $-38^{\circ}30'$ to -76° have been photographed. Measurements of the $-53^{\circ}30'$ zone have been reduced using Perth 70 catalogue places. The final solution awaits the SRS catalogues (W.H. Robertson).

B. CHILE

The Santiago Astrolabe (ESO/University of Chile) is completing a second catalogue containing 358 FK4 stars and 164 FK4 Suppl stars. The planet Uranus has been observed on 185 occasions, including 67 double transits. (F. Noël).

C. DENMARK

At Brorfelde the automation of the Meridian Circle continues. Computer-controlled setting to an accuracy of 2 arc seconds takes an average setting time of 10 seconds. The circle scanners have been fitted with stepping motors, and the electronics made more reliable. First stellar observations with the slit micrometer are expected in the autumn of 1978. Planning for moving the MC to La Palma is now well advanced, with a target date of 1980. 50,000 differential observations are being prepared for publication ($\varepsilon\alpha \cos\delta$ = 0''190, $\varepsilon\delta$ = 0''218). Observations of 1617 NPZT stars have been reported to H. Yasuda. Høyer has determined proper motions for 8000 GC stars in Perth 70, and confirmed that Perth 70 has negligible magnitude equation. Høg has constructed the multislit micrometer for the MC (see above) and made a study of aerodynamic and thermal conditions in a MC pavilion. L. Lindegren (Lund, Sweden) has reduced 530 planetary observations made by the Hamburg-Perth expedition. The phase corrections applied included geometrical and limb-darkening components. (Fogh Olsen, Høg).

References

Catalogue of 257 FK4 stars and 712 others: 1977 (041.025).
Theory, phase corrections of planets: A & A 57, 55, 1977.
Kinematic mounting: A & A 41, 107-109, 1975.
Positions, 54 southern circumpolar FK4 stars: Mitt. A.G. Nr. 38, 178-181, 1976.
Perth 70 Catalogue: Abh. Hamburg Sternw. 9, 40 + 334 pp, 1976.
Survey of Future: Highlights of Astronomy, 4, Part I, 361-367, 1977.
Photoelectric observations of planets: A & A Suppl. 30, 125-129, 1977.
Review: Colloquium on European Satellite Astrometry, Padova, 1978.

D. FEDERAL REPUBLIC OF GERMANY

At Hanover, Pilowski has studied the possibility of setting up an absolute fundamental catalogue based on observations of AGK3 stars with a transportable zenith camera. He assumes that geographical coordinates and polar motion are accurately known, and shows that absolute proper motions are only required if intervals of a century or more are in question. At Munich Observatory, observations of absolute declinations of FK4 stars and of planets were continued. Work on catalogues containing measurements dating from 1957 was completed. At Hamburg, the radio-source optical-position programme continues. Results for 17 sources have been published, and a catalogue of secondary reference stars for 34 fields is being prepared. The general rigorous block-adjustment computer program is being used in a study of the original AGK2 and AGK3 plate measurements. Error propagation in large block areas is under examination. Further investigations for the proposed new four-fold coverage of the northern hemisphere continue, and a decision on the realization of this project is expected in 1979. (Pilowski, Schmeidler, de Vegt).

References

Transportable zenith camera: Astr. Station Hannover Tech. Univ. Monograph, 1973
Supp I, No. 9, 1974 Supp II, No. 10, 1976.
Supp III in preparation.
Pilot investigation, 17 sources: A & A 67, 5, 1978.
Rigorous block-adjustment: MNRAS, 167, 169, 1974.
Completion: A & A Suppl. 32, 141, 1978.
Fourfold coverage (AGK4): Trans. IAU, XVIB, 181.

E. FRANCE

At Paris, the Astrolabe group continues to provide observational results for the BIH and the SIMP, 400 groups having been observed since September 1976. There is collaboration with CERGA and Nice Observatory for astrolabe observations of the Sun. A campaign of observing the radio-source β Persei has continued for some years. Variation of the instrumental zenith distance during observations can now be taken into account. An analysis of the first 18 years of Paris latitude observations has been published. Astrolabe observations of planets have continued at Paris, San Fernando, and CERGA (Mars, Jupiter and Saturn) and at Santiago de Chile (Uranus). Earlier campaigns, revised where necessary, are described in CERGA publications. A study of the Mars observations has revealed discrepancies in the different ephemerides. Results of five Saturn campaigns at Paris are in press. A general review has been published. (Chollet, Débarbat).

References

Paris astrolabe observations: A & A Suppl. 29, 241, 1977.
Sun observations with astrolabe: A & A 56, 207, 1977.
Control of astrolabe zenith distance variation: IAU Symp. 78, Kiev, 1977 and
IAU Symp. 82, San Fernando, 1978.
18 years, Paris latitude: Wiss. Z. Tech. Univ. Dresden, 25, H1, 1976.
Mars observations: A & A Suppl. 28, 305, 1977 and 32, 335, 1978.
Bol. Obs. San Fernando, Ser C, No. 79, 1977.
A & A Suppl. 32, 331, 1978.
A & A Suppl. 31, 171, 1978.
Saturn observations: A & A Suppl. 25, 365, 1976.
Bol. Obs. San Fernando, Ser C, No. 78, 1976, and No. 79, 1977.
A & A Suppl. 32, 323, 1978.
A & A Suppl. 31, 167, 1978.
Jupiter observations: Bol. Obs. San Fernando, Ser C, No. 79, 1977.
A & A Suppl. 31, 169, 1978.
Uranus observations: A & A Suppl. 30, 189, 1977.
Older observations: A & A Suppl. 32, 325 and 327, 1978.
Mars analysis: A & A 64, 281.
General synthesis: Vistas in Astronomy, 21, 93, 1977.

F. GREAT BRITAIN AND CAPE

The results of the Cape TC observations of the SRS zones have been sent to USNO Washington. Observed mean positions for stars on the FK4 and BS lists have been sent to AR-I Heidelberg and to Kiev. The planet observations await reduction.

At Herstmonceux, regular observations of Sun, Moon, major and minor planets, and fundamental stars are continuing. Nine more minor planets have been added to the observing list, which previously contained only the four brightest minor planets.

Observations of zodiacal stars and the NPZT stars are continuing, but both programmes are practically exhausted. Most of the results for the NPZT stars have been sent to Tokyo.

The investigation of the Herstmonceux instrumental system has now been completed. Corrections have been adopted for diurnal variation of various parameters, clamp effects and observer effects. New places and proper motions have been adopted for the azimuth and clock stars, independent of the periodic terms in FK4, but with FK4 equinox, to provide the basis of the local system of RA. The local declination system is based on revised values of flexure and latitude, and is independent of FK4. The reduction of the first sixteen years of Herstmonceux TC observations on the Herstmonceux system is now ready to proceed.

Progress with the modernisation of the Cooke TC continues to be disappointing. The promised Electronic Circle Reading system has not been delivered, and even further delay appears likely, because of staff changes at the laboratory where the equipment is being prepared. A stepping-motor drive has been installed for the RA micrometer, and a Helipot recording system on the declination micrometer. A new control panel has been installed, and new computer programs for data acquisition have been developed.

Work continues on the preparation of catalogues for publication, and on the computer handling of catalogue data used in determination of proper motions.

Detailed planning for the collaboration with the Copenhagen University Observatory in Meridian Astronomy on La Palma is now at an advanced stage, but further progress awaits the signature of the formal international agreement on the NHO. (Tucker).

G. JAPAN

At Tokyo, regular observations of the Moon and planets referred to the FK4 system have continued with the Gautier MC. There are seven minor planets on the list. Results for solar system bodies from 1974 to 1977 were published. The error in relative positions determined by the Tokyo MC was estimated by comparing with the photo-electric observations of the occultation of SAO 158687 by Uranus.

Systematic differences (Tokyo-FK4) were derived with an accuracy of ±0.″01. The Tokyo catalogue magnitude equation was demonstrated. The programme of OB stars was completed in the middle of 1978.

A preliminary NPZT catalogue is being compiled from results received from Abbadia, Bordeaux, Copenhagen, Herstmonceux, San Fernando, Belgrade and Tokyo. A catalogue of PZT stars for epoch 1958 is being compiled from Herstmonceux, Hamburg, Tokyo and Washington results.

At Mizusawa, ten years of astrolabe observations were used to correct FK4 and FK4 Supp star positions, including the systematic errors depending on RA. A general least-squares estimation process including star positions and motions and short-period latitude and longitude variations, was developed, and applied to past ILS data. Corrections to the declinations and proper motions are obtained with mean errors of 0.″01 and 0.″005/yr respectively. (Yasuda).

<u>References</u>

Solar System Obs. 1974-1977: Tokyo Astr. Bull. 252.
Error of Relative Positions, Tokyo MC: Publ. Astr. Soc. Japan, <u>29</u>, 631-637, 1977.
Magnitude equation in Tokyo Catalogue: Publ. Astr. Soc. Japan, <u>30</u>, 173-190, 1978.
Mizusawa - FK4 corrections: Publ. Int. Lat. Obs. Mizusawa, <u>10</u>, 2, 1976.
Star-place corrections theory: Publ. Int. Lat. Obs. Mizusawa, <u>11</u>, 2, 1977.

H. ROMANIA

At Bucharest, the results of the SRS observations were sent to USNO, Washington, and the BS results to Kiev. Publication of the corresponding catalogues has been delayed by organizational difficulties at the publishing establishment. The catalogues include notes on the FK4 reference star positions and comparisons with AGK3R and Perth 70. The NPZT observations are partly reduced. The observing list has been extended by including the Cagliari PZT stars. The RA and declination micrometer screws have been replaced by new screws, and the study of the screw errors is continuing. Considerable delays in all work continue as a consequence of the earthquake in 1977. (Marcus, Rusu).

I. SPAIN

At San Fernando, analysis of the astrolabe catalogues CASF1 and CASF2 has given the systematic departures of these catalogues from FK4 and FK4 Supp. A third catalogue has been published, and a fourth catalogue observation campaign is in progress. Planetary observations have continued, as reported above (section E).

The Meridian Circle has completed 82% of the NPZT observations. Results for two zones have been sent to Tokyo. Planet observations continue. New differential drives have been installed in the astrolabe micrometer and in the Meridian Circle RA micrometer.

The "Carte du Ciel" Astrograph is observing the 20 selected minor planets of the ITA Program. (Quijano, Sánchez).

References

Astrolabe catalogues CASF1 and CASF2: A & A Suppl. 25, 9, 1976.
Analysis of differences: A & A 60, 61, 1977.
Third catalogue: A & A 29, 245, 1977.

J. U.S.A.

At Washington the 7-inch TC is undergoing a rather thorough renovation. An image dissector photon-counting "micrometer" has been giving promising results on bench tests. The Automatic TC is having various parts re-worked at the factory. The 6-inch TC continues its current zodiacal observing program. A new electronics system for the Inductosyn circle-reading device is being evaluated. At Bloomington, Atkinson continues consideration of yielding and non-yielding Earth models in the study of nutation. A re-analysis of the Cookson Zenith Telescope results shows that the plates would be capable of giving better results if measured on a modern machine, but this is impossible because the plates have been destroyed. (Hughes, Klock, Atkinson).

References

Short program for aberration: MNRAS, 172, 469, 1975.
Use of C & D for reduction of radial velocities, MNRAS, 172, 465, 1975.

K. U.S.S.R.

At Pulkovo, the Zverev Photographic Vertical Circle is in operation. The original photographic circle reading system gives an accuracy of $0\overset{''}{.}05$ for a single reading. Differential determinations of declinations of Pulkovo Zenith telescope stars have begun. Reconstruction of the horizontal meridian circle and the Toepfer meridian circle continues. A highly sensitive eyepiece micrometer is being constructed. 4000 stars of the latitude and zenith-tube programme have been observed by the Ertel Vertical Circle. Absolute RA catalogues have been published for FK4 and FKSZ stars in northern and southern zones, using the Large Transit Instrument at Pulkovo and at Santiago de Chile. An absolute RA catalogue has been prepared for the Backlund-Hough stars observed at the Melbourne Observatory 1928-1940. An astrolabe catalogue of RA and Decl has been published. The Santiago declination observations are being reduced.

At Golosseyevo, the eyepiece micrometer of the Wanschaff Vertical Circle has been automated. A compilation catalogue using 15 RA and 14 Declination catalogues, 1939-1973, has been completed jointly with Pulkovo. A nine-year series of observations of Sun and planets shows considerable deviations of Mars from the ephemeris position. A new compilation of Washington, Greenwich and Nikolayev observatories has the aim of improving the FK4 equinox and equator corrections.

At Kazan (Engelhardt) the new refraction tables are completed. Latitude star observations are being reduced. Declination observations of PZT programme stars with the Meridian Circle have been completed.

At Kharkov, progress continues in automation of meridian observations and reductions. A digital signal storage device has been constructed. Minor planets (ITA programme) are regularly observed. Three differential RA catalogues have been published, together with corrections to RA of 372 FK4 stars.

At Kiev, two positional catalogues have been published, and a compilation catalogue of bright stars is in preparation.

At Leningrad, observations of 10 selected minor planets (1950-1974) are being reduced. Regular observations of 20 selected minor planets are being organized for 21 observatories, of which 10 are in USSR.

At Moscow, a device for studying personal pointing errors on disc-like objects has been built. A text book on General Astrometry by V.V Podobed and V.V. Nesterov has been published. Observations with the Sanders-Raimond method have been completed. An absolute RA catalogue, a Differential Catalogue of ZT and PZT stars, and a Compilation Catalogue of radio source positions have been published. Corrections to the FK4 equator and equinox have been found from Venus observations (1969-1972) amounting to $\Delta A = -0\overset{s}{.}003 \pm 0\overset{s}{.}001$ and $\Delta\delta o = +0\overset{''}{.}16 \pm 0\overset{''}{.}03$. Corrections $\Delta k = -0\overset{''}{.}08 \pm 0\overset{''}{.}16$ and $\Delta n = +0\overset{''}{.}37 \pm 0\overset{''}{.}14$ have been found from star proper motions relative to galaxies in 30 areas.

At Nikolayev, it is now possible to graduate metal circles with diameters up to 1200 mm. The Spitzbergen observations of absolute RA of stars are completed, and being reduced. Objects of the solar system continue to be observed with meridian and photographic instruments. One Declination catalogue and 2 RA catalogues have been published.

At Odessa, the Meridian Circle has been fitted with a recording measuring device. RA observations of major planets have been made. One RA catalogue and 2 positional catalogues have been published.

At Tashkent, regular solar observations and RA observations of inferior planets are made with the Meridian Circle. 2 Absolute RA catalogues, corrections to 36 FK4 stars, and RA of Sun, Venus and Mercury (1974-1975) have been published. Observations and reductions of RA observations of FK4 stars in various declinations are nearing completion. (Tavastsherna).

References

Catalogue FK4, FKSZ, RA: Trudy Gl. Astr. Obs. ANSSR. 82, 1977.
Automation, eyepiece of Wanschaff VC: A.C. No. 940, 1977.
New reduction, equinox to equator: A & A 31, 1977.
Differential RA Catalogue, FK4,: Vestnik Kharkov State Univ., 137, 1976.
FK4 (-20° to $+35^{\circ}$) Vestnik Kharkov State Univ. 160, 1974.
FKSZ (-20° to $+90^{\circ}$): Vestnik Kiev Univ. Astron. No. 20, 1978.
"General Astrometry" (text book): Podobed and Nesterov.
Absolute RA Catalogues, Tashkent: Circ.Astr. Inst. Uzbek. Acad. Sci. 65, 1976, and 67 (414), 1976.
Corrections to 76 FK4 stars: Circ. Astr. Inst. Uzbek Acad. Sci. 79, (426), 1977.
Correction of Fundl. RA.: Circ. Astr. Inst. Uzbek Acad. Sci. 44, 1973.

L. YUGOSLAVIA

At Belgrade, the Meridian Circle continues with observations of the NPZT stars. A preliminary catalogue of these observations shows mean errors of $\varepsilon\alpha = \pm 0\overset{s}{.}017 \sec\delta$ and $\varepsilon\delta = \pm 0\overset{''}{.}13$. Daytime observations of Sun, Mercury, Venus and stars are also made. Declination catalogues have been compiled for ILS stars and Latitude stars (GCLS, IKSZ). Systematic errors $\Delta\delta\alpha$ in the latitude observations are being studied.

The Transit Instrument and Vertical Circle are both engaged in observing BS stars of the north polar zone. The TI collimation and azimuth variations are controlled by vacuum meridian marks. The positional stability of the plane glass plate closing the vacuum tube has been demonstrated by a novel collimation method. A new method of calibrating and observing a bubble-level, using a collimator at 6 m distance, has been developed. The Vertical Circle has restarted observations after a successful reconstruction. A bibliography of star catalogues and a review paper on Fundamental Astrometry has been published. (Sadzakov, Teleki).

References

Decl and pm ILS stars: 1929-1972, Publ. Astr. Obs. Beograd, No. 21, 1975.
General Catalogue Latitude stars: Publ. Astr. Obs. Beograd, No. 24, 1978.
Reconstruction of Askania VC: Report of Commission 8 to XVI IAU.
Refraction table progress: Bull. Obs. Astr. Belgrade, 128, 7-10, 1977.
ILS and Refraction: Mitt. Lohrmann-Obs., Dresden, 33, 913-917, 1976.
Askania VC after reconstruction: Publ. Dept. Astr. Belgrade, 6, 123-124, 1976.
Astron Refraction: Proc. Int. Symp. Stockholm, Aug 1974, 5, No. 1, 1977.
Earth Rotation and Refraction: Bull. Obs. Astron. Belgrade, 128, 19-22, 1977.
Present and Future of Fund. Astrometry: Publ. Dept. Astr. Belgrade, 7, 31-67, 1978.
Bibliography of Catalogues: Publ. Dept. Astr. Belgrade, 7, 69-212, 1978.
Atmospheric Influences: Bull. Obs. Astron. Belgrade, 129, 1978.

3. Other Reports

A. WORKING GROUP ON ASTRONOMICAL REFRACTION (Chairman: G. Teleki)

The Group concentrated on two main tasks, the elaboration of the new international refraction tables, and the organization of the IAU Symposium No. 89 (co-sponsored by IAG of IGGU). Some progress has been made in the elaboration of the tables. A report of the Symposium follows.

B. IAU SYMPOSIUM NO. 89

"Refractional Influences in Astrometry and Geodesy", Uppsala, Sweden, 1-5 August 1978.

The Symposium was co-sponsored by the IAG of IUGG, with the aim of summarising results achieved in different fields, and of exchanging views and experiences of refraction.

The session topics were: astronomical refraction, general; astronomical refraction in astrometry; astronomical refraction, regional and local experiences; refraction effects in deep-space-network, VLBI and satellite geodesy, including the Moon; refraction effects in 3D-geodesy through meteorological and adjustment approaches; statistical and direct methods of determining refraction effects in 3D-geodesy; 3D geodetic network and stellar triangulation; instruments constructed for taking care of refraction effects; levelling and photogrammetric refraction; the use of dispersion methods for determining refraction effects in astronomy and geodesy; wave propagation; local dynamics from refraction-corrected measurements.

The Symposium approved the following Resolutions, incorporating 9 recommendations.

The Symposium on Refractional Influences in Astrometry and Geodesy considering the fact that astronomical and terrestrial refraction are still severe limitations on the attainable accuracy of measurements in Astrometry and Geodesy recommends:

1. that the existing astronomical refraction tables or computation methods should be improved so as to bring them nearer to reality,

2. that for higher accuracy demands the real state of the atmosphere should be taken into account. Especially, means should be found to determine the influence of the deviations of the atmospheric layers from the hypothetical shape,

3. that all possible preventive measures be taken by astrometric observers to minimize the effects of refraction. Careful site selection using the appropriate scientific criteria is of utmost importance. Attention must also be given to, e.g. the design of buildings and the organization of the observing programme,

4. that established observing sites be investigated by all means possible in order to determine the local refraction effects not corrected by the application of the general theories in use,

5. that since investigations to determine the influence of refraction on directions and distances using two or more wavelengths showed encouraging results, all efforts should be made to test these methods in the field and to bring them to a state where they can be applied in astrometric and geodetic practice,

6. that since after extensive researches into refraction effects in geometric and trigonometric levelling there are still problems with refraction in relation to the current demands for higher accuracy, more investigations of the refraction be encouraged using either classical means or new methods and techniques,

7. that since the meteorological approach to refraction problems has led to encouraging results, further investigations should be undertaken along these lines including the testing, development, and application of improved atmospheric models, and improved methods for probing and remote sensing of atmospheric parameters,

8. since both astronomers and geodesists are concerned with refraction problems, that the existing cooperation of the IAU and the IAG in this field be intensified, and that a special joint commission of the IAU and the IAG should be established with the participation of interested meteorologists and physicists and that this recommendation be transmitted to the Executive Committees of the IAU and the IAG,

9. that in view of the possibilities of using Space Lab facilities to investigate refractional effects in the atmosphere, that steps be taken by the IAU and the IAG to request the European Space Agency to consider such research.

The Symposium expresses its most cordial thanks to the IAU and the IAG for the sponsorship of the meeting, to the Geodetic Institute at Hällby, the Organizing Committees, and Uppsala University for the kind hospitality and for the most excellent arrangements made for a successful joint meeting.

C. IAU COLLOQUIUM NO. 48
"Modern Astrometry", Vienna, Austria, 12-14 September 1978.

The Scientific Organizing Committee included C.A. Murray (Chairman), W.F. van Altena, H.K. Eichhorn, B. Elsmore, L.W. Fredrick, W. Gliese, R.S. Harrington, E Høg, D.D. Polozhentsev, G. Teleki, R.H. Tucker, Chr de Vegt. The Chairman of the Local Organizing Committee was Professor J. Meurers. The Proceedings are being published by the University of Vienna with F.V. Prochazka and R.H. Tucker as Editors.

The Colloquium was attended by 132 participants from 16 countries. Topics covered included Astrophysical Astrometry, Astrometry of Radio Sources, New Developments in Ground-based Astrometry, Positional Astrometry and Impact of New Techniques in Astrometry. The Concluding Summary was given by Professor A. Blaauw, President of the Union.

The following Resolutions were passed:

1. Recognizing the need for an improvement in the Stellar Reference Frame over the whole sky, we urge the completion of current and proposed projects, including those in the southern hemisphere, for the determination of positions of stars, and we recommend that Commissions 8 and 24 set up a Joint Working Group to coordinate efforts in this field.

2. The participants at this Colloquium strongly support the independent and

complementary astrometric programmes proposed for the ESA Astrometry Satellite and the NASA Space Telescope.

D. ASTROLABE REPORTS.

These have been given above under the appropriate national headings of Chile (2B) and France (2E).

E. ASTROMETRY SATELLITE REPORT

The ESA study of an astrometry satellite has now demonstrated that it is feasible, and that the technological problems can be solved. The satellite containing a two-axis telescope as proposed by P. Lacroute will scan the stars from a geo-synchronous orbit. A number of about 100,000 preselected stars, most of them brighter than $m_B = 11$, will be observed. The observing program will need to be defined beforehand by astronomers according to the selected astrometric and astrophysical criteria. A rather uniform distribution of the stars on the sky is however required for technical reasons. All 60,000 stars of $m_B < 9$ may be included.

The predicted accuracy of the observed parallaxes, proper motions per year and positions is $\varepsilon = 0\rlap{.}''002$ for stars of $m_B < 11$, degrading to $\varepsilon = 0\rlap{.}''010$ at $m = 14$. This includes all sources of error: photon statistics, attitude instability, optical aberrations, thermal disturbances, etc.

It would be useful if astronomers express their views (as many have already done) on desirable projects based on data from such a satellite. The projects should assume an accidental accuracy of $\pm 0\rlap{.}''002$ per star and systematic errors less than $0\rlap{.}''001$

The estimated time schedule for the satellite if it is finally approved by ESA comprises a launch in 1984 followed by 2.5 years of operation. A number of preparations before launch is expected from the scientific community: definition of investigations, selection of stars, ground-based observations of radial velocities and photometric data, development of reduction procedures. ESA's responsibility will be the development and launching of the spacecraft as well as data acquisition, transmission to the ground and a first evaluation of the data. Final evaluation and application of the astrometric data will be the responsibility of astronomical institutes.

The report on the feasibility study is available from ESA as DP/PS(78)13 Paris 26 April 1978. A colloquium on European Satellite Astrometry took place in Padova 5-7 June 1978 with the participation of European and American astronomers.

4. Late Reports

M. GERMAN DEMOCRATIC REPUBLIC

At Dresden the pilot program for the "Lohrmann Program" is still under way. 90 plates $3\rlap{.}^\circ1 \times 3\rlap{.}^\circ1$ overlapped by $0\rlap{.}^\circ6$ in zone $\delta = 52\rlap{.}^\circ5$ were taken with the Tautenburg 2-m Schmidt. The latitude observations (Horrebow method) continue. A new astrograph 300/1500 cm was installed in Gönnsdorf. A program on minor planets (ITA Leningrad) was started. Photoelectric observations of occultations were continued.

At Potsdam, the PZT catalogue was published in 1976. A photo-electric device was tested on the Potsdam 100/1000 cm transit. Tests confirm the Dresden results. (Sandig).

References

Pilot Lohrmann Program: Wiss. Z. TU. Dresden, 28, No. 3.

Latitude observations: Circ. Lohrmann Obs., Nos. 77-93.
Minor planet program: Wiss. Z. TU. Dresden, 28.
Use of occultations in astronomy, geodesy & selenodesy: Mitt Lohrmann Obs. Nos. 38 & 39.
Potsdam PZT catalogue: Veröff. Zentralinst. f. Phys. Erde, No. 37.
Photoelectric device tests: Wiss. Z. TU. Dresden, 28, No. 3. (Potsdam)
Wiss. Z. TU. Dresden, 25, No. 4. (Dresden)

Dr. Syuzo Isobe has announced his retirement from the Commission.

N. ASTRONOMICAL RECHEN-INSTITUT, HEIDELBERG

The work on the FK5 has been continued on the lines described in the report of Commission 8 (Trans. IAU Vol. 16 A - Part 2, 9). Numerous catalogues of observation have been studied with the aim to find out whether they may serve for one of the following purposes: improvement of the FK4 system, correction of individual data of FK4 and FK4 Sup stars, extension of the FK4 to fainter stars (Fricke, Gliese). Suitable catalogues have been taken on magnetic tape (Lederle, Walter); the U.S. Naval Observatory, Washington, D.C., and the Centre de Données Stellaires at Strasbourg have contributed to this task.

Completed are studies and tests of various computerized methods for the determination of systematic differences between catalogues of observation and the FK4 (Bien, Fricke, Lederle, Schwan). In the work on the FK5 two of them will be applied simultaneously: a computerized version of the conventional numerical method and an analytical method. The analytical method uses products of Fourier series, Legendre polynomials and Hermite polynomials for describing systematic differences; it has evolved from a method developed by Brosche and Schwan (19.041.041) in which products of spherical harmonics and Hermite polynomials were used instead.

Some progress has been made in the work on the determination of the equinox and equator of the FK5 (Fricke, Scholl). Observations of the Sun and planets and lunar occultations confirm the existence of a non-precessional motion of the FK4 equinox, which has to be eliminated in the FK5. Furthermore, attention is drawn to published work related to the FK5; Lieske et al. (19.043.006) have developed the expressions for the precession quantities based upon the IAU (1976) System of Astronomical Constants, and Fricke (19.043.007) has presented the basic material on which the new value of the lunisolar precession is based, and has outlined the arguments in favour of the change in precession (19.043.002). Lederle (Bull. CDS, No. 14) has computed tables giving information on the accuracy of FK4 data at different epochs. Walter (20.041.026) has compiled a list of positions of radio stars which may serve for tying the extragalactic reference frame to the fundamental reference coordinate system. (W. Fricke, W. Gliese, T. Lederle and H.G. Walter).

R. H. TUCKER
President of the Commission

10. SOLAR ACTIVITY (ACTIVITÉ SOLAIRE)

PRESIDENT: G. Newkirk, Jr.
VICE PRESIDENT: V. Bumba
SECRETARY: A. Bruzek
ORGANIZING COMMITTEE: R. J. Bray, T. Hirayama, V. A. Krat, J.-L. Leroy, S. F. Smerd (deceased), N. V. Steshenko, M. Stix, P. Sturrock, J. Wilcox, C. Zwaan

This report summarizes research on solar activity over the three year period from September 1975 through September 1978. Every attempt has been made by the Organizing Committee to produce a comprehensive review of the highlights of the last three years within the limitations of available space. It is inevitable that some researches have not been described in great detail and others may have been inadvertantly omitted. For such oversights we apologize. The President wishes to acknowledge the effort of the authors who have assembled this report and our colleagues who have furnished material for inclusion in this summary of progress over the past three years.

It is our sad duty to report the death of S. F. Smerd, Member of the Organizing Committee. He served the IAU in a variety of capacities, and the loss of his astute scientific guidance and warm friendship is felt by us all.

At the 1976 General Assembly the IAU passed a resolution to encourage the Debrecen Observatory to undertake the responsibility of the continuation of the Greenwich Photoheliographic Program, which the Greenwich Observatory was forced to discontinue in January 1977. Dr. Dezső, Director of Debrecen, reports that the 10cm telescopes at Debrecen and at Gyula (separated by about 100km) have been producing white light photographs with a solar image of 10.5cm diameter each clear day since January 1977 to continue this program. (Pub. Debrecen Obs. 1, Nos. 11-14, 1978). The Greenwich system generally operated with an aperture of 10cm with a solar image of 19 cm. Dezso has visited Drs. Graham Smith and Stickland at RGO to assure that the methods of observing and analysis used at that institution for more than a century are maintained.

Since Debrecen has been carrying out a photoheliograph program for many years and has some 40,000 plates covering the last 22 years, ample material exists for comparing the RGO and Debrecen results on the same days before 1977 to calibrate such measured parameters as sunspot area. Debrecen is developing a television scanner to automate such measurements. Debrecen has to date secured the cooperation of the Pulkovo, Kislovodsk, Kodaikanal and Royal Greenwich Observatories in supplying data of similar quality to fill in the gaps in the Debrecen series.

The solar physics community owes Dr. Dezső and his colleagues at Debrecen and at the other collaborating stations a debt of gratitude for taking on the responsibility of maintaining continued homogeneity of these important data.

1. THEORY OF THE SOLAR CYCLE
(M. Stix)

Understanding the fundamental processes underlying the solar cycle still presents one of the principal challenges of modern astrophysics. The period of this report has been one of intense activity, some of which was undoubtedly stimulated by IAU Symposium No. 71, "Basic Mechanisms of Solar Activity," Prague, 25-29 August 1975. Most dynamo models which have been more or less successful in explaining the observed features of the solar cycle fall in the category of $\alpha\omega$-dynamos and are based on helical convection ("α"), a non-uniform mean angular velocity ("ω"), and turbulent diffusion of the mean magnetic field as an additional essential ingredient. Current research on each of these factors is described below.

A) The *angular velocity*, according to most of these models, should *increase* with depth in order to yield an equatorwards traveling mean field. However, most theories of the convection zone and differential rotation predict that the angular velocity should *decrease* with depth because the restraint of rotation tends to cause angular velocity to be constant on cylinders co-axial with the rotation axis. Observed differences of eastwards and westwards propagating solar p-modes indicate an increase with depth (Deubner *et al.*, *AA*, in press) in a shallow layer below the photosphere; however, the behavior of angular velocity in deeper layers is uncertain. An inwards increasing angular velocity is, however, predicted by calculations of Belvedere and Paternò (*SP* 54, 189) on the basis of a latitude dependent mean heat flow across the convection zone. This crucial question remains unresolved.

B) The α-effect employed in most models is scalar. It is, however, known that α is a tensor in general, i.e., the mean electric field arising from helical convection need not be exactly parallel to the mean magnetic field. Wälder (1978, Thesis, Univ. Göttingen) studied such an α, and Weisshaar (1978, Dipl, Univ. Göttingen) computed mean field dynamos on this basis. The behaviour of α at high electrical conductivity has been debated: Deinzer (*MAG* 40, 156) and Wälder (l.c.) find $\alpha \to 0$ in this limit, but Moffatt (*AAM* 16, 119; *Magnetic Field Generation in Electrically Conducting Fluids*, Cambridge Univ. Press) argues that a finite value is obtained if there exist contributions to the helicity spectrum at zero-frequency. Levy (*ApJ* 220, 325) finds a finite α for high conductivity, and Kraichnan's (*JFM* 75, 657; 77, 753) numerical simulation also indicates such behaviour.

C) Considering *turbulent electromagnetic diffusivity*, Knobloch (*JFM* 83, 129; *ApJ* in press) realized that the turbulent diffusion of a mean divergence-free vector field differs from that of a scalar, due to velocity correlations of higher than second order. Kraichnan (l.c.) found that helicity fluctuations influence the magnetic diffusivity, which can even be negative, while the diffusivity of a scalar remains positive. Nakano (*PASJ* 28, 451) introduced a wavenumber dependent turbulent diffusivity, and thus generalized the two-scale formalism which is normally adopted in mean field electrodynamics (e.g. Moffatt, 1978, l.c.).

MHD turbulence has been studied numerically by Pouquet *et al.* (*JFM* 77, 321) and by Pouquet and Patterson (*JFM* 85, 305). They find transfer of energy from kinetic into magnetic form together with a cascade of kinetic energy and an inverse cascade of magnetic energy. Since anisotropies, compressibility, a realistic geometry and boundary conditions are not considered in these calculations, application to the sun does not yet appear possible. But the non-linear interactions are studied in an exact fashion, even at moderately large magnetic Reynolds numbers.

Mean field models of the solar cycle predict well-defined phase relations between the toroidal and poloidal mean field components. Comparison of such models with Mt. Wilson magnetograms suggests that the rate of rotation increases with depth (Stix, *AA*, 47, 243). Yoshimura (*SP* 50, 3) argued that latitudinal shear is more important than radial shear in order to explain such observations and that the fields originate in the upper part of the convection zone. The latter result is in contrast to a conclusion of Parker that the fields would be much weaker than those observed if they were generated in the upper layers, and that, therefore, their origin must be deep. In fact, the solar dynamo seemed to be able to exist only below or at the base of convection zone. Unno and Ribes (*ApJ* 208, 222) and Schüssler (*AA* 56, 439) found that the buoyant magnetic fields could well originate from within the convection zone but only in its deeper layers. The rise of the buoyant field has been described in a linear model by Parker (*ApJ* 215, 370) as a mean upward flow with the consequence that the dynamo works with less ease. Yoshimura (*ApJ* 220, 692; *ApJ* in press) described the dynamic effects by a number of free parameters including a time delay in the build up of the field and its inhibition of the fluid motions in a non-linear model and was able to simulate numerically long term variations of the amplitude of the solar cycle. Non-linear models including an incompressible mean flow were calculated by Nelle (1977, Diplom, Univ. Göttingen); her results yield qualitatively correct butterfly diagrams, but the fields generally appear at too high latitudes. Application to the sun appears to be limited also because the Lorentz force arising from the second order correlations of the fluctuating field has been omitted. Compressible motions have been considered by Schlüsser (*AA*, in press) in a simplified geometry. Sector-type mean fields generated by the solar dynamo have been suggested to control the origin, rotation, and evolution of coronal holes (Stix, *AA* 59, 73).

The fore-mentioned models all use a mean field induction equation where the interaction of rotation with convection is described by the α-effect. More ambitious are the attempts to describe directly this interaction by a non-linear set of hydromagnetic equations. Kennet (*SAM* 55, 65) and Baker (*GAFD*, in press) used truncated modal expansions which are limited in applicability since, due to lack of convergence, they sometimes yield fallacious dynamos. Gilman has extended his spherical shell, thermal convection numerical model (*GAFD* 8, 93) to include the dynamo action of an incompressible fluid. Preliminary results indicate that convergence is obtained. These calculations reproduce oscillating fields and simulate the butterfly diagram. However, the dynamo period is too short by at least one order of magnitude. It is worth noting that this model maintains the dynamo with equatorward drift of the fields even though the mean angular velocity decreases with depth. The model appears to illustrate a basic difficulty in that the convective motions needed to drive the correct amplitude differential rotation have too much helicity to give the correct amplitude for the α-coefficient of the traditional $\alpha\omega$-dynamo. The effect of field concentration into small flux filaments might resolve this problem; but a numerical treatment of the effect in a global dynamo has not yet been done and its complexity will surely demand that the effect can be introduced only by means of non-physical parameterization.

Recent reviews on the dynamo problem have been published by Moffatt (1976, l.c.), Vaĭnsteĭn (1976, *SPU* 120, 613; *Engl.* 19, 987), Parker (*ARAA* 15, 45) and Stix (1978, *2ESM*). Moffatt (1978, l.c.) has contributed an entire monograph to the field.

2. EMPIRICAL ASPECTS OF THE SOLAR CYCLE
(V. A. Krat)

Detailed reviews of several aspects of the solar cycle have been published by Kuklin (*IAU* 71, 147) and have appeared in, "The Sun and the Atmosphere of the Earth," by Vitinsky, Ol' and Sazonov (Gidrometeorzdat, Moscow, 1976) and "The Solar Output and Its Variation," ed. by O. R. White (Colorado Associated University Press, 1977).

Combining various data of cycle 20 including the intensity of the corona at different latitudes, Gnevyshev (*SP* 51, 175) concluded that the 11 year cycle contains two maxima. The first coincides with the maximum in sunspot number and is accompanied by high activity at all latitudes. The second occurs 2-3 years later and appears predominantly at low (about 10°) latitudes. Constructing a "butterfly diagram" from the large scale magnetic fields for the period 1959-1974, Yoshimura *SP* 47, 581) has determined that the poloidal field breaks up into two zones in each hemisphere with opposite polarity in each zone. Svalgaard and Wilcox (*SP* 49, 177) have pointed out that the borders between large scale magnetic regions on the sun which give rise to interplanetary sector boundaries have different properties depending upon whether the change in magnetic polarity with longitude is opposite to or the same as that of sunspot groups in the same hemisphere. Near the latter, designated Hale boundaries, emission in the green coronal line and the magnitude of the non-spot magnetic field are a maximum while near the non-Hale boundaries both are a minimum. The connection of this phenomenon to the presence or absence of coronal streamers above these boundaries has not yet been investigated. These Hale borders appear to coincide with the position of active longitudes as shown by the presence of photospheric faculae for the 19th and 20th solar cycle (Kramynin, *SDB* 8, 60).

Karlinsky (*BAC* 28, 200) showed statistically a dependence of the displacement of active longitudes on the level of solar activity. Tuominen (*SP* 47, 541), Kozhevnikov (*AZAN* 53, 389), Antonucci *et al* (*SP* 53, 519), found that both the latitudinal drift of sunspot groups and the variation of rotation of the chromosphere and corona with latitude vary with the phase of the cycle. Using what data exist on the variation of solar activity, solar radius, and the solar constant, Vasiliev and Kandaurova (*SDB* 10, 71) suggest that the sun may pulsate with a main period of 22 years.

Several papers have dealt with the properties of the 80 year cycle (Vitinskij, Ol' and Sazonov, *The Sun and the Atmosphere of the Earth*, 173, Gidrometeoizdat, Moscow; Berdichevskaya, *AZAN* 53, 822; *AZAN* 53, 1046; Gleissberg, *VAIF* 59, 1977; Kopecky, *7CSP*, 249; Kuklin *IAU* 71, 147; Schröter, *SP* 50, 501; Yoshimura, *SP* 47, 581), Vitinskij (*SDB* 11, 59) and Vitinskij and Miletsky, (*SDB* 5, 1977) discussed the spatial and temporal distribution of various activity indices. Trellis (*CR* 284, 453) indicated that active longitudes of sunspot areas oscillate with a period of about 20-30 revolutions. A statistically significant North-South asymmetry in the frequency of large flares, of complex magnetic class sunspots, and of large sunspots has been detected in the data covering the last two solar cycles (Roy, *SP* 52, 53). A similar asymmetry appears in the distribution of white light flares reported since 1959.

Renewed interest has emerged concerning the topic of the so-called Maunder Minimum (XVII century). Eddy (*SC* 192, 1189) and "The Solar Output and Its Variation" (l.c) presented further evidence that solar activity as revealed by sunspots essentially disappeared between 1645 and 1700. Some investigators (Gleissberg, *SW* 16, 229; Link *AA* 54, 857; Vitinskij *SP*, in press) consider that this minimum in the overall envelope of solar activity is a natural consequence of the existence of a periodicity in solar activity with a period of several hundred years. The abundance of radioactive carbon, which reflects the flux level of galactic cosmic rays at the top of the atmosphere and thus, solar activity, in the rings of trees (Dergachev and Kocharov, *IANS* 41, 422; Damon, "The Solar Output and Its Variation", 429 (l.c.)) leaves little doubt that major variations in the level of solar activity have occurred over the last 5000 years although the presence of regular periodicity remains uncertain.

A new method of solar activity forecasts using geomagnetic data was outlined by Ol' (*SDB* 9, 73; *SDB* 12, 87). Forecasts of the properties of Cycle 21 were made also by Ramaswami (*N* 265, 713) and Vitinskij (*SDB* 11, 59). In connection with this problem paper by Kopecky (l.c), Dodson, Hedeman and Mohler (U) should be mentioned.

Several authors (Pystina, Shpitalnaya, and Vasilieva, *SDB* 3, 60; Romanchuck, *Astron. Obs., Kiev*, Preprint #10, 1975; Khlystov, *SDB* 10, 78; Kozhevnikov, *AZAN* 53, 389) have attempted to prove a dependence of solar activity on such external forces as planetary tides and the galactic magnetic field. That this idea is not widely accepted can be seen in the work of Gudzenko and Chertoprud (*TAI* 90, 154) and Smythe and Eddy (*N* 266, 434).

New indices of solar activity have been proposed by Xantakis and Poulakos (U) and Kopecky and Ruzichkova-Topolova (*BAC* 29, 65).

3. SOLAR ACTIVE REGIONS AND PHOTOSPHERIC FACULAE
(A. Bruzek)

A. *Active Regions*

During the three-year period covered by the present report, research on solar active regions has been carried out mainly on (1) birth and development as observed at optical wavelengths; (2) photospheric faculae; and (3) the upper (mainly coronal) part of AR's employing EUV and soft X-ray observations from space.

Solar active regions were the subject of the 8th Consultation of Solar Physicists in Irkutsk 1976. There, a review on ground-based observations, physical characteristics and processes involved in the birth and development of active regions was given by Stepanyan (*8CSP* 1, 3). Jäger (*8CSP* 1, 114) gave a review on the development of AR's and its prognosis. The birth and growth of AR's in the photosphere and in the corona were studied by Baranowsky and Stepanyan (*8CSP* 1, 27), Baranowsky *et al* (*8CSP* 1, 31), and Sykora (*8CSP* 1, 42). Alikaeva (*8CSP* 1, 21) found in an extended cooperative study that the development of AR's takes place in distinct pulses. Other papers dealt with the first phases of spotgroup development (Bumba *et al*, *8 CSP* 1, 59; Romanchuk and Krivodubskij, *8CSP* 2, 91) and with magnetic and velocity field development (Deszö *et al*, *8CSP* 1, 48; Grigoryev *et al*, *8 CSP* 1, 54; Ioshpa *8CSP* 1, 70).

The third of the ATM-Skylab Workshops sponsored by NASA and operated by the High Altitude Observatory was devoted to active regions with particular emphasis on problems of energy balance, the development of refined quantitative models of the conditions above active regions, and the mechanisms responsible for energy transport. The Workshops consisted of a series of three meetings spaced over several months with F. Q. Orrall as director and R. Noyes, G. van Hoven, and R. G. Athay as advisors. Proceedings of these meetings of about 40 researchers will be available.

I. *General*

The smallest bipolar magnetic regions are associated with coronal, soft X-ray bright points. Only a small fraction of bright points lives more than a day and develops into larger active regions (Golub *et al*, *SP* 53, 111). X-ray emission is detected within three hours of the first Hα brightening of AR's. Temperatures of 2.3-4 x 10^6K and densities 1.9-10 x $10^9 cm^{-3}$ were inferred from the analysis of soft X-ray flux in two bands (Little and Krieger *BAAS* 9, 341; Wolfson *et al*, *SP* 55, 181; Vorpahl *SP* 57, 297). Vorpahl found good agreement between the configuration of X-ray loops and the calculated potential magnetic field topology and concludes that the coronal magnetic field quickly attains a relaxed configuration. Svestka *et al*, *SP* 55, 359 detected dark gaps in the soft X-ray photographs of the later stages of AR development. These gaps are interpreted as areas of open magnetic field analogous to coronal holes. Large coronal loops or arches interconnecting neighboring AR's are commonly found on X-ray photographs and indicate a coronal connection by the magnetic field (Chase *et al*, *SR* 16, 917; Svestka *et al*, *SP* 52, 69; Howard and Svestka, *SP* 54, 65). Often several AR's are magnetically tied together to form a larger entity referred to as a "complex of activity". The speed with which these connections are established (Sheeley *et al*, *SP* 40, 103) suggests that some form of anomalous resistivity is present in the corona.

II. *Global Energy Balance*

The global energy losses of the atmosphere of an average AR above $T_e = 10^4$K were estimated by Evans *et al* (*SP* 55, 387), who found the total radiative output to be $\sim 2 \times 10^{19} Js^{-1}$ with the largest loss occurring as radiation from coronal material at $T_e > 10^6$K. Conduction losses as well as mass loss to the solar wind were estimated to be negligible. Jakimiec (*8CSP* 2, 5) has reviewed models of coronal AR's and their energy supply derived from XUV and X-ray observations obtained by the

Intercosmos satellites. Temperatures T $\geq 10^7$K in flare-free active regions were inferred by Silvester (*8CSP* 2, 24). The heating of hot coronal regions by chromospheric ejections (Vasiljev *et al*, *8CSP* 2, 66) and by shock waves (Vilkovsky *8CSP* 2, 70) was also investigated. Prime candidates for the source of energy above AR's remain Alfvén waves (Wentzel, *SP* 50, 343; *SP* 52, 163) and the dissipation of magnetic fields (Nolte *et al*, *SP* 55, 401; Rosner *et al*, *ApJ* in press) although the quantitative adequacy of either mechanism remains to be established. One aspect of the latter mechanism has been explored by noting that the emergence of the magnetic fields of new AR's necessarily leads to the formation of current sheets, where magnetic field reconnection and dissipation can take place (Syrovatskii, *SAL* 2, 35; Heyvarts and Priest, *SP* 47, 223; Wolfson *et al*, *SP* 55, 181) suggest that the AR magnetic field emerges in a stressed or complex configuration ready to release energy by field reconnection. Somov and Syrovatskii (*SP* 55, 393) discuss the possibility of quasi-steady energy release in current sheets as the main source of heating of AR atmospheres. They conclude the $\sim 10^{28}$ergs s^{-1} may be released by a steady current sheet over hours or even days as EUV radiation at T $\sim 8 \times 10^4$K, which heats the chromosphere. However, in order to account for the heating of the corona and its X-rays secondary processes must be considered.

III. *Loop Structure*

The presence of magnetically confined loops of plasma has emerged as a major factor in understanding the atmosphere above AR's. Active region loops seen in EUV lines formed at temperatures 10^4-2×10^6K were studied in detail by Foukal (*ApJ* 210, 575) and Levine and Withbroe (*SP* 51, 83), who found a coaxial shell structure with a low temperature core of a few hundred km radius surrounded by an extended sheath of coronal temperature and a constant density of $\sim 7 \times 10^8$cm^{-3}. Energy loss by radiation is of the order of 10^{-4}ergs cm^{-3}s^{-1} in the hot outer shell. Observations from rockets and OSO-8 suggest downward velocities of 30-100km s^{-1} in loops over spot umbrae. Foukal assumes that the radiation losses along the loops are balanced by the thermal energy and the gravitational energy of the falling material and that magnetic fields play no role in the support of the loops. Chiuderi, Giachetti and van Hoven (*ApJ* 213, 869; *SP* 54, 107; *SP* 55, 371) on the other hand, investigating the problem of MHD stability of loops, find that in order to support the pressure gradient a large departure from the force-free Lundquist field is necessary. This is achieved by the addition of the effects of cross-field currents.

Pye *et al* (*AA* 65, 123) constructed the first detailed temperature and density map of an AR at T > 10^6K from 4-23Å observations. The resulting model of the region contains a system of loops at temperatures 2.6-3.2 x 10^6K. They assume a homogeneously filled circular cross section at densities 1-6 x 10^9cm^{-3}. They conclude that the birth, stability and disappearance of a hot loop are controlled only by the energy input, which determines not only the temperature but also the scale height and density in the loop by mass interchange with the low atmosphere. Gerassimenko *et al* (*SP* 57, 103) presented observational evidence of continual heating in X-ray emitting coronal loops.

B. *Photospheric Faculae*

Considerable progress in the interpretation of photospheric faculae has been achieved by the introduction of non-LTE models which take into account the association of facular points with small-scale magnetic flux tubes. It has been known for several years that magnetic fields in facular areas and in the photospheric network are highly concentrated in flux tubes with field strengths of 1000-2000 G and cross sections of several hundred km. The flux tubes are confined because of reduced internal gas pressure, which implies reduced opacity and also gives rise to the Wilson depression. Thus, the interior of the flux tube will be surrounded by hot walls, and radiation form the walls adds to the intensity emerging from the tube. Non-LTE multidimensional radiative transfer calculations by Stenholm and

Stenflo (*AA* 58, 273) show that radiation channeling effects in small-scale flux tubes, indeed, can give rise to locally increased line intensities, which correspond to calculated temperature enhancements in LTE models while no true temperature increase is present. Spruit (*SP* 50, 269) also concludes that the contrasts observed in continuum faculae may be caused by radiation transfer effects and do not require additional heating. The "bright wall" effect may explain the center-limb variation of facular contrast and resolve the discrepancy between the temperatures derived from the CLV of continuum intensities (ΔT = 500-1000K) and from line profiles (ΔT = 100-300K), respectively.

Frazier (*AA* 64, 35) suggests another explanation of that discrepancy: His observations show that the shape of the contrast line profile of faculae, and therefore the LTE facular temperature structure, depend largely on the total magnetic flux, i.e., the size of the faculae. The ΔT discrepancy above may then be a selection effect in that continuum contrast observations select faculae bright in the continuum, i.e., small faculae. On the other hand, line profile observations select faculae bright in the line core, i.e., large faculae.

The most elaborate LTE facular models available so far have been presented by Chapman (*ApJ Sup* 33, 35). He achieves reasonable agreement between observed and calculated line profiles by taking into account the effects of scattered photospheric light and of magnetic fields. He derives field strengths of 1500-1900 G in faculae and temperature contrasts of up to 880K. if such temperature contrasts are real, some form of non-radiative heating is required. Less refined models have been given by Müller and Hirayama. Müller (*SP* 45, 105) employed high resolution continuum photographs of photospheric faculae with the 50cm refractor at Pic-du-Midi Observatory. In his model facular granules consist of hot clouds (ΔT = 750K) 100km above a locally cooler photosphere. Hirayama (*PASJ* 30 337) infers a temperature enhancement ΔT = 1000K for 2" diameter facular granules photographed with a 10cm balloon telescope. Koutchmy, employing a restoration technique, (*AA* 61, 397) determined the contrast of filigrees in the continuum at λ6441A to be $\geq$ 100%. Stellmacher and Wiehr (*AA* in press) constructed a facula model employing the high temperature filigree model (ΔT 1000K in the deeper layers) of Koutchmy and Stellmacher (*AA* 67, 93). They consider the observed facular emission as a composite--due to limited spatial resolution--of granular, intergranular and filigree emission. This model represents satisfactorily the central intensity and the wings of the lines as well as the decrease of faculae contrast in the red. It does not describe the CLV of the continuum contrast. Physical conditions in faculae were also discussed by Mitropolskaya and Sitnik (*8CSP* 1, 64) and a heating mechanism was proposed by Vainstein *et al* (*8CSP* 2, 95). Their molecular spectra have been studied by Dubey and Tripathi (*BAC* 22, 161).

Important for the interpretation of plages are the observations of Giovanelli and Brown (*SP* 52, 27) and Giovanelli and Slaughter (*SP* 57, 255), who found systematic downward motions in plage elements and in small EFR's with velocities increasing with depth in the atmosphere up to 0.4 km s^{-1}. Giovanelli (*SP* 52, 315) suggests that the downflowing material is replaced by neutral gas flowing into the closed flux tube near the temperature minimum under the influence of the gas pressure gradient between the outside and the inside of the magnetic tube.

4. SUNSPOTS
(C. Zwaan)

During 1976-78 considerable progress has been made in exploring the "sunspot problem": Why is the magnetic field concentrated in patches with large field strengths (spots, faculae, ...), and why sunspots have the structure we observe? Some problem items have been solved and remaining questions appeared in a new perspective. In this review the whole range of discrete magnetic structures is considered but we concentrate on sunspots.

A. *Magnetostatic Models*

The work by Deinzer and Yun before 1976, and by Spruit (*SP* 50, 269; 55, 3) shows that the main features of the range of magnetic structures, from sunspots down to facular points, can be explained by a set of magnetostatic flux tube models in the top of the convection zone. The underlying assumptions are that within the tubes the vertical heat flow is about one fifth of the heat flow in the normal convection zone, and that there is no heat exchange across the field except by electromagnetic radiation. In Spruit's models the transition from dark pores to bright facular points, and the center-to-limb variation of the facular contrast in continuum and line wings are explained as consequences of the Wilson depression, the cool interior and the warmer walls.

The objection that a sunspot flux tube would show up with an observable bright ring around it (Parker, *ApJ* 204, 259; Isenberg, *SP* 50, 49) is refuted by the rapid increase of the heat transport coefficient with depth (Spruit, *SP* 55, 3). The effect of an anisotropy in the heat conduction on the surface temperature distribution across an idealized sunspot was investigated by Eschrich and Kraus (*AN* 298, 1).

Staude (*BAC* 29, 71) computed magnetostatic umbral models fitting adopted magnetic field configurations and convection zone models. His analysis of the resulting thermodynamic stratifications indicated that a strong non-radiative flux is needed for heating the umbral atmosphere.

As for the dynamical stability Meyer, Schmidt and Weiss (*MN* 179, 741) put forward the argument that flux tubes in pressure equilibrium are stabilized against the exchange instability by the buoyancy of the tube overlying the denser plasma outside. They showed that vertical tubes with fluxes larger than about 10^{19}Mx are stable in the photosphere and in the very top of the convection zone. At larger depths an additional stabilizing mechanism is required.

B. *Energy Transport, Umbral Structure and Waves*

The very existence of tubes of intense magnetic field requires a reduction of the temperature inside the tube over several scale heights in the upper convection zone. Consequently, the energy flow heating the flux tube should be strongly reduced, but a consistent quantitative theory is still lacking. There is an argument about what mechanism is the most plausible: inhibition of convection by the magnetic field or cooling by emission of Alfvén waves generated by convective overstability. Parker (cf. *MN* 179, 93 P) had criticized the inhibition mechanism on the ground that whereever heat is dammed up beneath the spot the resulting pressure increase would disperse the magnetic field. Cowling (*MN* 177, 409) criticized Parker's simplification of the inhibition mechanism by adopting a thin inhibition layer beneath the spot and by treating the heat conductivity as nearly uniform and isotropic. Moreover, Cowling questioned the 80 per cent efficiency of the conversion of heat into Alfvén waves which would be required to explain the "missing flux".

As mentioned above, in the magnetostatic models inhibition of convective transport is assumed at least near the cylindrical wall of the flux tube in the top of the convection zone, and the problem of the absence of the bright ring around spots has been solved. As to the one-fifth of the normal photospheric flux radiated by umbrae, should we ask why it is so little or should we ask why it is so much?

Besides the papers already mentioned there are several recent papers dealing with the problem of energy transport in and near flux tubes. From a detailed linear analysis of overstable Alfvén modes of a Boussinesq fluid in and around a column of uniform magnetic field, Roberts (*ApJ* 204, 268) concluded that overstable Alfvén waves may cool spots. Boruta (*ApJ* 215, 364) found from a model for the inhibition of convection and for the cooling by Alfvén waves that some initial field concentration may increase and develop into a cool spot.

Roberts and Webb (*SP* 56, 5) investigated small amplitude motions in a vertical *slender* flux tube in temperature and pressure equilibrium with the surroundings, taking into account magnetic forces, buoyancy and *compressibility*, and discussed the pressure and velocity perturbations within the tube.

From line profiles from a large spot at various positions on the disk Beckers (*ApJ* 203, 739) concluded that the Alfvén flux in the umbral photosphere may amount to 20 to 50 per cent of the "missing" flux of electromagnetic radiation, provided that the non-thermal Doppler broadening is entirely attributed to pure Alfvén waves. However, line widths in coronal arches over a sunspot indicate that the flux of Alfvén waves leaking into the corona is extremely small (Beckers and Schneeberger, *ApJ* 215, 356, see also Evans *et al*, *SP* 55, 387). Using a simple but adequate temperature model Thomas (*ApJ* 225, 275) showed that any flux of Alfvén waves generated in the umbra is almost completely reflected downwards well before reaching the corona.

From power spectrum analysis of velocity oscillations measured in several spectral lines Schröter, Soltau and Wöhl (*AA* 49, 463; 50, 367) found three ranges of periods, viz. 470-300 s, 196-164 s, and 123-110 s; the latter short period oscillations had not been reported earlier.

Bumba *et al* (*BAC* 26, 315) and Hejna (*BAC* 28, 126) show and discuss the pattern of bright umbral dots, pointing out that the distribution of the structure sizes (i.e. the distances between adjacent dots) is very similar to that of the surrounding photospheric granulation. The reader examining the plates in the first mentioned paper may find once more that the dots in small or irregular spots are much brighter than in roundish large umbrae.

The Oslo observers noticed significantly different radiation temperatures in various spots (Maltby, *SP* 55, 335). Albregtsen and Maltby (*N* 274, 41) found a dependence of the umbral intensity with the phase of the solar cycle, in the sense that the radiation flux increases during the cycle. Clearly, this intriguing result requires further investigation.

C. *The Formation of Flux Tubes; Mass Flow Along the Tube*

The effects of vigorous convection on an initially weak and uniform magnetic field have been investigated in a Boussinesq fluid by Galloway *et al* (*N* 226, 686) and by Peckover and Weiss (*MN* 182, 189). The magnetic field is concentrated in ropes or sheets in which the field strength may exceed the equipartition value for suitable combinations of parameter values characterizing the viscous, thermal and magnetic diffusivities, the initial field strength and the Rayleigh number. Galloway *et al* applied these results to the solar atmosphere by assuming that the energy balance adjusts in such a way that the internal temperature and pressure reduce in step with the increasing magnetic pressure.

The idea that sunspots are concentrated by supergranulation from an initially dispersed field (Meyer *et al*, *MN* 169, 35) is not correct, however. Observational data indicate that in active regions magnetic field emerges in bunches of flux tubes which were already concentrated *before* penetrating in the atmosphere, and sunspots grow by coalescence of recently emerged tubes (cf. Piddington, *ASS* 34, 346; Zwaan, *SP* in press). Zwaan estimated that during the emergence the field strength amounts to several hundred gauss, which is close to the equipartition field strength in the top of the convection zone. Once the tubes have surfaced, radiative cooling sets in and then the cooled gas starts a convective downdraft carrying the cooling deeper down. New measurements in photospheric and low chromospheric lines confirmed strong downdrafts in emerging flux regions and particularly in growing pores and spots (Kawaguchi and Kitai, *SP* 46, 125; Bachmann, *BAC* 29, 180). The downdraft in spots stops when the spot ceases to grow.

Beckers (*ApJ* 213, 900) analysed line shifts in umbrae of several mature sunspots during their passage across the disk. Rotation rates from proper motions and from Doppler shifts agree within 1 per cent. The vertical flow is undetectable (< 25 m s^{-1}); there is no "convective blueshift" either. However, a systematic downdraft has been observed in the small magnetic elements in plages and network (cf. Harvey, *HA* 4 II, 223; Giovanelli and Slaughter, *SP* 57, 255).

Parker (*ApJ* 221, 368) found from calculations for slender flux tubes that a downdraft of only 100m s^{-1} would produce sufficient cooling with respect to the superadiabatic surroundings to concentrate magnetic fields up to 1500 gauss or more. Clearly, this effect may keep slender tubes concentrated as long as the downdraft lasts, and observations suggest that downdrafts are a persistent feature of thin flux tubes. This raises the question how the mass is supplied from above (cf. Giovanelli *SP* 52, 315). Spruit (*SP* in press) showed that slender flux tubes with surface field strengths less than 1300 gauss will collapse to an equilibrium state with higher field strengths ($\approx$ 1700 gauss) if triggered by some downdraft. However, the collapsed tube is not in thermal equilibrium with its surroundings, and Spruit argued that in the top part an overstable oscillation will be excited. He suggested that a brightness-velocity correlation may account for the (seemingly persistent) downdraft in the magnetic plage and network structure.

In order to explain the coherent motion of parallel flux tubes during the emergence and the coalescence into spots Parker (*ApJ* 222, 357) investigated hydrodynamic attraction by the effect of the wake of a preceding tube on a following one and by the Bernoulli effect on two tubes moving side by side. These attractions vanish once the rising of tubes stops, and breakup and decay sets in. In a new series of papers Parker (*ApJ* in press) investigated the dynamical stability of long-lived spots assuming a model consisting of many separate tubes held together by the buoyancy of the top parts and by a postulated downdraft in between the separate tubes well below the visible surface. Parker suggested that the individual tubes are cooled by the convective generation of Alfvén waves which are preferentially emitted downward because of the downdraft. Parker admitted that the postulated downdraft is not easily reconciled with the upwelling in the moat supergranule. This is a pressing problem in Parker's model since the moat supergranules are frequently observed around long-lived spots.

Zwaan (*SP* in press) proposed a tentative model for the amplification, ascent and emergence of intense flux ropes, which assumes, particularly in the deep convection zone, concentration by differential rotation, containment by the external turbulent pressure, and adequately efficient lateral heat exchange by the remaining internal turbulence. The crude model describes even the largest active regions as loops from toroidal ropes of a reasonably small diameter rooted near the bottom of the convective zone. After the cooling of the top parts by electromagnetic radiation and convective downdraft the conditions are created for the magnetostatic models heated by a reduced energy flow without requiring a blocking of heat somewhere within the tubes.

D. *Penumbrae*

Galloway (*SP* 44 409) proposed a simple model for the penumbral structure, based on the well known convective rolls parallel to the magnetic field which is strongly inclined with respect to the vertical. This model predicts that the magnetic field in the dark striae is stronger and more horizontal than in the bright striae. A large radial outflow is expected in the dark striae, and a smaller inflow in the bright striae. Note that in this model the penumbra is necessarily the highly structured and dynamic fringe of the largely magnetostatic structure of the whole spot. Its structural and dynamic features are in agreement with observational data. Recently Abdussamatov (*SP* 48, 117) confirmed from 2" resolution spectrograms that the field strengths in dark penumbral matter are significantly higher than in bright matter. Müller (*SP* 48, 101) confirmed his earlier discovery of the movement of bright penumbral grains towards the umbra, reaching the maximum speed of about 0.5km s^{-1} near the umbral border.

By means of a two-layer model Nye and Thomas (*ApJ* 204, 582) concluded that the lowest "plus" eigenmode of magneto-atmospheric waves would fit the running penumbral waves observed in Hα. This result suggests that the waves are a phenomenon in the photospheric part of the penumbra. Musman *et al* (*ApJL* 206, 175) have observed oscillations in a photospheric line with periods comparable to the Hα periods, but the photospheric waves are more intermittent and the horizontal phase velocity is much larger than in the Hα waves.

This review is focused on the structure of sunspots in the deeper parts of the atmosphere and on the mechanisms causing spots and smaller patches of concentrated magnetic field. Clearly, in this way only part of relevant sunspot investigations 1976-1978 have been covered. We briefly refer to some other groups of investigations below.

E. *Chromospheric and Coronal Structure over Sunspots*

Studies in the optical, radio and particularly in the EUV and X ranges have greatly increased our knowledge of the structure of the chromosphere, transition zone and corona as influenced by the magnetic fields of spots. Several of these papers will be referred to in other sections of this report.

F. *Sunspot Spectroscopy*

Exploitation of the sunspot spectrum for solving problems in solar or stellar spectroscopy is represented by two examples: Beckers' study (*ApJ* 213, 900) of Doppler shifts in umbrae observed center-to-limb revealed a redshift which is within the accuracy of 25 m s^{-1} identical with the predicted gravitational redshift of 636 m s^{-1}. Moreover, the interpretation of the limb effect in the photospheric Doppler shifts by an average convective blueshift is in agreement with the absence of the limb effect in umbral spectra.

Wing *et al* (*ApJ* 216, 659) identified bands of the FeH molecule in the near IR spectra of spots and by comparison also in M and S type stars. The vibration-rotation spectra of CO_2, NO, SiO, and TiO expected in sunspots has been studied (Pande and Gaur, *N* 253, 104; Gaur *SP* 46, 121; Gaur *et al*, *SP* 56, 67; and Sinha, *BASI* 5, 49).

5. PROMINENCES
(T. Hirayama)

Major progress in this period has been in the spectroscopic study, Skylab and OSO-9 (Bonnet *et al*, *ApJ* 221, 1032) observations, polarimetric observations and some theoretical works on magnetohydrodynamic treatments. But theory of formation of various types of prominences is still far from complete. Even the topology of the magnetic field in quiescent prominences is still under hot discussion. The IAU Colloquium No. 44 on "Physics of Solar Prominences" was held in Oslo (1978).

A. *Quiescent Prominences*

Remarkable progress has been made in understanding the spectra of quiescent prominences; see reviews by Hirayama (*IAU* 44) and Yakovkin and Zeldina (*SP* 45, 319). The kinetic temperature and non-thermal motions of the central part of quiescents is found to be 4500-8500K and 3-9km/s while in the outer part the extended wings of emission profiles indicate values up to 22000K and 15km/s (Landman *et al*, *ApJ* 218, 888). Other authors find the same values as in the central part (Mouradian and Leroy, *SP* 51, 103). Since it seems that these are real differences, effort should be directed to finding the physical conditions producing the difference. The excitation temperature inferred from the intensity ratios of Balmer lines and triplet helium lines was found on the average, to be 7200K and 4460K, respectively (Landman *et al*, *ApJ* 220, 666). It is not known whether the former temperature is the electron temperature or the radiation temperature. Optical depths (∿ten) in the Lyman continuum were obtained from Skylab observations (Orrall and Schmahl, *SP* 50, 365).

A large amount of data of the total intensities of H, He and calcium lines has been accumulated largely by Landman and his collaborators (*SP* 45, 339; *AA* 49, 277; *SP* 50, 383; *AA* 55, 103; *ApJ* 218, 888; *ApJ* 220, 666) and by Engvold (*SP* 56, 87) together with reports at IAU Colloq 44 by Stellmacher and Kubota. These data are generally in good agreement with the theoretical predictions from isobaric and isothermal models by Heasley and Mihalas (*ApJ* 205, 273) and by Heasley and Milkey (*ApJ* 210, 827; *ApJ* 221, 689), where the transfer equations have been solved for H, He, He^+ and Ca^+ but without assuming detailed balance in the Lyman lines in some cases. The result shows that the observed excitation temperature of the helium triplet can now be well reproduced. The controlling factor is the photospheric radiation. The helium singlet to triplet ratios are interpreted as due to the varying effect of optical depth in the helium resonance lines; but an appreciable discrepancy with observations still remains when brighter prominences are treated and it is interpreted as due to the neglect of the filametary structure in the models. In faint prominences, the interpretation of the observations led to a determination of helium to hydrogen number ratio of 10% ± 2.5% (Heasley and Milkey, *ApJ* 221, 689). Hirayama *et al (IAU Colloq* 44) on the other hand found 16% ± 4%. Serious studies of resonance line profiles of HI and MgII have just begun both on the observational side from OSO-8 (Vial *et al*, *SP* submitted) and Skylab and on the theoretical side on the effect of partial frequency redistribution (Milkey *et al*, *IAU* 44).

Energy balance in the cold part of prominences has been treated by Heasley and Mihalas (*ApJ* 205, 273). This work has made considerable progress over earlier theories in that they could, for the first time, reproduce the temperature variations inside the prominence in general agreement with the observations by treating radiative equilibrium with coupled transfer solutions for hydrogen and helium. The lowest temperature in the denser part of a prominence was found to be 4500K, which can be explained as a consequence of the irradiation by a gray body photosphere at an effective temperature of 5800K. In the outer regions 7000-8000K temperatures are expected as a result of the heating from incoming UV radiations. Lerche and Low (*SP* 53, 385) have treated the mechanical equilibrium by means of simplified energy balance relations.

The study of the fine structure and its motion is becoming an important part of prominence physics: Engvold (*SP* 49, 283) found from Hα pictures that the diameter of predominantly vertical threads increases from 400km at lower heights to 1500km at larger heights while Ramsey (*SP* 51, 307) measured the length of individual threads to be about 5000km. From a spectroscopic study the thread diameter was found to be ∿200km (Hirayama *et al* *IAU* 44) in agreement with the Hα pictures. Vertical motions were detected by Engvold with photo-subtracted K-line movies; but this wave-like motion may be only apparent. Maltby (*SP* 46, 149) and others have found no systematic motions of more than a few km/s on the disk. Oscillation with a period of 20-30 minutes was discovered by Landman *et al* (*ApJ* 218, 888) from D_3 observations. Zirin (*IAU* 44) has observed large scale motions in a part of a quiescent and also found that if the height of a quiescent exceeds a certain value, the quiescent will later erupt. On active region filaments Vial *et al* (*IAU* 44) and Mein (*SP* 52, 369) found a larger velocity shift near the edges of filaments than in the middle part. The lifetime of prominence-threads was estimated by Engvold (*SP* 49, 283), Engvold and Malville (*SP* 52, 369) and Maltby (*SP* 46, 149) to be about five to ten minutes while fainter threads seem to persist much longer. Non-thermal motions in the prominence-corona-transition zone was found to be small (2-7km/s) and independent of temperature in contrast to the transition zone in the quiet region (Feldman and Doschek, *ApJL* 216, 119). Although the interpretation of these motions as deviations from a force-free-field (Malville, *SP* 50, 79), Alfvén waves (Ramsey, *SP* 51, 307) or MHD oscillations (Landman *et al*, *ApJ* 218, 888) has been discussed, no comprehensive theory has yet been presented. Mercier and Heyvaerts (*AA* 61, 685) suggested that a general slow motion downward would be expected as the result of photospheric Joule dissipation occurring in separated tiny points.

Important results have emerged on the magnetic field of quiescents: Leroy, Bommier and Sahal-Bréchót (*AA* 54, 811; *AA* 59, 223; *AA* 60, 79; *AA* 64, 237; *IAU* 44) concluded from the magnetic Hanle effect that the active region prominences show the direction of the magnetic field to be parallel to the long axis of prominences, while the polar crown prominences make an angle of 30°-40° to the axis. Other interesting results are that the average field strength (4.5 gauss) might show a systematic variation over the solar cycle and that the changes of the field direction in the polar regions as inferred from polar crowns follow the 22 year cycle when compared with earlier measurements. Further, the general increase of the magnetic field with height has been confirmed (the average value $\sim 1.1 \times 10^{-4}$ gauss/km) but a smaller value was found for young filaments. Martres *et al* (*SP* 46, 137) showed from the magnetograph observation that the stable portions of filaments are often found in regions on the magnetic neutral line where the photospheric radial velocity is small ($\lesssim$ 0.3km/s). Magnetic field strengths measured by Fabry-Perot techniques give 10-30 gauss for semi-active prominences and 80 gauss for an active prominence (Den *et al*, *SP* 52, 35) in accordance with previous results. Koval' (*IKAO* 57, 133), however, has reported values up to 1000 gauss. Further progress is being made at Sacramento Peak-High Altitude Observatories for measuring the full Stokes parameters (Smartt and House, *IAU* 44).

Information on the prominence-corona interface is being deduced from EUV spectra and radio data (see review by Schmahl, *IAU* 44): Orrall and Schmahl (*SP* 50, 365) found the gas pressure to be ~ 0.03 dyne cm^{-2} at $T_e = 9 \times 10^4$K from Skylab data. Young *et al* (*SP* 45, 351) derived 0.5 dyne cm^{-2} from EUV eclipse observation. The latter value, however, is not consistent with the ground based observation by Nikolsky *et al* (*SP* 21, 332) for the same object.

Observations at radio wavelengths have been reported by Butz *et al* (*SP* 55, 161), Straka *et al* (*SP* 45, 131), Apushkinsky and Topchilo (*AZ* 53, 572), Apushkinsky *et al* (*AZ* 53, 1249) and Rao and Kundu (*SP* 55, 161; *IAU* 44) and the averaged spectra require a depression in temperature from ∿5% at 3mm to ∿40% at 6 cm compared with the quiet region (Raoult *et al*, *IAU* 44). Rao and Kundu (l.c.) found from radio spectra that little or no thermal energy is conducted into the main body of a filament. This conclusion is substantiated by EUV observations (Schmahl and also Maltby, *IAU* 44). The presence of the filament cavity is found to be important for the interpretation of radio spectra (Kundu *et al*, *AA* 62, 431).

Anzer (*IAU* 44) has given a critical review of the various magnetic field configurations required to sustain prominences. Although the Kippenhan-Schlüter model seems to satisfy observational data, it is hardly possible to reject the neutral sheet or a combination of K-S and neutral sheet models because of the lack of stability analyses for 3-dimensional models. The MHD equilibrium for the Kippenhahn-Schlüter model has been studied by Lerche and Low (*SP* 53, 385) and Bondal and Pande (*BAC* 27, 263). Uchida and Jockers (U) have calculated quasi-three dimensional MHD solutions in a neutral sheet, where the inclusion of the field component parallel to the X-type neutral sheet may be an essential element to understanding the equilibrium and the mechanisms of formation. Malville (*IAU* 44) stressed the importance of the total electric current in classifying various prominences.

Pikel'ner's siphon mechanism for the formation of the quiescents was examined by Engvold and Jensen (*SP* 52, 37). It is, however, not as yet clear whether the assumed stationary flow exists and whether the coronal heating mechanism does not affect the energy balance. Thermal instability in a current sheet was investigated by Smith and Priest (*SP* 53, 25), who showed that a slow compression process leads to cooling if the condensing magnetic tube is long enough to limit heat conduction. This process might be combined with the Uchida-Jockers (l.c.) configuration. Glencross (*ApJL* 207, 205) considered condensation after heating in sheets sheared due to sub-photospheric motions.

B. *Active Prominences*

Internal motions ($\leq$ 100km/s) in active, surge and eruptive prominences were investigated by Stoyanova (*SDB* 10, 93), Jockers and Engvold (*SP* 44, 429), Engvold *et al* (*SP* 48, 137), Malville (*SP* 50, 395), Ruzdjak and Kleczek (*BAC* 28, 193), Ciurla and Rompolt (*BAC* 28, 217) (see reviews by Zirin and by Malville; *IAU* 44). Engvold *et al* (l.c.) reported that vertical structures become twisted during eruption and indicate a field aligned electric current of $\sim 10^{10}$ amps. Others have stressed the importance of spiraling motions.

Although post-flare loops should be included in the category of flares, they are partly treated here: Kureizumi *et al* (*PASJ* 29, 129) analyzed the visible spectra of a post-flare loop and found T_k = 8000-9000K, n_e = 2-3 x $10^{12}cm^{-2}$, the ionization degree of hydrogen and neutral helium to be $\sim$0.9 and $\sim$0.2, respectively. The total effective thickness of the Balmer-line emitting region is 2-10km. The high amount of mass in post-flare loops was explained as a consequence of the trapping of coronal material by reconnecting field lines following the flare (Kopp and Pneuman, *SP* 50, 85), but the total energy in the soft X-ray flare (10^7K) which later would become Hα post-flare loops does not seem to be well predicted from their theory.

On surges, Maeda *et al* (*PASJ* 30, 533) confirmed the earlier result that the material motion is wholly governed by gravity with different initial velocities. They further found that the direction of motion is the same as the direction of the type III burst and that the root of the surge had n_e = $10^{13}cm^{-3}$. Roy (*SP* 48, 149) found n_e = $10^{12}cm^{-3}$ from another surge. X-ray filtergrams have shown that most of the surges do not heat the corona (Rust *et al*, *SP* 54, 53), though in one case a coronal response was seen as a temperature pulse of short duration (Smith *et al*, *SP* 52, 379). Zirin (*SP* 50, 399) reported the formation of a short lived filament apparently triggered by an occurrence of a surge.

Sakurai (*PASJ* 28, 177) has calculated non-linear, three dimensional motions of magnetic flux tubes after the kink instability sets in using the finite element method. The result shows that the smaller the ratio of the azimuthal to axial magnetic field in the initially straight prominence, the lower the final height attained and the smaller the overall twisting motions, in accordance with observations. This work is a first great step towards full understandings of eruptive prominences, coronal transients and flares and further work should be carried out using more realistic boundary conditions. Other theoretical treatments include the work by Birm *et al* (*SP* 57, 89).

6. SOLAR FLARES AND ENERGETIC PARTICLES
(P. A. Sturrock)

A. *Activities*

The last three years have brought substantial advances in our understanding of solar flares and the clarification of key problems requiring observational and theoretical attack. The OSO-8 Spacecraft was launched on June 21, 1975 and has begun to yield new information on the X-ray emission from flares as reported in the OSO-8 Workshop held in Boulder, Colorado, 7-10 November 1977.

An international symposium on the Solar Atmosphere including a discussion of solar flares was held in London, 14-15 January 1975, and the proceedings have been published (*PTRSL, A* 281, 293). Observational data concerning the flares of August 1972 have been reviewed by Nakagawa (*SSR* 19, 459) and flare models by Somov *et al* (*SPU* 19, 813). Solar flares are also reviewed by Rust in the volume "Solar System Plasma Physics" (Kennel *et al* North-Holland, in press), based on a study commissioned by the National Academy of Sciences of the USA. Another notable compilation of information about solar flares has been provided by Svestka (*Solar Flares*, Reidel). During 1977, a Workshop on Solar Flares, based on Skylab data, was held at High Altitude Observatory, Boulder, Colorado; P. A. Sturrock served as director. The output of this Workshop will be published in 1979 by Colorado Associated University Press. This Workshop involved over 70 scientists, mainly from the US but also from Argentina, Australia, the Federal Republic of Germany, Ireland, Japan, the Netherlands, the United Kingdon, and the USSR. Participants, including not only space scientists but also optical and radio observers and theorists, organized themselves into seven teams, each attacking a specific aspect of the flare problem. The following sections are based in part on the proceedings of this Workshop.

There are several major initiatives for future research concerning solar flares. One of these is the Flare Buildup Study organized under the aegis of SCOSTEP by a steering committee chaired by C. deJager. Another is the Solar Maximum Year initiated by an IAU Commission 10 Working Group chaired by David Rust. The next major space mission to be devoted to solar flares is NASA's Solar Maximum Mission which will fly experiments built by European and United States scientists. This spacecraft, due to be launched in October 1979, will carry the following instrumentation: (a) Coronagraph/Polarimeter; (b) UV Spectrometer and Polarimeter; (c) Soft X-Ray Polychrometer; (d) Hard X-Ray Imaging Spectrometer; (e) whole sun Hard X-Ray Spectrometer; (f) whole sun Gamma Ray Spectrometer; and (g) whole sun Solar Constant Monitoring Package.

B. *Significant Problems*

I. *Preflare State and Activity*

a. *Coronal transients.* One of the major discoveries of the Skylab mission was that of coronal mass-ejection transients, seen as bright, outward moving structures above $2R_{\odot}$ (Gosling *et al*, *SP* 48, 389). For all events, the average speed within the field of view of the experiment (1.75 to 6 solar radii) was 470km s^{-1}. Typically, flare associated events (importance 1 or greater) traveled faster (775km s^{-1}) than events associated with eruptive prominences (330km s^{-1}); no flare-associated event had a speed less than 360km s^{-1}. Metric wavelength type II and IV radio bursts are associated only with events moving faster than about 400km s^{-1}; all but two events moving faster than 500km s^{-1} produced either a type II or IV radio burst or both. In another article (Gosling *et al*, *SP* 40, 439), the group described the direct observation of a flare-related coronal and solar wind disturbance. This showed a direct association of a coronagraph observed mass ejection, which followed a 2B flare, with a large interplanetary shock wave disturbance observed at 1 AU. Estimates of the mass (2.4×10^{16} grams) and energy content (1.1×10^{32} erg) of the coronal disturbance are in reasonably good agreement with estimates of the mass and energy content of the solar wind disturbance at 1 AU.

The nature of coronal transients is not yet completely understood. Photographs certainly indicate that magnetic field is essential to these structures. However, it is not yet clear whether the driving force is a pressure gradient, driving out the material as a bullet from a gun, or whether the stress is magnetic, in which case the stress would appear to be magnetic pressure rather than magnetic tension.

b. *Forerunners*. More recently, it has been found (Jackson and Hildner, *SP* in press) that these coronal transients are typically rimmed by broad halo-like regions termed "forerunners." Of over 20 transients studied, no exceptions have yet been found in the Skylab coronagraph data. Forerunners rimmed transients of all speeds and comprised from 10 to 20 percent of the underlying transient mass. They can extend the transient disturbance to heights greater than $2R_o$ above what had heretofore been considered the transient leading edge. The nature of the "forerunner" phenomenon has yet to be determined: it is not yet clear whether it is to be regarded as an outward flow of mass or a wave propagating through the corona.

c. *Preflare magnetic field geometry*. It is generally agreed that the energy released in a solar flare is the excess magnetic energy released when a stressed magnetic field relaxes to an unstressed or less stressed state with the same boundary conditions (normal component of magnetic field) at the photosphere. In the case that the plasma stresses (pressure and gravitational force) are negligible, the pre-flare field configuration will have the form of a force-free field, for which $j \times B = 0$, with the possible addition of discrete current sheets. There is therefore considerable interest in calculating force-free field configurations as well as attempting to study the stability of such configurations.

In the case of "linear" force-free fields, for which κ = constant where $j = \kappa B$, the relevant equation is a Helmholtz equation. Chiu and Hilton (*ApJ* 212, 873) and Barbosa (*SP* 56, 55) have computed solutions of this equation using Green's function methods.

The more difficult "nonlinear" force-free field problem, for which κ is not constant, has been studied by Low (*ApJ* 212, 234), and Birn *et al* (*SP* in press), using an iterative technique.

The above procedures for computing force-free fields all involve simplifying assumptions. It is important to develop a technique, for instance the development of the relaxation procedure (Barnes *et al*, *ApJ* 174, 659), with more rapid convergence properties, which could be used as a general technique for computing field configurations in active regions.

II. *Magnetic Energy Release*

It has been recognized for some time that the free energy of a stressed magnetic field can be released if "reconnection" can occur. This topic has been reviewed by Priest (*SP* 47, 41). Heyvaerts *et al* (*ApJ* 216, 123) have investigated reconnection in the context of an "emerging flux model," according to which, as more and more flux emerges, reconnection occurs, to produce some preflare heating. If the current density in the sheet should exceed a critical value, the impulsive phase of the flare is triggered.

A modification of the emerging flux model has been proposed by Uchida and Sakurai (*SP* 51, 413). They point out that a large-scale configuration including a current sheet may be unstable against the MHD interchange instability. This can develop explosively, leading to fine-scale structure which permits rapid reconnection, which they identify with the explosive phase of a flare.

In contrast to the previous model, which tends to produce a "magnetic arcade," a model has been analyzed by Spicer (*SP* 53, 305) which involves only a single twisted flux tube. The principal flare instability in this model is the resistive kink instability which plays the role of thermalizing some of the field energy in the arch and generating x-type neutral points which may be sites for particle acceleration.

Since the plasma processes involved in the release of magnetic energy are so complicated, it is important that this phenomenon should, if possible, be studied experimentally in the laboratory. Such experiments have been done by Syrovatskii (*PLI* 74, 1) and by Baum and Bratenahl (*SP* 47, 331). The difficulty is in setting up experiments for which the magnetic Reynold's number is much larger than unity. It is important that further experiments of this type be carried out, since they have much to teach us about the plasma processes involved in the release of magnetic energy in solar flares.

III. *Energetic Particles*

a. *Abundances.* The properties of energetic particles in flares can be inferred from observations made in space. Measurements made with the IMP 7 and 8 spacecraft (Armstrong *et al*, *SP* 49, 395) have shown that the Fe/O ratio is highly variable for high-energy particles produced by a solar flare. In addition, the charged states of energetic iron were measured directly for the first time in a solar particle event of 1974 May 14-15 (Gloeckler *et al*, *ApJ L* 209, 93). The iron was not fully stripped but has a mean ionization state if 11.6, remarkably similar to the mean ionization state of iron in the quiet solar wind. The ionization states of carbon and oxygen also have been measured for a number of flare events (Sciambi, *ApJ* 214, 316). The ionization states were found to be surprisingly constant, despite great variations in other event parameters and were similar to the respective ionization states in the solar wind. This new information strongly suggests that acceleration of ions to high energies occurs in the corona sufficiently rapidly that the ionization states are not changed from those characteristic of the normal corona.

A remarkable property of particles accelerated during flares is the great variation in the abundance of ^{3}He (Anglin, *ApJ* 198, 733). In the range 1-20 meV/neucleon, the ^{3}He/^{4}He ratio varies from about 10^{-2} to more than 1. However, the enhancement in ^{3}He is not accompanied by a similar enhancement in either ^{2}H or ^{3}H. These large enhancements usually accompany small proton events, and they are always associated with enhancements of Fe nuclei (although the converse is not true) (Anglin *et al*, *15 ICCR* 5, 43). The ^{3}He problem has been addressed by Colgate *et al* (*ApJ* 213, 849). They attribute the production of ^{3}He to spallation of ^{4}He by protons and explain the comparative absence of ^{2}H and ^{3}H as the reslut of thermonuclear reactions in filaments of very hot (200 keV), very dense (3×10^{15}cm^{-3})plasma. Fisk (*ApJ* in press) has proposed that the enhancement of ^{3}He is due to selective acceleration by electrostatic ion cyclotron waves in a plasma in which ^{4}He/^{1}H $\geq$ 0.2.

b. *Acceleration.* The data concerning charged states and abundances of ions accelerated in solar flares provides even more stringent boundary conditions on any successful theory of acceleration. The general requirements and models have been thoroughly reviewed by Svestka (*Solar Flares*, Reidel, 1976). The current view appears to be that there are two distinct stages of acceleration. The first stage preferentially accelerates electrons to mildly relativistic energies (Hoyng, *AA* 55, 23; 55 31). Smith (*ApJ* 212, 891; 217, 644) attributes this stage to strong plasma turbulence such as might arise in the region of field-line reconnection. A method for distinguishing between alternate mechanisms of first stage acceleration has been proposed by Hoyng *et al (SP* 58, 139).

Second-stage acceleration, which preferentially accelerates ions, is usually attributed to shock waves on the basis of their close association with type II radio bursts. A detailed theory for the acceleration of particles in MHD turbulence, such as might develop behind a collision-free shock wave, had been developed by La-Combe (*AA* 54, 1).

c. *Reverse current.* Colgate *et al (ApJ* 213, 849) and others have drawn attention to a problem related to impulsive X-ray bursts, which are sometimes attributed to streams of electrons flowing from the corona down to the chromosphere (Hoyng *et al, SP* 48, 197). If taken at face value, the currents associated with these electron fluxes are so high that they would produce very strong magnetic fields, the energy of which would greatly exceed the total energy of the flare. This problem has been addressed by Knight *et al (ApJ* 218, 306) and by Hoyng *et al (SP* 48, 197). They find that sudden acceleration of electrons, such as would produce an electron stream, will lead to space-charge separation developing an electric field which will then drive a "reverse current," neutralizing the primary current. This mechanism may have implications for the sudden heating of the corona during the impulsive phase of a solar flare. The propagation of an electron beam in an ionized plasma has also been re-examined by Hoyng and Melrose (*ApJ* 218, 866).

IV. *Impulsive Phase*

a. *X-ray spectra.* Space scientists tend to analyze X-ray spectra in terms of simple models. The simplest are power-law spectra and thermal bremsstrahlung. The former is usually termed a "nonthermal" spectrum and may be produced by a beam of electrons with power-law spectrum. Craig *et al (N* 264, 340) have argued that observational data typically do not allow one to discriminate between these possibilities. However, Gabriel (*N* 267, 410) expressed a dissenting opinion. It seems clear that future observational data should be made with as high accuracy as possible over as wide a range of energy as possible if one is to attempt to discriminate between thermal and non-thermal models of the X-ray source (e.g. Boehme *et al, SP* 53, 139).

b. *Location and possible beaming of X-ray source.* A beam model would tend to produce X-ray emission low in the sun's atmosphere, where collisions become important. On the other hand, thermal emission must occur in a low-density region, high above the photosphere. Hence the nature of the emitting mechanism is coupled with the question of the height at which hard X-ray emission occurs. Since hard X-ray detectors have, to date, had no angular resolution, we have no direct information on this point. However, the hard X-ray detector on SMM should provide valuable information.

Roy *et al (SP* 40, 165) have analyzed a number of events which were attributed to flares occurring over the limb. Their data extended over the energy range 5-100keV. Their conclusion was that disk and over-the-limb bursts were similar in spectrum and they inferred that hard X-ray emission is produced by regions extending to great heights.

If hard X-rays are produced by a high-energy beam of electrons propagating towards the photosphere, the X-rays also would tend to be beamed. Brown *et al (AA* 41, 395) and Bai *et al (ApJ* 219, 705) have argued that, in this case, one could obtain information about the height of the source and the degree of beaming by observing the "albedo patch," i.e. the area of the sun's surface from which X-rays are scattered.

It has also been proposed that evidence for or against beaming could be obtained by studying the longitude distribution of impulsive X-ray bursts. Savenko *et al (SP*

50, 447), analyzing data from the Prognoz and Explorer satellites, found that the X-ray burst frequency (normalized to the Hα flare frequency) increases towards the solar limb. This would argue in favor of a beam model. On the other hand, Datlowe *et al* (*ApJ* 212, 561), analyzing 148 bursts observed on the OSO-7 satellites, found no center-to-limb variation. They concluded that their result rules out downward-streaming thick-target models for hard X-ray production. On the other hand, Bai *et al* (*ApJ* 219, 705) and Langer *et al* (*ApJ* 215, 666) have made detailed calculations of the angular dependence of X-rays which would be produced from a beaming source, taking into account scattering of X-rays by the sun's atmosphere, and conclude that no center-to-limb effect should be apparent.

It is clear that new data is required to help distinguish between thermal and non-thermal models for hard X-ray production during the impulsive phase of flares. Hopefully, data to be obtained from the SMM spacecraft will provide valuable new information relevant to this question.

c. *Time variation of hard X-ray emission.* The hard X-ray emission may fluctuate rapidly during the impulsive phase. Hurley *et al* (*SP* 52, 107) observed the solar flare on 1974 July 4 in hard X-rays using a balloon-borne detector. When analyzed with a time resolution of 100ms, four 2-second long spikes were observed which were correlated with decimetric emission. Crannell *et al* (*ApJ* 223, 620) have studied a number of simple impulsive solar flares in the OSO-5 hard X-ray data and in microwave and meter-wave radio data. They find that the emission typically has the form of a simple, fairly symmetric, spike with a duration of a fraction of a minute. It is possible that the X-ray emission from large flares comprises a sequence of such elementary spikes. These data suggest that the basic time constant for electron acceleration could be only a few seconds. (Type III radio burst data suggests that it may be less than 0.1 second.)

It has been known for some time (Svestka, *Solar Flares*, Reidel) that microwave emission from flares tends to occur in pulsations which may be quasi-periodic. There has been a hint that hard X-ray radiation from flares also contain periodic components. Lipa (*SP* 57, 191) has analyzed the hard X-ray emission from 28 large solar events and found periodicity in 26 of them with periods in the range 10-100 seconds. It is clearly an important problem to try to understand the origin and significance of these periodicities. The data analysis of Crannell *et al* and of Lipa are both suggestive of betatron-type acceleration occurring in an "electron trap," as suggested by Brown *et al* (*ApJ* 200, 734; *SP* 49, 329), but it is likely that alternative models will be proposed in the future.

V. *Energy and Mass Flow*

a. *Lower atmosphere.* Although the primary energy release is believed to be the conversion of magnetic energy into high-energy particles and/or high-temperature plasma, much of the flare energy leaves the sun in the form of electromagnetic radiation, and a large part of this energy is in the form of lines characteristic of the "lower atmosphere," i.e. the transition region and chromosphere. Hence an important problem concerning solar flares is to identify the mechanism whereby energy propagates from coronal heights to chromospheric heights where it is absorbed and radiated.

Since Hα kernels brighten in coincidence with hard X-ray emission, it appears possible that the chromosphere is heated by electron bombardment, but Roy (*SP* 48, 265), examining data for the flare of 1972 August 2, concludes that chromospheric heating cannot be attributed to the hard X-ray producing electrons.

Orrall and Zirker (*ApJ* 208, 618) and Lin and Hudson (*SP* 50, 153) have considered the possibility that the lower atmosphere may be heated by beams of non-thermal protons. It appears that the least ambiguous evidence for the presence of such non-thermal protons would be the asymmetry in the wings of Lyα predicted by Orrall and

Zirker. Canfield and Cook (*ApJ* in press) have searched for this effect without success.

Somov (*SP* 42, 235), Henoux and Nakagawa (*AA* 57, 105) and Machado (*SP* in press) have examined theoretically the possible heating of the lower atmosphere by soft X-ray flare radiation from the corona. It appears that soft X-ray heating probably plays a significant, though not dominant, role in the chromospheric flare energy budget.

Machado and Linsky (*SP* 42, 395) have taken a different approach and constructed an array of photosphere-chromosphere models designed to fit the CaII lines in a variety of flares. This model implies that the temperature may increase by about 100K during a flare in the region of temperature minimum. Machado and Noyes (*SP* in press) have also examined the SiI data of Cook and Brueckner (*ApJ* in press) and this again leads to the inference of a temperature increase near the temperature minimum. Machado and his colleagues find it difficult to assign a mechanism to heating in this region.

b. *Thermal X-ray plasma.* From the impulsive phase on, flares typically emit soft X-rays which may be attributed to bremsstrahlung from thermal plasmas at temperatures of order 10^7K. This "thermal X-ray plasma" may also provide energy to the lower atmosphere by heat transfer or by X-rays. Moore and Datlowe (*SP* 43, 189) have analyzed a number of flares using Hα and OSO-7 X-ray data. Of the possible cooling mechanisms of the thermal plasma, they find that heat conduction tends to be more important than radiation. However, Underwood *et al* (*ApJ* 224, 1017) have examined EUV and soft X-ray data of a highly compact flare in which they find that the coronal density is sufficiently high ($>10^{11}$cm^{-3}) that radiation may dominate the cooling.

The density of the thermal X-ray plasma is much higher than that of the corona in active regions. It is now generally believed that this mass "evaporates" from the chromosphere as a result of an energy flux from above. Kostyuk and Pikel'ner (*SA* 18, 590) and Kostyuk (*SA* 19, 458) have modeled the interaction of the chromosphere, transition region and corona in a hydrodynamic framework following impulsive heating by an electron beam. Craig *et al* (*SP* 50, 133) have also examined the response of a flare filament to a localized injection of energy. Antiochos *et al* (*ApJ* 220, 1137) have examined the evolution of the plasma in a closed flux tube when the primary energy release is over. They find that initially the hot plasma cools by heat conduction but this heat conduction drives a mass flux from the chromosphere into the corona. When the density is sufficiently high, radiation losses will become dominant, evaporation will cease, and mass will eventually flow back to the chromosphere.

An important question concerning large flares is whether the primary energy release is limited to the impulsive phase. Pallavacini *et al* (*SP* 45, 411) and Vorpahl (*SP* 47, 147), studying the flares of 1973 June 15 and 1973 September 5, respectively, find evidence for continued heating during the "decay" phase of these flares. If this conclusion proves correct, it will be important to understand the mechanism for this energy release and why it is slow compared with energy release during the impulsive phase.

7. RADIO PHYSICS
(S. F. Smerd and D. J. McLean)

A. *General*

A detailed summary of the observations, interpretations and theories of 'Solar Noise Storms' has been presented in book form by Elgaroy (Pergamon Press, Oxford, 1977); particular attention is paid to the type I, i.e. the high-frequency portion, of storm bursts and continua. The author concludes that before solar storms can be properly understood there is need: (i) to elucidate the relations between the metric (the type I) and the dekametric (the type III) components of storms; (ii) to obtain much more precise quantitative information, for instance, on storm spectra and source positions [the author notes "It is a sad state of affairs that most solar interferometers operate in one dimension and with inadequate resolution." The effects of this inadequacy spread, of course, right across solar radio astronomy.]; and (iii) to know much more about the complex physical conditions in the storm centres so as to allow meaningful applications of plasma physics to the generation of storm radiation. One such physical quantity is the magnetic field. However, because of the current uncertainties in a profusion of type I theories, Dulk and McLean (*SP* 57, 279) excluded type I bursts (as also fundamental type III bursts and moving type IV bursts) from the observational evidence which they used in their derivation of average 'Coronal Magnetic Fields' in active centres where solar bursts originate. These authors conclude that "observation of harmonic type III polarization promises to be one of the best ways of estimating coronal magnetic field strengths."; they suggest that the empirical formula

$$B = 0.5\ (R/R_o - 1)^{-1.5}\ G, \qquad (1.02 \lesssim R/R_o \lesssim 10)$$

fits all the data presented to about a factor of three, *provided the fields are associated with electron densities only twice those of the equatorial corona at sunspot minimum.* These densities are lower than any other derived from solar-burst observations; [see section on type III bursts].

Two specifically solar radio astronomy reviews were presented by Rosenberg (*PTRSL* 281, 461) and by Smerd (*ISSTP* ed. D. J. Williams, I, 193). Rosenberg pays special attention to a variety of fine structures in such type IV bursts as occur near the plasma level. The suggested interpretations usually involve non-linear coupling between plasma waves and other wave modes, for instance whistlers or Bernstein waves. Rosenberg also examines the radio evidence for clues on acceleration processes in flares. He concludes that during the flash phase of a flare (but also in type I sources) low-frequency plasma-turbulence is generated by m.h.d. instabilities, for instance the tearing-mode instability; the growth rate is thought to be large enough to accelerate electrons to (10-100) keV in $\sim$1 s. Further acceleration to higher energies is again attributed to Fermi acceleration.

Smerd concentrates on macroscopic radio structures and the corresponding solar disturbances. The most notable recent addition to slowly-varying structures are 'coronal holes'; co-operative studies (e.g. Dulk *et al* *SP* 49, 369) show that, on current solar-atmosphere modelling, the EUV-observations yield much higher electron densities than do the radio observations. An 'over abundance' of heavy elements might have to be invoked, but also explained. Smerd uses dynamic burst spectra, source positions and polarizations--in particular as observed in 2 dimensions at 43, 80 and 160 MHz with the Culgoora radioheliograph--to identify possible disturbances and radiation mechanisms. Spectra and heliograms confirm three distinct disturbance speeds, namely: $\sim 10^5$km/s for electron streams; $\sim 10^3$km/s for m.h.d. and at times shock waves; $\sim 10^2$km/s for clouds of magneto-plasma. We note that the moving disturbances can result in moving or in stationary burst sources; and that the observed circular polarization does not help to distinguish between harmonic plasma radiation and synchrotron radiation as the emission mechanism of certain stationary type IV continuum bursts.

On a far less specialized scale and seen through the eyes of a non-radio solar physicist, there is a substantial survey of solar radio astronomy included in Svestka's book 'Solar Flares', although it includes only a small part of the evidence represented.

Another, but highly-specialized, survey of USSR and of Australian radio-astronomy developments was reported at the USSR/Australia Solar Symposium at Pulkovo in September 1976; the invited papers have been published in Russian in *Izv. VUZ Radiofiz.* 20, 9 and in English in *Radiophys. Quantum Electron.* 20, 9.

B. *Type I Storms and Bursts*

The relationship of type I storms to changes of the photospheric magnetic field structures are stressed by Yurovskaya (*IKAO* 53, 139), Zanelli and Zlobec (*SP* 53, 497) and Heyvaerts *et al* (*AA* 66, 81).

A number of authors have studied fine structure in type I storms, including Dröge (*AA* 57, 285), Chernov (*AZ* 54, 1081) and Kuijpers and Slottje (*SP* 46, 247). A fairly detailed emission theory has been developed by Mangeney and Veltri (*AA* 47, 165; *AA* 47, 181) (this is well covered in Elgaroy's book). A review on continuum storms by Sakurai (*ASS* 42, 349) also includes a short theoretical section. Simultaneous observations from a space probe and earth indicate beaming of type I emission; Bougeret and Steinberg (*AA* 61, 777) have devised a fibrous model of coronal fine structure to explain these observations.

At longer wavelengths, de la Noë and Gergely (*SP* 55, 195) have found displacements between the sources of continuum, type I, type III and drift pairs.

C. *Type II Bursts*

Interesting observations of type II bursts due to "behind-the-limb" flares were reported by Gergely and Kundu (*SP* 48, 357) and McLean and Nelson (*R* 20, 359); both show similar features although the interpretations offered are different. The structure of type II burst sources and its relation to loops visible in soft X-rays is discussed by Stewart (*PASA* 3, 157).

Zaitsev and Ledenev (*SAL* 2, 172) and Ledenev (*SAL* 3, 144) have developed a theory for type II bursts which does not assume perpendicular shocks. It is interesting to contrast this theory for which emission occurs ahead of the shock with the conclusion drawn by Davis and Feynman (*JGR* 82, 4699) and Chertok and Fomichev (*PLSC* 24, 459), both of whom re-interpreted the IMP-6 observations of a type II burst at kilometric wavelengths, and concluded that the radiation is emitted from behind the interplanetary shock front. A different theoretical approach by Lampe and Papadopoulos (*ApJ* 212, 886) concentrates on the acceleration of electrons in the wake of a shock front.

D. *Type III, Type V and Inverted-U Bursts*

The Proceedings of the Workshop on Mechanisms for Solar Type III Radio Bursts held in May 1975 at Space Sciences Laboratory, Berkeley, California, were published in *Solar Phys.* 46, 433, edited by Lin. These Proceedings contain a number of observational and theoretical papers on type III bursts as well as abstracts of other papers which have since been published in full.

The theory of type III burst emission is now supported by the detection *in situ* from Helios I of Langmuir waves in the plasma at the time of a type III burst (Gurnett and Anderson, *SC* 194, 1159; *JGR* 82, 632).

There is disagreement about the electron density law appropriate to type III bursts; to interpret the observed source heights, Stewart (*SP* 50, 437) requires that type III sources lie in structures 8 to 10 times more dense than the quiet corona whereas Hoang *et al* (*AA* 56, 283) require that the sources lie in low-density parts of the corona, to explain the results of the STEREO experiment. Leblanc and de la Noë (*SP* 52, 133) compared positions and K-corona data and concluded that type

III sources lie in low-density structures. [We note however that this last paper was criticized by Mercier (*SP* 57, 423)]. Dulk and McLean (*SP* 57, 279) found the best agreement between different estimates of the coronal magnetic fields by adopting a low-density model, more nearly consistent with those deduced from optical observations.

Further advances in the study of burst trajectories in interplanetary space were made by Baumbach *et al* (*SP* 48, 361) and Weber *et al* (*SP* 54, 431). Both experiments used simultaneous data from two spacecraft to study trajectories using fewer assumptions than in previous work.

The relative simplicity of the type III phenomenon has continued to attract the attention of the theoreticians; Bespalov (*FP* 3, 1118; *SJPP* 3, 619), Magelssen and Smith (*SP* 55, 211), and Esconde and Genouillac (*AA* 68, 405) have studied propagation of electron streams; Nicholson *et al* (*ApJ* 223, 605) and Robinson (*ApJ* 222, 696) stressed the nature of the Langmuir wave spectrum or its evolution, while Melrose *et al* (*AA* 66, 315) predicted the polarization to be expected for fundamental and harmonic components: this theory provides the key to one of the most promising indicators of coronal magnetic field strengths. Smith (*ApJL* 216, 53) discussed the theoretical consequences of the *in situ* detection of Langmuir waves quoted above. Pilipp and Benz (*AA* 56, 39) have explained type V bursts as resulting from resonant scattering of the electrons generating a type III burst, whereas Robinson (*SP* 56, 405) attributes type V bursts to long electron streams.

Suzuki (*SP* 57, 415) measured opposite polarization for the two legs of some U-burst in the range 25 to 220 MHz, but at 237 MHz Benz *et al* (*AA* 56, 123) find the same polarization for the two legs. Suzuki points out that the discrepancy may be due to different propagation effects for different observing frequencies.

E. *Type IV Bursts*

Nelson (*PASA* 3, 159) has recorded two-dimensional images at three frequencies of a moving type IV and finds new difficulties of interpretation. Kai (*SP* 56, 417) has found an unexpected relation between the observed polarization of moving type IV bursts and the polarity of the photospheric magnetic field. A metre-wave stationary type IV burst was singled out for intensive study by Dulk *et al* (*SP* 49, 369) because of the coronagraph observations of an associated white-light transient. A fairly detailed model, including magnetic field strengths, emerged from this study.

F. *Microwave*

A variety of instruments have been used to obtain high-resolution data on the emission from active regions and burst sources: the new RATAN-600 (Parijskij *et al*, *SA* 20, 577), the Stanford interferometer (Falchi *et al*, *SP* 48, 59; Falchi *et al*, *SP* 56, 355; Felli *et al*, *SP* 51, 65), Westerbork (Kundu *et al*, *ApJ* 213, 278; Allissandrakis and Kundu, *ApJ* 222, 342; Chiuderi *et al*, *AA* 61, 79), the German 100-m dish (Kundu *et al*, *AA* 62, 431) and the US NRAO interferometer (Kundu *et al*, *SP* 50, 429). Some interesting features such as very small regions with high brightness temperatures are revealed by these studies.

An interesting aspect of some microwave bursts is their association with hard X-ray bursts (Neidig, *SP* 57, 385; Mätzler *et al*, *ApJ* 223, 1058); Mätzler (*AA* 70, 181) has proposed a thermal model in which the temperature is 10^8K, which contrasts with the earlier, continuous injection model of Mätzler (*SP* 49, 117).

Also a thermal model for impulsive microwave bursts has been described by Uralov and Nefed'ev (*SA* 29, 438; *SA* 20, 590; *SA* 21, 749). In a different theoretical approach to the explanation of hard X-ray and impulsive microwave bursts, Böhme *et al* (*SP* 53, 139) use a two-component model. Neidig (*SP* 54, 165) has used optical estimates of magnetic field to test Takakura's earlier model, and found them consistent.

G. *Quiet Sun*

The quiet Sun is still the object of intensive study over a very wide range of wavelengths from dekametric to sub-millimetric. At metre and decimetre wavelengths, renewed interest arises from the recognition of coronal holes (Dulk *et al*, *SP* 52, 349; Smerd, *RQE* 20, 917; Trottet and Lantos, *AA* 70, 245; Covington, *SP* 54, 393). In the discussion of the results a discrepancy between the radio and EUV results arises, which has not yet been resolved. The arguments involved rely heavily on accurate measurements of the quiet Sun flux such as given by Kundu *et al* (*SP* 53, 489) and Erickson *et al* (*SP* 54, 57). Chambe (*AA* 70, 255) has drawn attention to the instability of some of the mathematical processes involved. At centimetre and millimetre wavelengths interest is in active regions (Righini-Cohen and Simon, *ApJ* 217, 999; see also Microwave Section), filament cavities (Kundu and Lantos, *SP* 52, 393; Rao and Kundu, *SP* 55, 161) and the centre-to-limb distribution with its information about chromospheric structure (Kislyakov and Kuznetsova, *SDB* 8, 71; Bachurin *et al*, *IKAO* 58, 40; Kuseki and Swanson, *SP* 48, 41; Labrum *et al*, *SP* in press; Labrum, *PASA* 3, #3).

8. CORONA
(J.-L. Leroy)

A. *The Slowly Variable and Active Region Corona*
I. *Coronal active regions*

a. *General structure.* The determination of accurate values of local physical parameters from observed quantities integrated along the line-of-sight is a major problem in the case of coronal active regions where the inhomogeneous structure is striking (Kahler and Krieger, *SP* 56, 351); uncertainties about coronal abundances make the problem harder (Davis *et al*, *SP* 45, 393; Withbroe *SP* 45, 301). Results are much safer if one has some knowledge of the geometrical structure of the condensation under study provided for instance by good quality X-ray images or EUV spectroheliograms (Jordan *IAU* 68, 1974; Pye *et al AA* 65, 123). In the general case, the derivation of the line-of-sight X-ray measure can lead to erroneous interpretations if there are noticeable line-of-sight temperature gradients (Underwood and McKenzie, *SP* 53, 417).

Many determinations of N_e and T_e in coronal condensations have been published (see for instance Bromboscz and Jakimiec *RZ* 51, 528; Feldman *et al*, *ApJ* 219, 304; Gelfrejkh and Korzhavin *RZ* 51, 509; Gerassimenko *SP* 48, 121). The general temperature structure of denser regions may be variable: in a condensation observed at the 1970 March 7 eclipse Gabriel and Jordan (*MN* 173, 397) have found that the density increased with increasing temperatures and that the low, central parts of the active region contained more material at high temperature than the neighbouring loop system however, even the central core must have a loop structure (Kreiger, 1977, *OSO-8 Workshop*). On the other hand, Kurokawa (*SP* 43, 385) has described a condensation observed during the 1973 June 30 eclipse as containing a cool and dense core ($T = 10^6$K; $N = 6 \times 10^9 cm^{-3}$) at the lower part of the condensation. An attempt to provide a four parameter model of active regions has been published by Landini *et al* (*SP* 44, 69): the parameters are the maximum temperature ($2.2 < T_M < 2.8 \times 10^6$K), the maximum density ($2 < N_M < 9 \times 10^9 cm^{-3}$), the width of the pressure distribution σ_P and the width of the temperature distribution $\sigma_T = \alpha^{1/2}\sigma_P$ ($2 < \alpha < 7$). The maximum temperature may be larger ($\sim 5 \times 10^6$K) according to Pye *et al* (*AA*, 65, 123). The height structure of X-ray condensations has been studied by Livshits and Bulatov *SA* 21, 1977); the height of the high temperature core can also be estimated from radio centimetric observations (Donati-Falchi *et al*, *SP* 56, 335) and has been found in the range 10000-40000Km.

b. *Energy balance.* Comprehensive reviews on this topic have been given by Withbroe (*IAU* 36, 263) and by Orrall and Kopp (*IAU* 36, 457). The need for continual energy deposition in active regions loops has been proved by Gerassimenko *et al* (*SP* 57, 103) and Levine and Withbroe (*SP* 51, 83). The evaluation of various losses shows that generally radiative losses are larger than conductive ones (up to 15 times according to Neupert *et al*, *SP* 43, 359). Solar wind losses are generally estimated to be unimportant. Teske and Mayfield (*ApJL* 210, 153) have studied the thermal energy content of a solar active region and shown that it was not in energy equilibrium at the time of observation; the time rate of change of thermal energy content was comparable to the rate of energy loss by radiation. In another global investigation of the energy balance in an active region, Evans *et al* (*SP*, in press) find that the region under study did not radiate and almost certainly did not receive more than a very small part of the "missing flux" of the spot group. Beckers and Schneeberger (*ApJ* 215, 356) reached a similar conclusion (the corona above a sunspot receives an Alfvén wave flux of less than 0.08% of the missing flux) after an analysis of the profile of the 5303Å line. In nearly all studies it is assumed that the energy deposition is of mechanical origin (Landini and Monsignori Fossi, *AA* 42, 213); Compagno (*OMOAA* 104, 74) and Jordan (*RSL* 281, 391) have provided an evaluation of the rate of change of the mechanical energy deposition as a function of height. However,

Rosner *et al (ApJ* 222, 317) have proposed recently a process of coronal plasma heating by anomalous current dissipation which takes into account the inhomogeneous structure of the lower corona and its links with fine scale magnetic fields (Krieger *OSO-8 Workshop*).

II. *Streamers*

a. *Three-dimensional structure.* The model of a streamer as a coronal blade has been advocated again by Dollfus and Martres (*SP* 53, 449) in an analysis of K-corona measurements. On the other hand, after a study of a streamer observed over 2 months from Skylab, Poland (*SP* 57, 141) finds that the most probable structure is that of several radial cylinders of enhanced density extended in longitude and overlying both filaments and active regions.

b. *Dynamical models.* Weber (*SP* 53, 59) has developed a dynamical model of the interaction between a magnetic field emerging from the photosphere and the expanding solar atmosphere which leads to the formation of the streamer. One advantage of this model is that it does not require an extraordinary low electrical conductivity to obtain a reasonable thickness. The resulting structures are to a great extent nearly identical to those found in stationary models but there is a transverse field considerably larger than in stationary models. It will be noted that the initial magnetic configuration in this model is that of a bipolar region and it is expected that the model would fit better the so-called helmet streamers than the coronal blades which have been considered above. The long term studies of Poland (*SP* 57, 141), which depict the slow variation and mass changes of a typical coronal streamer provides us with good data for further modeling of streamers which appear to be persistent rather than stationary according to Poland.

B. *Coronal Magnetic Fields*

I. *Measurements*

a. *Optical methods.* High quality measurements of the polarization of coronal forbidden lines are now available (Arnaud, *ROL* 12, 137; Querfeld, *ROL* 12, 109) although puzzling discrepancies with eclipse measurements are still present (Sykora *et al*, *BAC* 28, 1; Picat *et al*, *AA* in press). Thus it should be possible to get good information about the orientation of the magnetic field in the lower corona; however, the difficult problem of inversion of the line-of-sight integrated measure of polarization is not yet solved.

II. *The computed coronal field*

a. *Improvement of numerical models.* Previous results have been discussed in several papers (Newkirk, *Proc. Symp. SSIMT*, GSFC, 1976; Schatten *RGSP* 13, 589; Vasilev and Kutakov, *AV* 10, 138; Roy *JRASC* 70, 292). One of the limitations of the current-free calculations has been insufficient spatial resolution by comparison with small scale arch structures, which are visible on X-ray images, in global calculations for the whole Sun. A first improvement has been to introduce as primary data the measurements obtained at Kitt Peak with a 2.5" or 1" resolution. Next an improvement of the spherical harmonics method, with a maximum index of 90 (instead of 9) has been achieved by Altschuler *et al (SP* 51, 345) and allows a spatial resolution of about 2° heliographic. A similar result has been obtained by Adams and Pneuman (*SP* 46, 185) by a different method (fixed mesh solution to Laplace's equation using line-of-sight boundary conditions). In both cases the loop structures in the lower corona become much more conspicuous than in previous computations. Altshculer *et al* however stress that the spherical harmonic expansion technique provides essential information about the contribution of different scales of magnetic structures. Still a different approach has been chosen by Sakurai and Uchida (*SP* 52, 397) to compute the coronal field above active regions: the photospheric field is represented by an aggregate of solenoidal current loops located below the photosphere at the position of sunspots. The authors expect that discrepancies between computed lines of force and observed EUV structures will denote the presence of coronal current sheets about which they make some computations (energy stress storage). The next approximation after the current free case is that of force free fields which up to now has been treated essentially in the case of constant α ($\nabla xB = \alpha B$) (Levine *SP* 44, 365 ; Barbosa *AA* 62, 267; Barbosa *SP* 56, 55; Nakagawa *et al*, *AA* in press). Finally the case of non-force-free solar magnetic fields in magnetohydrostatic equilibrium has been considered in an axisymmetric situation by Comfort *et al* (*ApJ* in press).

b. *Comparison with XUV structures.* While interesting comparisons can be made between coronal structures as seen on XUV images and the photospheric magnetic field (Teuber *et al*, *SP* 53, 97) the lack of measured values of the coronal magnetic field has led many authors to compare coronal structures and computed lines of force. First comparisons have shown an overall correspondence between X-ray structures and the computed potential field (Poletto *et al*, *SP* 44, 83; Sakurai and Uchida, *SP* 52, 397; Poletto *et al*, *OMOAA* 104, 175) although significant discrepancies could be found in active regions (Krieger *et al*, *SP* 47, 117). McGuire *et al* (*SP* 52, 91) and Nakagawa *et al* (*AA* in press) have found that constant α force-free lines of force gave a better agreement than potential field lines in the case of some stable arches visible either on X-ray images or on K-corona pictures. Levine (*SP* 46, 159) has found it possible to fit nearby loops above an active region by different sets of force-free computed field lines corresponding to opposite values of α, which the author interprets as being due to currents flowing in opposite directions. However, some doubt about the validity of such a procedure has been raised by Schmidt (*IAU* 36, 321) and it is not sure that the analysis of discrepancies between observed coronal structures and computed lines of force will be sure enough to provide quantitative data on coronal currents.

III. *Evolution of magnetic structures during the life of active regions*

a. *Arch structures inside active regions.* Howard and Svestka (*SP* 54, 65) have followed the evolution of the magnetic structures (identified with X-ray arches) of a coronal complex of activity. The basic components of the activity complex were permanently interconnected but the visibility of individual loops was greatly variable and typically shorter than one day. Only loops connecting active regions to the remnants of old fields could be seen with about the same shape for many days. Loop brightenings often occurred at the time of flare occurrence, triggered by the emergence of new magnetic flux, but it was also possible to observe growing and brightening of X-ray loops in old regions without flares. Although loop brightening seems, according to Howard and Svestka, to be related to a variation of the photospheric magnetic field at their foot point, it is wourth remembering (Krieger *et al*, *SP* 47, 117) that temperature and perhaps density variations are likely to explain much of the visibility change of X-ray structures.

b. *Large-scale structures.* The final decay of the complex of activity studied by Howard and Svestka (*SP* 54, 65) was accompanied by the penetration of a coronal hole into the region where the complex existed before. However, open field configurations can be found there even during the life of active regions: according to Svestka *et al* (*SP* 55, 359) X-ray observations confirm the prediction of potential field computations (Levine *et al*, *ApJ* 215, 636) that some of the dark gaps seen between interconnecting loops and inner cores of active regions may be the loci of open fields. That would occur only in a later stage of the active region development.

A long-lived coronal arch system has been followed during several rotations by McGuire *et al* (*SP* 52, 91). Svestka *et al* (*SP* 52, 69) have described the life history of a transequatorial loop system which connected a new-born active region and an older one; magnetic field variations in the young center seemed to cause striking brightenings and twisting of the loop system. Finally, the evolution of the equatorial K-corona radiance all along the Skylab mission indicates (MacQueen and Poland, *SP* 55, 143) a rapid response of equatorial outer structures to abrupt changes in the global surface field structure.

c. *Possible occurrence of reconnection of magnetic lines of force.* Examples of possible reconnections of magnetic lines of force have been pointed out by Howard and Svestka (*SP* 54, 64) but it is difficult to distinguish changes in the temperature distribution of matter and true changes in the general structure of the field (Krieger *et al*, *SP* 47, 117). This problem has been investigated in detail by Nolte *et al* (*SP* 55, 401): even if one can discard apparent configuration changes of magnetic structures due to variations of emission, it is still necessary to have observational data to decide whether reconnection has occurred or if one has observed a large-scale reconfiguration of a "frozen-in" coronal magnetic field. According to the authors, existing observations of emitting loops and magnetic fields are rarely sufficient to make the distinction.

Possible effects of reconnection are, nevertheless, considered in several publications: Sheeley (*SP* 47, 173) after a comparison between coronal spectroheliograms and photospheric magnetograms suggests that field line reconnection is a mechanism which takes place usually when coronal magnetic fields interact and prevents the building up of magnetic stresses in the lower corona. Consequences of magnetic reconnection (which would happen due to the differential rotation) for the global coronal field and for the location of quiescent prominences have been examined by Hansen and Hansen (*SP* 44, 503; *SP* 51, 169). Finally, the possible role of reconnection processes during the relaxation phase of transient events, which could lead, for instance, to the formation of loop prominences, has been studied by Kopp and Pneuman (*SP* 50, 85; see also Rust and Webb, *SP* 54, 403).

C. *Coronal transients*

I. *Frequency and relation to solar activity*

Although the structure of active regions has been described in the previous paragraphs as if it was only "slowly variable", it is well known that there exists also fast events; and one of the most spectacular results of Skylab observations has been to observe them as well in X-ray in the lower corona as in the outer K-corona thanks to the coronagraph (Mac Queen *et al*, *RSL* 281, 405; MacQueen and Poland, *SP* 55, 143).

Hildner *et al* (*SP* 48, 127) have shown that transients observed in the K-corona are not rare (110 events observed in 227 days during the Skylab mission) and that they occur more often above strong photospheric and chromospheric magnetic field regions, which is consistent with the idea that magnetic forces may play a dominant role in the expulsion of coronal material. Transients may be detected in the lower corona using ground-based observations (Fisher, *SP* 55, 135) but the association with other solar phenomena is more easily seen on the disk, with the help of X-ray images; Kahler (*ApJ* 214, 891) has shown that X-ray events with long decay times are associated either with long-lived active regions, or with Hα prominence activations or with white light outer coronal transients. Rust and Webb (*SP* 54, 403) have found that out of 156 large-scale enhancements of X-ray emission in active regions, 44% were correlated with flares and 81% with filament activity. As flare related events and prominence activations will be described in other sections of this Report, only a brief account of the low corona counterpart of these transients will be given here.

II. *Transient phenomena in the lower corona*

a. *X-ray observations of coronal brightenings in active regions.* They are in most cases associated with Hα filament activity and are observed generally as X-ray loops or arcades of loops which can last several hours (Rust and Webb, *SP* 54, 403). These structures can probably be identified with classical Hα loop systems and the authors suggest that loops outline magnetic lines of force which are reconnecting after filament eruptions (see also Kopp and Pneuman, *SP* 50, 85). A well observed and particularly energetic event (1973 August 13) has been described by Vorpahl *et al* (*ApJ* 212, 550) who derive a temperature as high as 7×10^6K for the top of loops and a maximum density of $3 \times 10^9 cm^{-3}$. The correlation with a white light transient observed much higher in the corona has been established for this particular example by Rust and Hildner (*SP* 48, 381). Other examples of such associations are presented by Smith *et al* (*De Feiter Memorial Symposium*, in press). The life-time of the loops which have been formed can be very long (Sheeley *et al*, *SP* 45, 377); their decay has been studied by Krieger (*SP* 56, 107), who finds it longer than that computed for conductive cooling and thus asks for an additional energy source. In the case of the 13 August 1973 event Vorpahl *et al* (*ApJ* 212, 550) suggest that energy and possibly new material was added at the top of the loops as the phenomenon evolved.

b. *Coronal changes associated with disappearing filaments.* This class of transients which has been described by Webb *et al* (*SP* 48, 159) happens outside active regions: concomitantly with the Hα filament disappearance, X-ray emitting structures appear near the filament location; the peak temperature is of the order of 2.5×10^6K and the peak density of $10^9 cm^{-3}$, the total mass of the coronal X-ray emitting material is about 10% of the pre-existing filament. Kahler (*ApJ* 214, 891), in a study of X-ray events with a long decay time (according to the 1-8 Å flux), has suggested that these phenomena are similar to the X-ray enhancements near filament cavities. Other observations of X-ray and EUV emissions associated with the eruption of initially quiescent prominences have been described by Smith *et al.* (*SP* 52, 379) and Schmahl and Hildner (SP *55*, 473), who have been able to follow the phenomenon high into the corona and to study its link with the white light transient. The association of X-ray coronal transients outside active regions with disappearing filaments may be considered as a peculiar although very frequent case in the more general association of transients with neutral lines or more precisely with those which border growing equatorial coronal holes (Webb *et al*, *SP* in press).

III. *The "white light" transients*

a. *Kinematics* A lot of detailed data on the outward speed of mass ejections as observed in white light has been published (Hildner *et al*, *SP* 45, 363; Schmahl and Hildner, *SP* 55, 473). Gosling *et al* (*SP* 48, 389) have shown that while the overall scatter of velocities is between 100Km/s and 1200Km/s, flare associated events travel typically faster (775Km/s) than eruptive prominence associated events (360Km/s). This result is fully consistent with the long known difference (Valnicek, *BAC* 15, 207) between the kinematics of flare sprays and that of eruptive prominences. Type II and IV radio bursts (Kosugi, *SP* 48, 339) are associated only with events moving faster than 400Km/s (Gosling *et al*, *SP* 48, 389).

b. *Origin of ejected material.* The fact that the white light transient, although accompanying an eruptive prominence, is hot coronal material is proved by the polarimetric analysis of wide band pictures (Poland and Munro, *ApJ* 209, 927). Moreover the good coincidence between Hα and HeII images of the prominence shows that its material is not significantly heated and evaporated into the corona (Hildner *et al*, *SP* 45, 363). Therefore, the white light transient, which can be observed 1 $R_\odot$ higher than the top of ascending prominences (Hildner *et al*, *SP* 45, 363), must be coronal material originating from the low corona overlying the prominence. In the 1973 December 19 event studied by Schmahl and Hildner (*SP* 55, 473), the mass of the prominence material was about 2×10^{14}g while it reached 2×10^{15}g for the coronal transient. In other cases the mass of the prominence material was larger (Hildner *et al*, *SP* 45, 363) but the figure of 10^{15}g for the coronal transient was found in several examples (Hildner *et al*, *SP* 45, 363; Smith *et al*, *SP* 52, 379; Rust and Hildner,

SP 48, 381). The density in the transient is roughly 10 times greater than the normal neighbouring corona (Poland and Munro, *ApJ* 209, 927; Dulk *et al*, *SP* 49, 369). An analysis of the time evolution of the kinetic and potential energy (Schmahl and Hildner, *SP* 55, 473) leads to the conclusion that in the case of the 1973 December 19 event, the total energy increased 10 times during the 4 hours of observation implying the existence of some accelerating mechanism.

c. *The role of the magnetic field.* The simultaneous observation of a white light transient from Skylab and of the associated continuum radio burst (originating above the outward-moving white light loop in a compressed medium headed by a bow wave) at Culgoora has allowed Dulk *et al* (*SP* 49, 369) to determine the magnetic energy density. It was found to be more than 10 times the thermal energy density and larger than the kinetic energy density even in the fastest moving parts of the transient. Therefore, the plasma was magnetically controlled and it is likely that magnetic forces must provide the principal mechanism for acceleration of the transient material from the Sun.

d. *Interpretations.* One important point upon which present knowledge is not sufficient is the three dimensional structure of the transient; and the theoretical explanation of the phenomenon will be different for a "loop" than for a "shell" transient (Benz, *2ESM* in press). The loop structure has been proved for one example (Munro, U) and has been investigated by Mouschovias and Poland (*ApJ* 220, 675) and by Anzer (*SP* 57, 111). The white light transient is considered as a magnetic flux tube originating from the lower corona which expands in the background corona and magnetic field. A longitudinal current is assumed to flow along the loop to produce an azimuthal magnetic field, which can account for the expansion of the loop. The velocity versus height diagram is consistent with observed data (Anzer, *SP* 57, 111).

The alternative to the expanding loop is something like a piston driven expanding structure if the transient has a shell shape (Uralov and Kasinskii, *IANS* 53, 1264; Steinolfson and Nakagawa, *ApJ* 215, 345; Smith *et al*, *SP* 52, 379). Computations of Nakagawa *et al*, (*ApJ* 219, 314), Wu *et al* (*ApJ* 219, 324), Steinolfson *et al* (*ApJ* in press) show that is is possible to reproduce the observed transient velocities. The configuration of magnetic field is important; and it is found that only radial field topologies allow the propagation of transients as observed.

It can be thought (Benz, *2ESM* in press) that both loop and shell transients exist, the first corresponding to eruptive prominences and the latter to flares.

9. SOLAR WIND – SOLAR TERRESTRIAL
(John M. Wilcox)

The Skylab Workshop on coronal holes (reviewed by Zirker, *CHHSWS*; *RGSP* 15, 257) established that all large near-equatorial coronal holes seen during the Skylab period were associated with high-velocity solar wind streams observed at 1 AU (Krieger *et al*, *SP* 46, 303; Bell and Noci *OMOAA* 104, 111; Watanabe, *PASJ* 27, 385; Kovalenko and Molodykh, *AZAN* 54, 859; Rickett *et al*, *JGR* 81, 3845) but it is important to note that the converse does not hold, since there are considerably more high-speed streams than there are coronal holes. Coronal holes were established as sources of recurrent geomagnetic disturbances (Sheeley *et al*, *SP* 49, 271), but observations into the rising phase of sunspot cycle 21 showed that the solar atmosphere evolved from a structure having a few, large, long-lived, low-latitude coronal holes to one having numerous small, short-lived, high-latitude holes (Sheeley and Harvey, *CHSWS*, Naval Research Laboratory Skylab/Atm. Preprint, 1978). The high-latitude holes recurred with a synodic rotation period of 28-29 days instead of the 27-day period characteristic of low-latitude holes. The high coronal structure of high velocity solar wind stream sources was examined in Nolte *et al*. (*SP* 51, 459) and coronal holes were shown to be sources of open magnetic field lines, which extend from the solar photosphere to interplanetary space, as traced in a current-free (potential field) approximation using measured photospheric fields as a boundary condition (Levine *et al*, *JGR* 82, 1061). The holes magnetically open to the interplanetary medium having "toward" field polarity tended to connect to the south polar cap and those with "away" polarity tended to connect to the north polar cap (Wagner *AJ* 206, 583) during the lase cycle. The polar holes have persisted throughout the interval of observations (Sheeley and Harvey, l.c.). The magnetic polarity of low-latitude holes agrees with the polarity of the interplanetary magnetic field in the corresponding high-speed streams of the solar wind (Sheeley *et al*, *SP* 49, 271; Sheeley and Harvey, l.c.). Theoretical discussions of the effects of coronal hole structure on the solar wind are given by Fahr *et al* (*ASS* 43, 19), Dyer and Steinolfson (*JGR* 81, 5413) Holzer (*JGR* 82, 23), and Richter and Suess (*JGR* 82, 593). The diverging magnetic field geometry is found to be of key importance.

The three-dimensional structure of the solar wind (i.e. including north-south variations) received increasing attention and stimulated a proposed Solar Polar Spacecraft Mission to investigate the interplanetary regions above the polar regions. A large-scale latitude gradient in solar wind velocity of approximately 13km/s per degree of latitude was reported (Rhodes and Smith, *JGR* 81, 2123; Rhodes and Smith, *JGR* 81, 5833), but other workers have found intermittent (Bame *et al*, *JGR* 82, 173) or no (Diodato and Moreno, *ASS* 39, 409) latitude gradients of velocity. Interplanetary scintillation observations indicate a mean solar wind speed gradient of 2km/s per degree of latitude (Coles and Rickett, *JGR* 81, 4797), while analysis of the orientations of ionic comet tails gives no support for a higher speed near the poles than at the equator (Brandt, *Proc. Symp. SSIMT*, 95). These conflicting claims may be resolved (Hundhausen, *CHHSWS*, 225) by consideration of the longitudinal average which many of these techniques impose on the latitudinal gradient away from a warped current sheet in the heliosphere (Svalgaard and Wilcox, *N* 262, 766). When Pioneer II was at heliographic latitude of 16°N near sunspot minimum, it observed for several months a nearly continuous "away" interplanetary field polarity indicating that the warp in the equatorial current sheet at the time was less than 16° (Smith *et al*, *JGR* 83, 717). The large-scale heliospheric magnetic field is, at the present phase of the sunspot cycle, directed away from the sun northward of this warped "equatorial" current sheet and directed toward the sun southward of the current sheet. A theoretical discussion of the latitude dependence of solar wind speed as influenced by the coronal magnetic field geometry was given in Pneuman (*JGR* 81, 5049) and the subject was reviewed in Dobrowolny and Moreno (*SSR* 18, 685). Evidence for a solar equatorial belt from which few protons are emitted is given in Vladimirsky and Levitsky (*N* 260, 27). Observations with a solar probe near 0.3 AU indicated that the northern boundary of the high-speed stream associated with an equatorial extension of the south polar coronal hole was quite narrow--less than about 10° in latitude thick. The local latitudinal gradient in flow speed was at least 30km/s per degree of latitude (Schwenn *et al*, *JGR* 83, 1011). These observations appear to be consistent with the considerations described by Hundhausen (*CHHSWS*, 225).

After more than a century of controversy the possible influence of solar activity on terrestrial weather may be moving toward scientific respectability (Wilcox, *SC* 192, 745). Structures related to the warped current sheet described above appear to influence the size of wintertime low-pressure troughs (Wilcox *et al*, *JAS* 34, 382). Related investigations are described in Loginov *et al*. (*TGIM* 23, 43), and Knight and Sturrock (*N* 264, 239). Atmospheric pressure at the surface of the earth was found to change in an ordered manner after geomagnetic disturbances attributed to solar flares (Mustel *et al*, *AZAN* 53, 1060; Mustel *et al*, *AZAN* 54, 682; Chertoprud, *AT* 883, 4). A solar-generated quasi-biannual geomagnetic variation was described in Sugiura and Poros (*JGR* 82, 5621), and possibly related meteorological variations were desribed in Yakoleva (*TGGO* 355, 94). Large-scale reductions in the ozone content of the middle and upper stratosphere over the polar cap regions were associated with major solar proton event (Heath *et al*, *SC* 197, 886). Stratospheric circulation at 10 mb was associated with the sun's rotation (Ebel and Batz, *T* 29, 41). An effort was made to improve regular daily observations of the magnetosphere, ionosphere and atmosphere to aid in understanding the physical mechanisms involved in sun-weather influences (Wilcox, *JATP* 39, 173). An examination of several possible physical mechanisms was given by Dickinson (*BAMS* 56, 1240) and by Schneider and Mass (*SC* 190, 741) and the energetics of such relationships was discussed by Willis (*JATP* 38, 685). Further theoretical discussions of possible mechanisms were given by Volland (*N* 269, 400), Cole (*N* 260, 229), and Schuurmans, (*Z* 3.Jaarg, 358). Possible influences of cosmic rays on the terrestrial atmosphere were discussed by Chamberlain (*JAS* 34, 737) and Schelegel (*JATP* 39, 101). A wide-ranging review of many possible climate effects of solar phenomena was given by King (*ASS* 57, 209) and some specific relationships were discussed in Gerety *et al*, (*JAS* 34, 673) and Rubashev *et al*, (*SDB* 10, 98). The Maunder minimum in solar activity may have been associated with the little ice age, and similar relationships may have occurred in earlier centuries (Eddy, *SC* 193, 1189). There is some evidence that solar activity may have a significant influence on the accuracy of weather forecasting (Larsen and Kelley, *GRL* 4, 337) but much more work on this point is needed. The Proceedings of a recent conference provide a good summary of the current state of the field (Proceedings, Symposium/Workshop on *Solar Terrestrial Influences on Weather and Climate*, Ohio State University, July 1978, in press). The overall subject remains one of considerable controversy (Pittock, *RGSP* 16, 400) and agreement on which effects are real and accepted physical mechanisms has not yet emerged.

Streams of protons with energies a few Mev observed in the solar wind near earth were attributed to an unknown solar acceleration process (Roelof and Krimigis, *STIP*, 343), but as spacecraft observations between one and five AU became available it was observed that the intensity of such proton streams was larger at a few AU than at the earth. Thus most of the acceleration is attributed to interplanetary processes which are not yet well understood (Barnes and Simpson, *ApJL* 210, 96; Smith and Wolfe, *STIP* 227; Pesses *et al*, *JGR* 83, 553; Palmer and Gosling, *JGR* 83, 2037).

The solar wind emission is highly structured in latitude and longitude by the solar magnetic field, including the effects of coronal holes discussed above. It had been thought that low speed solar wind represented the theoretical "uniform sun" conditions, but recent studies have indicated that the high-speed solar wind is more uniform and probably should be used in any comparisons with isotropic theory (Bame *et al*, *JGR* 82, 1487; Gosling, *JGR* 81, 5054). Observations from 1964-1975 showed that high- and low-speed solar wind flows originated from preferred solar longitudes, with a modulation in solar wind speed of 20% associated with the synodic rotation period of 27.025 days (Gosling, *et al*, *JGR* 82, 2371). An internal structure deep in the sun may be postulated to explain these effects, which appear to rotate rigidly, or they may be a consequence of long lived circulation patterns. Radio techniques were increasingly used to probe the solar wind near the sun and out of the ecliptic (Coles and Rickett, *JGR* 81, 4794; Rickett *et al*, *JGR* 81, 3845; Watanabe, *UK* 19, 80; Woo *et al*, *ApJ* 210, 568; Rickett, *SP* 43, 237; Fitzenreiter *et al*, *SP* 52, 477; Woo *et al*, *ApJ* 218, 557; Coles and Harmon, *JGR* 83, 1413). Most of the observers found the solar wind speed to increase at higher latitudes. Yearly averages of solar wind speed observed with spacecraft were found to be rather constant with solar cycle except for broad high-speed streams observed during about two years before solar minimum (Bame *et al*, *ApJ* 207, 977; Gosling *et al*, *JGR* 81, 5061; Feldman *et al*, *JGR* 83, 2177; Sheeley *et al*, *SP* 52, 485). A study of geomagnetic records from 1868-1975 indicated that the broad high-speed streams observed in the 1973-1975 era may have been unusually large (Gosling, *et al*, *JGR* 82, 3311). Observations just *after* sunspot minimum found solar wind streams of lower amplitude, lower maximum speed and shorter duration than those observed in the preceding years (Nolte, *et al*, *GRL* 4, 291) and that the solar wind-coronal hole relationship inferred from the Skylab period may change in other parts of the solar cycle.

In order to reduce the discrepancy between solar wind measurements at 1 AU and the corresponding predictions of the basic two-fluid model, a nonthermal heating of solar wind protons in the outer corona up to distances of about 0.1 AU has been postulated (Auer and Rosenbauer, *JGR* 82, 1503). In this connection a number of discussions have been given of the momentum and energy deposited by various wave modes in the solar wind (Wentzel, *JGR* 82, 714; Jacques, *ApJ* 215, 942; Abraham-Shrauner and Feldman, *JGR* 82, 618). Two-fluid models continue to be popular for describing solar wind flow conditions (Nerney and Barnes, *JGR* 82, 3213; Singer and Roxburgh, *JGR* 82, 2677; Summers, *PLSC* 24, 799; Benz and Gold, *AA* 55, 229). When an azimuthal velocity dependence is included in a two-region model of the solar wind, it is found that the amount of energy converted into kinetic energy in the solar wind is only a small fraction of the total expansion energy flux and has little effect upon the final radial expansion velocity (Acuna and Whang, *ApJ* 203, 720). The effects of rapidly diverging flow may produce an increased conductive energy supply to the solar wind (Holzer, *JGR* 82, 23). It was found that including the Archimedes spiral field geometry did not have a large influence on the results obtained from the two-fluid solar wind equations (Nerney and Barnes, *JGR* 83, 3729).

A sheet-current approach was introduced to describe the force balance between the magnetic field and the gas pressure in the coronal-interplanetary space (Yeh and Pneuman, *SP* 54, 419). Dynamic modeling of coronal and interplanetary responses to solar events within the contexts of hydrodynamics and magneto-hydrodynamics has received considerable attention (Dryer *et al*, *JGR* 83, 532; Steinolfson and Dryer, *JGR* 83, 1576; Wu *et al*, *STIP* 43). A review of the observations and theoretical ideas concerning the role of kinetic processes in the solar wind was given in Dobrowolny and Moreno (*SSR* 20, 577).

Analysis of the number of geomagnetic storms showed that if the component of the large-scale solar magnetic field in an active region or in a flare region is directed southward then that region and that flare produce a geomagnetic storm (Pudovkin and Chertkov, *SP* 50, 213; Pudovkin *et al*, *SP* 54, 155). A correlation of the toward polarity interplanetary magnetic field sectors with the inward polarity main body of the solar magnetic "super giant" structures was found (Bumba, *BAC* 27, 153). A study of the interplanetary magnetic field in 1969-1974 showed a clear reversal of the sense of the Rosenberg-Coleman dominant polarity effect in the 1970-71 interval and that on the average sector boundaries near the earth were inclined approximately 12° to the solar equator (Hedgecock, *SP* 44, 204). Sector boundary distortion by a possible meridional velocity gradient in the solar wind was discussed (Suess and Feynman, *JGR* 82, 2405). Coronal streamers can be followed into interplanetary space as a curved surface that is accompanied by a change of polarity of the interplanetary magnetic field (Korzhov, *SP* 55, 505). Studies of long-term interplanetary magnetic field variations in 1963-74 showed that the yearly average magnitudes of all field sectors had virtually no solar cycle variation (King, *JGR* 81, 653; Mariani *et al*, *SP* 45, 241; Kovalenko and Malyshkin, *SDB* 9, 95). Since the large-scale solar magnetic field changes considerably as a function of latitude through the sunspot cycle some compensating effects that are not well understood at the present must exist in order to produce a constant magnitude of the interplanetary magnetic field. The solar wind electric field near the earth rises sharply at each sector boundary from a pre-boundary value of approximately 1 mV/m to peak values between 3-6 mV/m, resulting in several phenomena in the polar cap ionosphere (Bahnsen and D'Angelo, *JGR* 81, 683).

10. GROUND BASED OPTICAL SOLAR INSTRUMENTATION
(R. J. Bray)

A. *New Telescopes*

A 60 cm domeless solar telescope is due for installation at the Hida Observatory of the University of Kyoto in 1978. Carried by an altazimuth mounting on a 20 m tower, the telescope was constructed by Carl Zeiss, Oberkochen, (Kühne, *ZI* 84). It is a modified Gregorian and most of the optical path is in vacuo.

Carl Zeiss, Jena, is also active in solar telescope design and construction. Artus (*JR* 3, 138) has described a design concept for a large, sophisticated vacuum tower telescope fed by an altazimuth coelostat and incorporating a large range of computer-controlled auxiliary instrumentation. Zeiss is currently building five horizontal telescopes with apertures of 50 cm for the Czechoslovak Academy of Sciences, equipped with 10 m Czerny-turner spectrographs.

Several member countries of the Joint Organization for Solar Observations plan to erect telescopes at sites on the Canary Islands. Details of the various instruments under construction are to be found in *JOSO*; progress in advancing the organization's aims is also described in *JOSO*.

Two vacuum telescopes of 40 cm aperture are due for completion at Kanzelhöhe over the period 1979-80 (Pettauer, private communication).

SibIZMIR (Irkutsk) is building a large telescope at Lake Baikal in Siberia. The two-lens objective has a diameter of 76 cm and a focal length of 40 m; the first images are expected by the fall of 1979. A brief description will appear in *NTA*, Nauka (in press).

There are three solar observatories currently operating in China - the Purple Mountain Observatory near Nanking, Yunnan Observatory near Kunming, and Peking Observatory. The largest solar telescope is a Chinese made Cassegrain coudè with an aperture of 60 cm located at Peking.

High Altitude Observatory is completing a new K-coronameter ('Mark III') to be installed at the Mauna Loa Observing Station during the spring of 1979. The new instrument, of 20 cm diameter aperture, features improved spatial and temporal resolution over its predecessors and is designed particularly to permit inner coronal observations of coronal changes and mass ejection events on the sun. A companion instrument to observe prominence material far from the solar limb and determine the velocity of ejecta in Hα will be brought into operation at the same site in the summer of 1979, prior to the launch of NASA's Solar Maximum Mission satellite.

B. *Filters*

There has been vigorous development of computer-controlled filters with narrow bandwidths ($\lesssim$ 0.1 Å) capable of rapid tuning to a number of Fraunhofer lines ('universal' filters). At Lockheed, Title (*AO* 15, 2871) has made a birefringent filter which, by using partial polarizers, achieves a significantly higher transmission than conventional filters. At λ5324 the bandwidth is 0.09 Å. The filter can be tuned to any wavelength in the range 4500-8500 Å by a number of stepping motors commanded by a minicomputer.

The universal birefringent filter designed several year ago by Beckers (SPO) and Zeiss (Oberkochen) is now available commercially. It may be tuned to any wavelength in the range 4200-7000 Å with a bandwidth varying from 0.09 Å to 0.28 Å, depending on the wavelength. At the Big Bear Solar Observatory such a Zeiss universal filter mounted on the 65 cm vacuum telescope operates under the control of a PDP-11 computer (Zirin, private communication).

Narrow-band birefringent filters are also being developed at SibIZMIR (Skomorovsky, *SP* 41, 254; 45, 2; Aleksandrovich *et al* , *NTA* 5, 34).

A well-recognized need exists for universal filters having much narrower bandwidths than present-day filters of the birefringent type. At SCLERA (Tucson, Arizona) Smolka and Hill (*JOSA* 66, 1089) are constructing a filter consisting of ten Michelson interferometers in series. Tuning will be carried out by means of piezo-electric transducers controlled by a minicomputer. The expected bandwidth ranges from 0.02 Å at 4000 Å to 0.09 Å at 8000 Å.

At Sydney a filter consisting of three servo- and computer-controlled Fabry-Perot interferometers is being assembled (Bray, *R* 20, 1318). To avoid parasitic images the interferometers must be tilted slightly; nevertheless, the calculated instrumental profile shows little degradation over a field 3 x 4 arcmin in extent (Loughhead *et al*, *AO* 17, 415). At Hγ (4340 Å) and Hα (6563 Å) the calculated bandwidths are 0.02 Å and 0.05 Å respectively.

A digitally controlled Fabry-perot interferometer for use in the infrared is being developed at Kanzelhöhe (Pettauer, private communication).

C. *Magnetographs*

Increasing effort is being applied to the measurement of vector magnetic fields as distinct from their purely longitudinal component. Details of performance figures and calibration procedures have been published by Baur and House (*PSPIE* 112) for the HAO/SPO Stokes Polarimeter. The instrument is mounted on the 40 cm coronagraph at Sacramento Peak and simultaneously measures all four Stokes parameters as a function of wavelength across the profile of any line in the visible spectrum. It is used to determine vector magnetic fields by means of the Zeeman and Hanle effects in both quiet and active regions on the disk and prominences beyond the limb. For longitudinal fields its sensitivity is typically 1 gauss, and for transverse fields, 20 gauss. The spatial resolution is typically 6 arcsec while the best achieved is 2 arcsec.

At Kitt Peak a Stokes polarimeter is expected to be available by the end of 1978. It will operate at the McMath telescope in conjunction with a two-dimensional diode-array detector (Harvey, private communication).

A new method of measuring vector fields which, like the Stokes polarimeter technique, is free of the saturation effects encountered in the conventional magnetograph, has been proposed by Loughhead and Bray (*SP* 50, 297; *PASP* 90, 230). Using a tunable narrow-band filter or spectroheliograph, they propose to determine the contrast profile of the point under examination in a suitable Zeeman triplet. The method is confined to strong (kilogauss) fields but is intrinsically capable of high spatial resolution (λ 1 arcsec) (see also Frazier, *AA* 64, 351).

A videomagnetograph for rapid reconnaissance of longitudinal fields built by Leighton in 1970 is now installed on the 25 cm vacuum refractor at Big Bear (Zirin, private communication). A PDP-11 is used to control this system as well as a universal birefingent filter and its associated cameras. Thus, one can carry out rapid wavelength scanning programs with the universal filter interleaved with magnetograms of the same region. The Big Bear Observatory has the capability of obtaining nearly simultaneous, real time magnetic and velocity measurements, filtergrams in a variety of spectral lines, and photographs in white light. The three vacuum telescopes employ apertures of 65, 25, and 22 cm and produce both high-resolution and patrol-type observations.

Modern methods of magnetic field measurement have recently been reviewed by Grigorev (*IKAO* 56, 166). This article is valuable for its account of vector magnetographs, the properties of various kinds of linear and two-dimensional solid-state photodetector arrays, and recent developments in the Soviet Union. A CSIRO translation is available.

Magnetographs in use in the Soviet Union have also been described by Klochek *et al*, *NTA* 5, 25) and Stepanov *et al*, (*IGAFS* 37, 147).

D. *Miscellaneous*

The intensity and linear polarization of certain coronal emission lines can be used to infer the direction, but not the strength, of coronal magnetic fields. Coronal emission line polarimeters have been placed in operation at Pic du Midi by Arnoud (*ROL* 12, 137) and at SPO by Querfeld (*PSPIE* 112, 200). The 15 cm coronameter at Pic du Midi observes the FeXIV 5303 Å line to $1.4R_o$ while the joint HAO/SPO 40 cm coronameter uses the FeXIII 10747 Å line out to $2.2R_o$.

The Haleakala Stokes polarimeter (Mickey, *ROL* 12, 81) is now equipped with a compact Czerny-Turner monochromator so that spectral profiles may be obtained with .030 Å at 6000 Å resolution in all four Stokes parameters. Since silicon diode detectors are used, the instrument has a spectral range of 3500-11000 Å. The effective spatial resolution is limited to 5 arc seconds by spar pointing stability.

A two-dimensional diode-array detector is now available for imaging and spectroscopic observations at the McMath telescope, Kitt Peak National Observatory. With the large infrared spectrometer it is now possible to build up spectroheliograms in the infrared (1 - 10 μm) using a single detector. The 1 m Fourier transform spectrometer is now in full operation (Harvey, private communication).

ABBREVIATIONS

AA	= Astronomy and Astrophysics
AAM	= Adv. Appl. Mech.
AASup	= Astronomy and Astrophysics Supplement
AiA	= Astrometrija i Astrofizika
AIAA	= Bulletin of the American Institute of Aeronautics and Astronautics
AIAAJ	= Journal of the American Institute of Aeronautics and Astronautics
AJ	= Astronomical Journal
AL	= Astrophysical Letters
AN	= Astronomische Nachrichten
AO	= Applied Optics
ApJ	= Astrophysical Journal
ApJL	= Astrophysical Journal Letters
ApJSup	= Astrophysical Journal Supplement
ARAA	= Annual Review of Astronomy and Astrophysics
ASS	= Astrophysics and Space Science
AT	= Astronomischeskij Tsirkular
AUP	= Australian Journal of Physics
AUPA	= Australian Journal of Physics, Astrophysical Supplement
AUW	= Acta Universitatis Wratislaviennis
AV	= Astro. Vestro
AZ	= Astronomicheskij Zhurnal
AZAN	= Astron. Zhurn. Akada. Nauk
BAC	= Bulletin of the Astronomical Institutes of Czechoslovakia
BAAS	= Bulletin of the American Astronomical Society
BAMS	= Bulletin Amer. Meteor. Soc.
BASI	= Bulletin Astron. Soc. India
CA	= Center for Astrophysics
CALR	= Univ. of Calif. at Riverside
CAPS	= Comments on Astrophysics and Space Physics
CE	= Cosmic Electrodynamics
CHHSWS	= Coronal Holes and High Speed Wind Streams, Colo. Assoc. Press

CHSWS	= Coronal Holes, Solar Wind Streams, and Geomagnetic Activity During the New Sunspot Cycle, Naval Research Laboratory Preprint
COSP	= COSPAR Symposium
CR	= Comptes Rendus de l'Academie des Sciences, Paris
CSP	= Proceedings Consultation of Solar Physicists from Socialist Countries, Irkutsk 1976
ECSPI	= First European Conference on Solar Physics, Florence
EOS	= EOS, American Geophysical Union
ESM	= 2nd European Solar Meeting
FP	= Fiz. Plazmy
GAER	= Geomagnetizm i Aeronomiya
GAFD	= Geophys. and Astrophys. Fluid Dyn.
GRL	= Geophysical Research Letters
HA	= Highlights of Astronomy
IANS	= Izv. Akad. Nauk SSSR
IAU	= IAU Symposium
ICCR	= International Cosmic Ray Conference
IGAFS	= Issl po Geomagn Aer i Fiz. Solntsa
IKAO	= Izvestiya Krymskoj Astrofizicheskoj Observatorii
ISSTP	= International Symp. on Solar Terrestrial Physics
ISSTR	= International Symposium on Solar Terrestrial Relations
JATP	= Journal Atmos. Terr. Physics
JAS	= Journal of Atmospheric Sciences
JFI	= Journal of Franklin Institute
JFM	= Journal of Fluid Mechanics
JGR	= Journal of Geophysical Research
JICR	= Journal of Interdisciplinary Cycle Research
JOSO	= JOSO Annual Report
JPP	= Journal of Plasma Physics
JR	= Jena Reviews
JRASC	= Journal of the Royal Astron. Soc. of Canada
KB	= Kleinheubacher Berichte
KI	= Kosmicheskie Issledovaniya
LPI	= P. N. Lebedev Physical Institute
MAG	= Mittelungen der Astronomischen Gesellschaft
MGYR	= Magnetnaya Gidrodynamika
MN	= Monthly Notices of the Royal Astronomical Society
N	= Nature
NPS	= Nature Physical Science
NRO	= Nouvelle Revue d'Optique
NTA	= Novaya Tekhn v Astron
OE	= Optical Engineering
OLI	= Ord. Lenin Inst. Priki, Mat. Acad.
OMOAA	= Oss. e Mem Oss Arcetri
PAAS	= Publications of the American Astronautical Society
PASA	= Proceedings of the Astronomical Society of Australia
PASJ	= Publications of the Astronomical Society of Japan
PASP	= Publications of the Astronomical Society of the Pacific
PCRC	= Proceedings of the International Cosmic Ray Conference
PIAFE	= Proceedings of the IAFE Flare Conference
PLI	= Proceedings Lebedev Institute
PLSC	= Planetary and Space Science
PRL	= Physical Review Letters
PSL	= Proceedings of the Physical Society, London
PSPIE	= Proceedings of the Society of Photo-Optical Instr. Eng.
PTRSL	= Philosophical Transactions of the Royal Society, London
R	= Radiofizika
RCSP	= Regional Consultation on Solar Physics

RGEO	= Reviews of Geophysics
ROL	= Reports Observatory of Lund
RQE	= Radiophysics and Quantum Electronics
RSL	= Royal Society London
RZ	= Referativnyj Zhurnal
SA	= Soviet Astronomy
SAL	= Soviet Astronomy Letters
SAM	= Studies in Appl Math.
SC	= Science
SDB	= Solnechnye Dannye Byulleten
SJPP	= Soviet Journal Plasma Physics
SP	= Solar Physics
SPO	= Sacramento Peak Observatory
SPU	= Soviet Phys. Usp
SR	= Space Research
SSI	= Space Science Instrumentation
SSIMT	= Proc. Symp. on Study of the Sun and Interplanetary Medium in 3-D
SSL	= Space Science Laboratory
STIP	= Study Travelling Interplanetary Phen.
SW	= Sterne und Weltraum
T	= Tellus
TAI	= Trudy Astrofizicheskogo Instituta
TAGU	= Transactions American Geophysical Union
TAOR	= Tokyo Astronomical Observatory Reports
TGIM	= Trudy VNII gidrometeorol inform. mirovojtsentr dannykh
TGGO	= Trudy Glav. geofiz. observ.
U	= unpublished
UCR	= UCAR preprint
UK	= Uchusen, Kenkyu
USP	= Uspekhi Fiz. Nauk
VAIF	= Veroeffentlichungen Astr. Inst. Frankfurt
Z	= Zenit
ZA	= Zeitschrift fur Astrophysik
ZI	= Zeiss Inf.

G. NEWKIRK
President of the Commission

12. RADIATION AND STRUCTURE OF THE SOLAR ATMOSPHERE
(RADIATION ET STRUCTURE DE L'ATMOSPHERE SOLAIRE)

PRESIDENT: M.K.V. Bappu
VICE-PRESIDENT: Y. Uchida
ORGANIZING COMMITTEE: J. M. Beckers, G. E. Brueckner, A. N. Cox, R. G. Giovanelli, C. Jordan, W. Mattig, S. I. Gopasyuk, V. M. Sobolev, P. Souffrin.

1. INTRODUCTION

The study of the solar atmosphere has witnessed a major expansion in recent years. A considerable improvement in spatial and temporal aspects of solar research from both ground based telescopes and space vehicles together with progress in theoretical interpretation have provided us with a much improved understanding of phenomena associated with the star closest to us. To report on these numerous contributions would be a near-impossible task even if space restrictions did not exist; this report, therefore, treats in some detail only a few subjects of topical interest. They are (1) Large scale circulations on the Sun (Dr. Gilman); (2) Five-minute oscillations on the Sun (Dr. Deubner); (3) Small scale solar magnetic fields (Drs. Livingston and Stenflo); (4) Alfvén waves in the solar atmosphere (Dr. Uchida); (5) Energy flow in the chromosphere and corona (Dr. Athay); (6) Coronal holes (Dr. Zirker); (7) Fine structure of the corona and the origin of the solar wind (Dr. Brueckner).

2. LARGE SCALE CIRCULATION ON THE SUN
(P. Gilman)

Introduction

In this review the author will comment from his own perspective on progress in observing, modelling, and understanding the global circulation of the sun. As implied in the title he will restrict himself to global features, and not, for example, attempt to discuss supergranulation. Howard (Rev. Geophys. and Space Phys. 1978, in press) has recently completed a broad review of the solar rotation problem. The most recent meeting on this problem is the Catania Workshop on Solar Rotation held in September 1978.

Observations

The effort to measure global velocities on the sun has risen dramatically in the last few years. At least four observatories in the United States (Hale, Kitt Peak, Sacramento Peak, Stanford) now make regular Doppler velocity measurements over the entire solar disk. Extensive measurements are also made at Locarno, Switzerland, and the Astronomical Observatory, Belgrade. Several other groups both in the United States and Europe, and the Soviet Union are active in analyzing a great variety of tracers of global circulation. There is renewed interest in sunspots as tracers, but the calcium network, coronal features, X-ray bright points, photospheric magnetic field patterns have also been used. It is becoming increasingly feasible to compare observations using the same techniques from different observatories, as well as different kinds of measures of global circulation for the same time period. Some of these comparisons are beginning to illustrate disquieting but potentially exciting differences.

From Doppler measurements, Howard (18.081.051) has reported long term changes in solar rotation, including a speeding up of the rotation at all latitudes from 1967 to 1976 of 3-4% as well as small changes in the amount of differential rotation between low and high latitudes. The 1976 rotation rate is within 1-2% of the classical long term average rate from long lived spots reported by Newton and Nunn. Howard also sees large short term ($\sim$1 week) fluctuations in the rotation rate of up to ten percent, which are highly correlated in latitude from about 50^0N to 50^0S, as pointed out by Gilman (Coronal Holes and High Speed Wind Streams Chapter 8, University of Colorado Press, 1977). These intriguing features, which might be interpreted as evidence of global east west convective motions, have unfortunately not been seen in the Stanford measurements for the same time period. Duvall at Stanford reports fluctuations of only 1-2 percent which are not nearly as well correlated in latitude. The relatively coarse resolution used at Stanford might account for some of the differences, but certainly not all.

Beckers (Sac Peak), Duvall (Stanford), and Howard (Hale) are all independently reporting evidence of axisymmetric meridional flow toward the poles in each hemisphere of magnitude 10-40 meters/sec. A suggestion of this feature is also seen in recent Belgrade data of Karabin and Dubicela (1978 preprint). Because of the lack of an absolute wavelength reference for the Doppler shift, and possible confusion of a meridional flow with various limb shifts, caution is urged in the interpretation of these measurements. Beckers (Ap. J.224, L143) has looked carefully at the rotation rate of the poles, and found no evidence of a polar vortex there, contrary to theoretical inferences from a spherical shell convection model made earlier by Gilman (Geophys. and Astrophys. Fluid Dyn., 8,93). It has so far proved extremely difficult to find Doppler shift evidence of other global circulation, such as giant cell patterns, probably because of obscuration by smaller scale motions, as well as instrumental and limb shift effects. In addition, some previously reported evidence of anticorrelation of mid and high latitude rotation rates has now been shown by Stenflo (20.080.053) and by Duvall and Svalgaard (Solar Phys.56,463) to be an artifact of the curve fitting used.

Virtually all tracers of global circulation are connected in important ways to the magnetic field. A variety of new measurements have been made recently, but the various results still present a rather complex picture. Foukal (17.071.004) has determined that the local Doppler rotation rate in the quiet photosphere increases with increasing magnetic field strength, such that regions with the strongest flux (> 10 gauss, for a 5 x 5 arc sec aperture) rotate more than 40m/sec or about 2%, faster than the regions of weakest neighbouring fields. Thus the strong field regions are moving through the weak field areas. Beckers (19.072.012) has shown that the Doppler and tracer rotation rates of sunspots agree within 1%. Since the plasma rotation outside sunspots is a few percent slower, this represents additional evidence that the field and plasma inside the spot are moving through the surrounding medium. Related evidence for rotation differences among magnetic structures come from compact X-ray emission features found in the low corona. Golub and Vaiana (Ap.J.219, L55) have found that the smallest features, lasting two days or less, rotate at about the rate of the photospheric gas, while the largest, longest lived features rotate up to 5% faster, a rate consistent with sunspots. They postulate this is further evidence of rotation increasing with depth, since the magnetic field of the longer lived features is thought to be rooted more deeply in the convection zone. Changes in differential rotation in latitude have been seen by Stenflo (20.080.053). He showed that photospheric magnetic field patterns show regions of larger than average differential rotation on the poleward side of the sunspot zones. These regions migrate toward the equator in pace with the migration of the sunspots zones.

Several groups have looked for evidence of giant convective cells in tracer velocity patterns. Schröter and Wöhl (18.080.103) report a four cell pattern in Ca^+ network seen in the summer of 1975, with rms velocities of 80 m/sec and substantial flow across the equator. However, they were unable to find corresponding patterns in the Doppler measurements for the same period from Mt. Wilson. Two new estimates of the latitudinal transport of angular momentum by global velocities have also been made by Schröter and Wöhl (18.080.103), and by Belvedere, et al. (18.080.021) using faculae. They both find transport toward the equator (the correct sense to maintain the equatorial acceleration), which is several times larger than previously reported. The data sample is apparently too small to estimate the variation of this transport with latitude, but these results do give some observational support to the theoretical prediction that Reynolds stresses in giant convective cells are responsible for maintaining the solar differential rotation.

Interest has been renewed in using past records of sunspots to deduce changes in solar rotation. Motivated by the rediscovery of the Maunder minimum by Eddy (17.072.049), Eddy, Gilman and Trotter (18.080.019; 20.080.055) have reconstructed rotation rates for two periods before the Maunder minimum in the 17th century. They find the equatorial rotation just before the Maunder minimum onset (1642-44) was faster by 3-5%, while the differential rotation between the equator and 20^0N was enhanced by a factor of 3. In contrast, the period twenty years earlier, when there were more spots, was much more like the modern average. Eddy et al (Bull. Am.Astron.Soc.10,400) have since reanalyzed the Greenwich sunspot record for the period 1888-1964 and found a long term decline in the average rotation of about 1%. This drop occurred through a period when successive sunspot cycles were rising in amplitude, and so is consistent with the 17th century results.

Promising new techniques for measuring global circulation are currently being developed and tested. Rhodes, Deubner and Ulrich are using frequency shifts of power bands in the K- ω diagram for 5 minute oscillations to deduce angular velocity changes with depth. Early results show an increase with depth of several percent in the first $15x10^3$ km, consistent with the larger reported differences between sunspot and Doppler surface rotation values. Beckers and Brown are currently developing a Fourier tachometer at Sacramento Peak, which uses a Michelson interferometer to measure the Doppler shift. Some progress is also being made on development of spectral wavelength references against which to measure the Doppler shift. The influence of scattered light on the Doppler differential rotation signal is also receiving renewed emphasis.

Theory

During 1976-1978, development of theory for global circulation of the sun has proceeded on at least 3 fronts; compressible convection models in which symmetry about the axis of rotation is assumed; Boussinesq convection models which include departures from axisymmetry; and applications of turbulence theory.

The most highly developed compressible axisymmetric model is that of Belvedere and Paterno (20.080.068). Starting from the assumption that a weak influence of rotation upon convection leads to a small variation in latitude of the rate of radial heat transport, the model generates a pattern of meridional circulation consisting of three cells extending from pole to equator, with typical flow velocities of 20cm/sec. The variation of radial heat flux is fixed by fitting the model to the observed surface differential rotation. There results a differential rotation which increases weakly with depth, and a surface heat flux variation with latitude as small as 10^{-4}, consistent with observations. This model has attractive features, but also problems. In particular, it depends critically on the assumption of weak influence of rotation upon convection. However, most estimates of convective turn-over times for the deep convection zone imply that the rotational influence is rather strong there. Furthermore, the form of the heat transport variation with latitude is ad hoc (though not unreasonable). If compressible convection in a deep slowly rotating spherical shell were allowed to develop in its own way, (including departures from axisymmetry) would the result be the same ? The author's own experience with Boussinesq spherical shell convection models suggests that something quite different would happen, namely that an equatorial deceleration would form when the rotational influence is weak. This possibility remains to be tested in the compressible case.

Boussinesq spherical shell convection calculations have been carried out primarily by the author. He has found (Gilman, Geophys. Astrophys. Fluid Dyn.8,93) that in this model, equatorial acceleration is generated only if the influence of rotation upon the convection is strong. As this influence is weakened, by increasing the thermal driving, the profile changes over to equatorial deceleration and high latitude acceleration. The model also generates large heat flux differences in latitude at the surface which have no counterpart on the sun. These are reduced but not eliminated if the model is constrained to have constant radial heat flux at the bottom (Gilman, Geophys. Astrophys. Fluid Dyn.10, in press). As a secondary effect, the model also generates an axisymmetric meridional circulation, at least an order of magnitude larger than seen in axisymmetric models of Belvedere and Paterno and earlier workers. Near the outer boundary, the flow is toward the poles, consistent with observational results described above. Large as it is, however, this circulation is not primarily responsible for the model differential rotation, contrary to the axisymmetric models, since Reynolds stresses still overpower it to drive the equatorial acceleration.

The Boussinesq model has proved useful in adressing other questions concerning the sun's circulation. Gilman and Foukal (1978 preprint) have calculated models of convection in a thin shell weakly influenced by rotation, to simulate the momentum transport effects in supergranules. They found that in such a shell the angular velocity decreased outwards at close to the rate expected if radially moving fluid particles conserve their angular momentum. This effect has been previously postulated by Foukal and Jokipii (14.080.002) as being responsible for the difference in rotational velocity traced by sunspots as compared to the Doppler values. Based on this and earlier model calculations, Gilman and Foukal have suggested that the sun's latitude gradient of angular velocity is formed deep in the convection zone where the rotational influence is strong and where there is also a strong tendency for the angular velocity to be constant on cylinders. The latitudinal gradient is transmitted to the photosphere by supergranulation, which also changes the sign of the radial gradient near the surface.

Gilman has also shown with the model that the polar vortex, characteristic of convecting layers of depth 20% of the radius, disappears when layers as deep as 40% are adopted. This result favours a deep solar convection zone, and is consistent with the observations of Beckers (Ap.J.224, L143).

Clearly all conclusion concerning the sun drawn from Boussinesq models must be tested again with a compressible model. In the future modelling efforts are expected to focus on the problem of compressible, nonaxisymmetric convection in a deep rotating spherical shell. Basic advances in the theory of compressible convection in cartesian geometry without rotation, recently made by many workers in several countries, suggest that the time is ripe for this. Greater sophistication in the parameterization of small scale turbulent transports will be needed.

3. FIVE MINUTE OSCILLATIONS ON THE SUN
(F. L. Deubner)

The 5-min oscillations made their first appearance in solar physics through the well known work of Leighton et al and have ever since been studied by many authors, both observationally and theoretically. Beckers' and Canfield's review (18.080.066) of 1975 gives an extensive account of what had become accepted wisdom about the 5-min oscillations at the beginning of the period to be covered by this report.

By that time, the display of the observed data in the format of diagnostic (k, ω) diagrams had become common practice. This display technique gives conveniently insight into the dispersion relations governing the observed fluid motions, and is therefore ideally suited for comparison with the relevant theoretical models.

The rapid progress which was achieved in 1975 with the analysis of several new sets of observations obtained by Deubner (14.071.042) marked a kind of renaissance in the study of solar oscillations. The new observations, soon after being confirmed by Rhodes et al.(20.080.050), clearly identified the 5-min.oscillations as solar "p-modes", as first predicted by Ulrich (04.071.060). The observed p-modes are standing acoustic waves, modified by the stratification of the atmosphere, trapped in a subphotospheric cavity formed by the temperature and density structure of the solar convection zone. Through the photosphere and the temperature minimum where the acoustic cut-off frequency is too high to permit running waves at the observed frequencies, the p-modes leak in the form of evanescent waves, wave energy decreasing rapidly with height. Beyond this layer, they couple eventually to a second cavity in the chromosphere when the temperature has risen enough to permit propagation of acoustic waves in this frequency range again.

Simultaneously with the observational progress, Ando and Osaki (14.080.058) have studied the stability of the solar convection zone against these modes, and found that these modes were all overstable or just marginally stable in the range of frequencies occupied by the photospheric 5-min. oscillations. Ando and Osaki predicted as a result of their spherical, nonradial, nonadiabatic calculations an extensive list of eigenfrequencies as function of radial mode number and horizontal wave number, which could easily be compared with the observations. Deubner has demonstrated with such a comparison, not only the p-mode nature of the oscillations, but also the very close agreement of the predicted and the observed eigenfrequencies.

Further observational progress, leading to an increasingly better definition of the p-mode ridges in the k,ω diagram within a very short time, was made possible by extended scans across a substantial fraction of the solar disk (enhancing the resolution in spatial wavenumber k), very long observing runs making use of most of the available daylight time (thereby enhancing frequency resolution), application of the "long slit" technique (introduced by Rhodes et al. which permits to observe (k_x,ω) spectra in any preselected direction with respect to the solar axis, e.g. parallel to the equator), and extremely careful tune-ups of the existing observing equipment.

As a result, it became possible to resolve observationally an increasing number of p-mode ridges in the k,ω plane (>10), and at the same time, by independent observations (20.080.050) and by Deubner, Ulrich and Rhodes, to establish a small but significant difference between the observed eigenfrequencies, and those derived from standard solar models. This difference which increases with both horizontal wavenumber and radial wavenumber, is established well enough to warrant theoretical work aimed at an improvement of standard solar models. Such work is already in progress, and first results have been published. Gough (IAU Coll.36,3,1976), and Ulrich and Rhodes (20.080.035) show that better agreement can be reached by increasing the fractional thickness of the shell occupied by the convection zone, which is equivalent to increasing the mixing length parameter l/H. However, these authors have also shown that a perfect agreement may not quite be reached in this way due to the fact that other problems become aggravated by the existence of an extended convection zone (like the neutrino problem and the Lithium abundance). Ulrich (pr.comm.) suggests to study corrections to the equation of state for further improvement of the theory. Clearly, there is still a need for a more detailed modelling of the solar convection zone.

By virtue of the rapidly improving resolution in the observed k, ω plane it has become possible to study the coherence of the p-modes in space and/or time. The width of the p-mode ridges was investigated by Deubner et al. who found that it could barely be distinguished from the "instrumental profile" determined by the data window and the numerical techniques applied. In particular, Deubner (Catania Workshop on Solar Rotation, 1978, in press) found that Q (i.e.the ratio between the decay time and the period of a given mode) rises from about 80 for high wavenumber surface modes to infinity for low wavenumber modes penetrating more deeply into the convection zone. Whether the finite number observed at high k is a consequence of the perturbing influence of random surface velocity fields (such as supergranules or granules) disrupting the coherence of the p-mode velocity field, of atmospheric seeing, imperfect guiding, or of any other instrumental or solar effects, still remains to be shown.

Most sunspots seem to participate in the p-mode oscillations (Giovanelli, Livingston and Harvey, Solar Phys., in press). At any rate, the spatial/temporal coherence was found large enough to introduce the p-modes as a new kind of tracer of large scale horizontal flows in, or below the photosphere (Rhodes, Deubner and Ulrich (Ap.J., in press) and Deubner, Ulrich and Rhodes, Astron. Astrophys., in press). Normal modes of the sun, running in opposite senses parallel to the direction of the flow, may be separated in the Fourier plane according to their direction of propagation. A non-zero flow velocity then causes a frequency shift among each pair of eigenfunctions which is proportional to the flow velocity. This method seems ideally suited for large scale flows and solar rotation in particular, because it suffers only very little from spectrographic drift, atmospheric seeing or stray light as most of the spectrographic methods do which are being used to determine large scale velocities. Neither are the results derived from the observed "splitting" of the modes sensitive to "random walk" and evolutionary effects typical for small scale tracers.

Since the radial dependence of the p-mode eigenfunction depends on the geometry of the actual observed mode (Wolff 10.080.035 and Ulrich et al., Ap.J., in press) filtering the results of a mode-splitting analysis with respect to radial and horizontal wavenumber allows a probing of horizontal velocities in a given range of depths in the convection zone. A preliminary result of such an analysis concerned with radial differential rotation (Deubner et al., Astron. Astrophys., in press) has revealed a negative gradient $d\omega/dr$ with the average velocity at a geometric depth of 15 Mm below $\tau(5000) = 1$ being about 80 ms^{-1} faster than the surface rotational velocity. The implications of such a finding with respect to solar cycle models call for corroborative measurements and serious efforts to study the inversion problems connected with a non- uniform and potentially non-monotonic law of differential rotation as a function of depth (Gough, Catania Workshop on Solar Rotation, 1978, in press).

The excitation mechanism for the photospheric p-modes is not yet fully understood. The χ - mechanism (Ando and Osaki 14.080.058) could well be responsible for most of the observed power in these modes. The very low amplitude of the oscillation ($\Delta r/r$ equals a few 10^{-5}) seems to contradict self excitation, whereas the very low noise level found in the "valleys" in between the p-mode ridges on top of the instrumental noise appears to be an argument in favour of it. Careful studies of the dissipation of the p-modes in the higher layers of the solar atmosphere in comparison with the predicted growth rates are needed for an understanding of these observed facts.

Observational efforts as well as theoretical work has already been aimed at a better coverage of the resonant eigenmodes in the chromosphere/transition region (Ando and Osaki 20.080.006, Ulrich and Rhodes 20.080.035). Theory predicts that the leakage of p-mode energy into the chromosphere and corona should not seriously change the stability of the modes. The power of the photospheric ridges would be shifted further towards higher frequencies corresponding to wave modes becoming progressive already further down in the atmosphere. The chromospheric cavity would produce additional trapped modes of which the two lowest ones appear in the k, ω diagram at frequencies of about $0.027s^{-1}$ and $0.036s^{-1}$, nearly independent of the horizontal wavenumber.

The observational picture of this region of the solar atmosphere looks very complex, and still rather confusing, although a lot of effort has been put into these studies.

The existence of the new chromospheric p-mode ridges has been looked for, but they have not been found in the relevant k, ω diagrams (Rhodes et al. Proc. OSO-8 Workshop, 1977, Deubner et al, Astron. Astrophys. in press). However, the periods near 240s and 180s are well known from earlier high resolution spectroscopic work to stand out as resonances in the frequency power spectra of motions observed in chromospheric lines such as CaII K or H (Orrall, Ap.J.,143,917,1966, Deubner, 12.071.026). Possibly, these resonances can build up only in the interior of supergranulation cells, where the structure of the atmosphere is sufficiently homogeneous and comparatively little disturbed by the presence of magnetic fields, spicules, etc., but do not develop any kind of large scale coherence like the photospheric modes.

The continuing interest in chromospheric p-modes is of course related to the question of how much energy is carried by this wave field into the higher layers of the atmosphere. Until today, of the many processes which have been suggested as being capable of heating the upper part of the solar atmosphere, not a single one has yet been proven by observation to be of considerable importance.

With the p-modes and related oscillations the difficulty comes mainly from the fact that we happen to observe these motions in a layer of the atmosphere, where they change from evanescent to progressive waves. Very accurate phase measurements between velocity and intensity fluctuations in spectral lines, and between velocity fluctutations at several wavelength intervals corresponding to various levels in the atmosphere are needed to determine the character of the observed wave phenomenon. An important prerequisite of such observations is high statistical stability of the data which gradually seems to become available (Lites and Chipman, preprint). For the interpretation of the relative phase relations it is vital to assess very carefully the height of formation in the atmosphere of the spectral quantities being observed. One dimensional weighting functions may not be adequate, as pointed out recently by Keil (preprint). High spatial resolution may well be important to detect the energy flux, one is looking for, (e.g. see Cram et al. 19.073.033, Deubner 18.080.015), and it appears advisable then to take advantage of the technique of separating the various wave modes with different propagation characteristics in the k, ω plane, rather than deduce the phase and amplitude relations from frequency spectra alone.

It may be because in none of the existing observations of chromospheric wave motions all of these requirements have been met at the same time, that a consistent picture of the wave field has not yet emerged. There is agreement among observers that in the photosphere the low frequency waves (3 to 5 mHz) are evanescent (Schmieder 18.071.52), with some dissipation. In chromospheric lines (MgI 5172.7Å. CaII K), where higher frequencies (5 to 8 mHz) become more prominent, some authors (Schmieder, preprint. Mein 20.073.066) maintain that there is little energy propagation detectable, perhaps from narrow isolated structures only (Mein, Solar Phys., in press, Carm et al. 19.073.033; a dynamic nonLTE model for bright K grains was proposed by Carm 18.073.001). On the other hand there is evidence in recent observations by Lites and Chipman (preprint), supported by an analysis by Musman (preprint), for preferentially upward propagating acoustic waves at about 6mHz. These authors point out, however, that the energy flux at a height of about 900 km above $\tau = 1$ is too low by nearly two orders of magnitude to explain the heating of the middle chromosphere to corona. The same conclusion was reached by Athay and White (preprints) who have in a series of papers analysed extensive data obtained with the OSO-8 satellite in SiII and CIV lines in the upper chromosphere to lower transition region.

The important role of radiation relaxation varying considerably with height throughout the upper atmosphere was recognized and studied in detail (see e.g. Schmieder 18.071.052, Giovanelli, Solar Phys., in press). A minimum of the relaxation time ($\sim$ 1.5 min.) in the lower chromosphere due to cooling by emission in spectral lines may enhance wave dissipation of the p-modes in this region.

A feature of the chromospheric p-modes which is not easily anticipated from the model analysis must also be considered in the analysis of observational data, according to Leibacher (OSO-8 Workshop, 1977). The response at high chromospheric levels (> 1000 km) to a low amplitude pulse in the sub-photosphere is far from linear, producing steep pulses at regular intervals of nearly 200 seconds (corresponding to the acoustic cut-off frequency at the temperature minimum), with pressure variations of > 50% and particle excurions, at a height of 1800 km, of the order of 600 km. Thus, the following part of the wave train will interact with an atmosphere which deviates strongly from a mean static equilibrium chromosphere.

Radio emission fluctutations have been observed at cm wavelengths, corresponding to the upper chromosphere. The presence of brightness fluctuations of solar origin in the frequency range of the p-modes has been clearly demonstrated (among many others by Kundu and Alissandrakis 14.077.008, Graf et al, Ap. J., in press). However, the power spectra exhibit the rather smooth character of a power law with a frequency index of -2, rather than significant peaks in the frequency range of 3 to 5 mHz. This is in accordance with earlier EUV measurements by Vernazza et al. (14.074.005).

Graf et al. have recently obtained the first k, ω spectrum of solar radio brightness fluctuations. One is tempted to look for the signature of the photospheric p-modes in this diagram although it is clear from the foregoing discussion that the ridge structure must have disappeared at this level of the atmosphere, and the instrumental resolution in frequency and wave-number would not be sufficient to detect the photospheric mode structure. Yet, the degree of redistribution of power among other frequencies and spatial scales, compared with the photospheric input, should contain important information on the processes of wave propagation and dissipation in the upper chromosphere.

4. SMALL SCALE SOLAR MAGNETIC FIELDS: NEW OBSERVATIONS AND PHYSICAL INTERPRETATIONS

(W. C. Livingston and J. O. Stenflo)

Introduction

During the last decade, observations have revealed remarkable and previously unsuspected properties of solar magnetic fields. We now believe that most of the photospheric magnetic flux observed with a spatial resolution of a few arc-sec, or larger, actually resides in discrete flux patches at field strengths of 0.1 - 0.2T (1000 - 2000 G). The physical conditions in these flux elements appear to be very much the same whether we examine quiet regions such as the "network", where the large-scale average field is below 1mT, or active regions (excluding sunspots and pores), where the flux concentration may be two orders of magnitude larger. These unique properties of fluxtubes over all the sun allow us to treat solar magnetic fields in a unified way. Fluxtube models should apply to a large variety of situations on the sun. The main parameter that distinguishes large sunspots from tiny network elements is then the fluxtube diameter. In this report we cover observational and theoretical programs within 1976-1978. For earlier work see the reviews of Stenflo (18.080.076; Rep. Prog. Phys. 41.865) and Harvey (20.071.014).

New techniques and instruments

To better exploit the line-ratio method for the indirect deduction of field strengths, a 3-line magnetograph was installed at the Locarno station of the University of Göttingen. The results for the temperature insensitive lines of Fe 6302, 6337, and 6408Å lays to rest any questions arising from the usually employed ground state Fe 5250/5247 pair. Field strengths in the range 0.12 to 0.22 T were found.

Giovanelli, Livingston and Harvey were able to measure the true oscillatory velocities within fluxtubes by a novel technique using the Kitt Peak multi-channel magnetograph. The cross-over point of the V-Stokes parameter was tracked, using the output of the lock-in amplifier as the error signal for a tipping glass-plate Doppler servo. Oscillations within the fluxtubes turned out to be similar in amplitude and phase to the surrounding non-magnetic region, setting new model constraints.

The possibility that large amounts of "hidden" flux exists, i.e. flux so tangled that opposite-polarities cancel, has been explored along several avenues. Harvey and Stenflo made line profile surveys of large spectral regions using the 1-m FTS to look for a correlation between line width and Lande g (see also Stenflo and Lindegren, 20.071.002). An even more sensitive test employs the Hanle effect, according to which polarization caused by coherent scattering in Fraunhofer lines is destroyed by the presence of magnetic fields. From near limb Hanle effect observations, Stenflo proposes an upper limit of 0.1 mT to the field strength in any hidden flux.

The direct measurement of sub-arc-sec fields appears best attempted by narrow-band birefringent filters (18.080.123; 19.080.044). Observations by Trabell et al. with a 0.13 Å passband on Fe 6302, using the Sacramento Peak Vacuum Tower, attain a flux limit of 2 (10^9) Wb for 0.5 arc-sec seeing. Yet they fail to detect the "inner-network" fields regularly seen at lower resolution, but with enhanced signal to noise ratio at the McMath Telescope with the 512-magnetograph. As the inner-network flux limit is $\sim 5(10^8)$ Wb, they suggest that if this flux is more concentrated, they should see it; and therefore, these flux patches may be intrinsically diffuse.

Finally, it is of interest that a high resolution magnetograph experiment has been accepted for Spacelab Mission 2 of the U.S. Shuttle program. A 30 cm Cassegrain telescope will feed a 60mÅ filter and CID array camera. Magnetograms having, hopefully, 0.5 arc-sec resolution will be acquired in such lines as Fe 5250, 6302, etc. While this resolution remains insufficient, it should be constant with time and thus allow the accurate study of how magnetic fields emerge, evolve, and dissipate. Space lab 2 is scheduled to be launched in April 1981.

Other Observations

Koutchmy and Stellmacher (Astron.Astrophys.67, 93) have obtained spectra of magnetic elements under exceptional seeing with the SPO Vacuum Tower, They confirm the existence of high field strengths in the network and show that the fields extend beyond the filigree structures. Elste and Teske show that the hyperfine structure of Mn 5394 makes this line insensitive to microturbulent and magnetic broadening (in the outer wings), and hence can be used for temperature studies of the properties of fluxtubes. A number of facular model constraints follow from their work.

Kotov (19.080.048) demonstrated that the network fields are the principal source of the solar mean field ("the sun as a star"). Duvall and Scherrer from Stanford mean field observations, support this conclusion by showing that the line ratio given by Fe 5250/5247 agrees with the value observed on the disk at high resolution. Since Livingston, Harvey and Heasley have found that the discrepancy factor 5250/5247 varies somewhat from region to region, approaching unity for what is apparently aged flux, it will be of interest to see if the mean field 5250/5247 ratio varies with the solar cycle indicating the periodic magnetic refreshment of the sun.

The topological arrangement of fluxes and polarities in the photosphere determines much of the physics in the atmosphere above. Although most of the flux that makes up the supergranulation network seems to occur in largely unipolar regions, it is also very common with flux patches of opposite polarity in close proximity of each other, corresponding to closed bipolar configurations. These so-called ephemeral active regions have a median life-time of about 12 hr. and are closely correlated with X-ray bright points (20.076.013). They are not confined to the usual zones of solar activity although they probably represent the small-scale portion of the size spectrum of bipolar magnetic regions. Larger bipolar regions are then the normal active regions with sunspots. This size spectrum shows large variations with the solar cycle (19.076.010).

Interpretative problems

The interpretation of the observations is complicated by the circumstance that the basic magnetic elements are far from resolved. While their inferred diameters may be of the order of 100 km, the magnetic-field recordings can rarely resolve structures below 1000 km. The physical conditions in fluxtubes are drastically different from the surrounding conditions. Thus the gas in a fluxtube should be partially evacuated to allow field confinement (see below), there is heating in the higher layers and a downflow of matter that increases with depth. Due to the partial evacuation, there is a depression of the isodensity surfaces (Wilson depression) in the fluxtube. When the fluxtube is viewed at an angle, we will see more of the hot walls of the depression as compared when viewing from the vertical direction. This may explain some of the center-to-limb variation of the facular contrast (18.071.100). Caccin and Severino have studied the effect of geometry on the continuum diagnostics of faculae. The detailed fluxtube geometry, however, also affects the interpretation of disk-center observations in spectral lines. Photons from the hot walls may be scattered in the vertical direction and thus add to the core intensity of spectral lines in the interior of the fluxtubes. This contribution to the 'line-gap' effect is decoupled from the local temperature enhancement in the fluxtube unless a proper multi-dimensional transfer treatment of the radiative-transfer problem is performed. The methods of treating this type of multi-dimensional transfer of the full stokes vector of polarized light have been developed by Stenholm and Stenflo (Astron. Astrophys.58,273; 67,33) and applied to the case of idealized cylindrically-shaped fluxtubes. The relative magnitudes of the effects of radiation channelling and heating depend however, on the details of the fluxtube model, how much the fluxtube interior is shielded from the hot-wall radiation, the magnitude of the Wilson depression, and the height divergence of the field. A realistic quantitative study of these effects has still to be made.

The principal method to avoid these problems and still be able to deduce information on the unresolved properties of the magnetic elements has been the 'line-ratio' method (18.080.076). By making simultaneous magnetograph recordings in two or more spectral lines, which belong to the same multiplet and are formed in the same way apart from the difference in Landé factor, it has been possible to isolate the magnetic-field effects from all the other coupled physical parameters and deduce the true field strengths. The interpretation of the results depends however, on the assumed shape of the magnetic cross section of the fluxtube. The smallest field strength, 0.1 T, is deduced assuming a rectangular (single-valued) cross section. Any other cross section gives higher values for the peak field strength in the element. Thus a Gaussian shape gives about 0.2 T or slightly more.

Fluxtube models

Various attempts to construct empirical models of the temperature and density stratification in fluxtubes from continuum and line-profile data (facular models) have been made. The most detailed LTE model of this kind is that of Chapman (19.072.013). It is based on line-profile data at disk center and shows a temperature enhancement that increases steeply with height. As noted above, however, part of the derived temperature enhancement may be an artifact of the LTE treatment, since multi-dimensional transfer effects will tend to modify the line profiles in the same direction as an increased temperature. Elste and Teske find a less steep gradient than Chapman. Further, as pointed out by Frazier (20.072.032; Astron. Astrophys. 64,351), there is nothing like a unique facular model. But we should rather have a one-parameter family of models, the varying parameter being the fluxtube diameter. The line profiles and thus also the temperature-density stratification vary continuously with fluxtube size. Small fluxtubes (network elements) are brighter in both the continuum and the lines, whereas larger elements (small sunspots) are darker in the continuum but brighter in the line cores. Facular models based on some average or typical observed facular contrasts may thus be severely contaminated by observational selection.

The continuous variation of fluxtube properties from the smallest bright elements in the quiet network to plage elements, pores and large sunspots, has been discussed by Spruit (18.071.100; 19.080.011) and Stenflo (I.A.U. Colloquium 36.143). Spruit, Beckers and Zwaan have tried to determine the relative size distribution of fluxtubes in an active region by measuring the bright points on high resolution Mg b_1 pictures. Unfortunately, much of the interesting portion of the distribution occurs at or below the spatial resolution. The distribution is also expected to vary greatly from one region to the next on the sun (enhancement of the small-scale end when we go from active to quiet regions).

Spruit (18.071.100; 19.080.011) has investigated the pressure balance and energy transport in fluxtubes and calculated a family of theoretical magnetostatic fluxtube models characterized by flux-tube radius, Wilson depression and heat flux. A major effect of the magnetic field is to inhibit convective energy transport across the field lines. For small fluxtubes the lateral inflow of radiation can cause the structures to appear bright. The effect of this inward flux is reduced in larger fluxtubes due to the increased lateral optical depth. Spruit's family of models seems to account for the major observed characteristics of fluxtubes at least qualitatively, but a detailed quantitative comparison with observed spectral features has still to be done.

The concentrated photospheric fields have important implications for the structure and energy balance of the chromosphere and corona. A model for the extension of the fluxtubes into the corona has been worked out by Gabriel (18.073.076) including the effects of the magnetic field on the temperature and density structure.

Formation, confinement and decay

The explanation of the concentration of most of the photospheric magnetic flux into 0.1 - 0.2 T fields is a major plasma physics problem. Parker has treated the basic physics of this problem in a series of papers in the Astrophysical Journal. He concludes (17.080.015) that the confinement of the field cannot be accomplished by the dynamic forces of the plasma motions nor by twisting (electromagnetic confinement, force-free fields). The only possible way to achieve confinement is by having a larger external gas pressure that balances the magnetic pressure in the tube. The gas pressure must then be reduced by evacuation or cooling. In sunspots, Parker believes that most of the cooling is achieved by the convective generation of Alfvén waves emitted preferentially downwards. In the case of the small fluxtubes in the network and faculae, Parker (Astrophys.J. 221,368) suggests that the main cooling mechanism is the adiabaticity of the downdrafts within the fluxtubes. As the magnetic field inhibits convective energy exchange between fluxtube interior and exterior, the interior is efficiently insulated from the surroundings, with the consequence that the downdrafts will be adiabatic within the fluxtube. Due to the superadiabaticity of the exterior convection zone, the fluxtubes will be cooler than their surroundings. The cooling tends to enhance the downdrafts further leading to a partial evacuation of the fluxtubes and the establishment of a Wilson depression. This effect was originally proposed by Nordlund. Unno and Ando have treated this downdraft effect as a convective instability. Spruit has gone a step further to explain the formation of the fluxtubes. Using the equations for the 'thin tube approximation' (Roberts and Webb , Solar Phys.56,6) he finds that fluxtubes with field strengths above about 0.13 T are stable against this convective instability. If the field strength is smaller, the instability will spontaneously concentrate the tube through a downdraft until the field strength reaches 0.13 T, or an updraft will dilute the field and try to make it infinitely weak. This may explain the unique properties of the observed fluxtubes.

In Spruit's model the downdrafts occur only during the stage of formation. Observations (Giovanelli and Slaughter, Solar Phys. 56,255; 20.071.014) and those made recently by Frazier and Stenflo show however, that downdrafts do occur inside the fluxtubes during most of their life time. Thus, unless the fluxtube decays as soon as it has been formed, the downdraft material has to be supplied from the exterior, through field line crossing. Giovanelli (20.071.033) has proposed that this is accomplished by an inflow of neutral gas near the temperature minimum. Unno and Ribes propose a steady state polytropic model, although their calculated downflows are somewhat higher than observed.

For alternative mechanisms of flux concentration, see Galloway et al.(19.080.014) and the review by Weiss (20.080.032).

Once formed, the fluxtubes may be subject to the flute or interchange instability and decay by fragmentation into smaller elements. Observations indicate that such fragmentation occurs at the surprisingly fast rate of 10^7 Wb s^{-1}(Stenflo, 18.080.076), although it is not clear whether this also applies to the sub-arc-sec fluxtubes. Meyer et al.(19.072.086) have shown that the buoyancy force has a stabilizing effect on the flute instability in the upper portions of the fluxtubes, where the field diverges rapidly with height. Deeper down, however, the flute instability may take over; this has lead Parker (1978b) to propose a sunspot model, in which the sunspot flux splits up into many separate fluxtubes somewhere within 1000 km below the visible surface. Piddington (18.080.064; Astrophys. Space Sci.55,401) has argued that the fluxtubes are stable against the flute instability because they are strongly twisted. In a series of papers in the journal Astrophysics and Space Science he has presented a flux-rope-fiber model of solar magnetic fields. He concludes that the fields cannot be concentrated at the surface but must emerge in the form of intense, twisted fluxropes. Gokhale (Kodaikanal Obs.Bull.Ser.A.2,10; 2,19) from a similar viewpoint has attempted to explain the creation of thin intense flux tubes in the deep convection zone and their subsequent upward motion. The stability of twisted flux tubes has been studied by Wilson (19.062.047; 20.062.073). Nakagawa and Stenflo find that pure rotational twisting motions in the photosphere or below can lead to a concentration of magnetic flux to the axis of the tube in the force-free region, accompanied by a density enhancement in the chromosphere and corona.

Some outstanding observational problems

As the fluxtubes in the network or in plages have not been spatially resolved, their properties have been inferred using more or less indirect methods, which are generally statistical in nature. We know almost nothing about the morphology and evolution of the basic magnetic elements. We do not know in what form the flux elements emerge, what are their life-times as cohesive entities, or how they decay. The necessary spatial resolution cannot be obtained by groundbased observations; so space magnetographs such as provided by Shuttle facilities are urgently needed. We need to be able to follow the interrelated evolutions of the magnetic elements, downdrafts, vorticity and azimuthal fields, oscillations, brightness enhancements, and the granulation pattern with much better resolution than has so far been achieved, in order to learn about the basic fluxtube physics. We need to determine the size distribution of fluxtubes in various active and quiet regions on the sun, in particular the latitude and time (solar-cycle) variation of the spectrum of bipolar magnetic regions. More stringent limits on the uniqueness of fluxtube properties are needed, i.e., we want to know how uniquely the fluxtube structure is determined by the fluxtube diameter.

It is also important to study the non-network weak-field fluxes with high spatial resolution, to learn about the evolution of their morphology and their relation to the strong fields. The Hanle effect in combination with high spatial resolution seems to offer promising possibilities for the study of the weak fields.

There are presently practically no observational limits on the height gradient of the field in fluxtubes. Simultaneous observations in different spectral lines may solve the problem, but a proper derivation of the height gradient requires the use of proper fluxtube models. The construction of empirical models is complicated when the observations do not resolve the structures, and the choice of fluxtube geometry affects the interpretation. However, the present rapid progress in instrumental and interpretative techniques provides hope that many of these problems may be solved during the next decade. An understanding of the basic physics of the magnetic elements in the solar atmosphere appears essential for understanding the mechanisms of the solar cycle.

Finally, since small scale magnetic structure is the dominant source of the sun's mean field, observations of the temporal variation of the fields of stars of spectral types similar to the sun may provide magnified, or clarifying, examples of solar cycle effects. Particularly if we could apply the line-ratio technique, observe stars pole-on, or simply monitor solar sector analogs on other stars, the underlying mechanism of the magnetic cycle might be deduced from models of stellar evolution.

5. ALFVEN WAVES IN THE SOLAR ATMOSPHERE

(Y. Uchide)

Introduction

In the last three years, the problem of Alfvén waves (here we do not confine ourselves to the pure Alfvén mode waves in the strict sense of MHD mode analysis in an infinite medium) has attracted attention in several aspects of solar physics. We confine this discussion to the following four major aspects among them. These are (i) the role of the Alfvén waves in the sunspot cooling mechanism whereby the cooling is effected by the escape of the excited overstable Alfvén waves rather than by the inhibition of the convective energy transport in the presence of the magnetic field, (ii) the problem of oscillations in sunspot umbrae and small magnetic knots, and of running penumbral waves, all attributed directly or indirectly to Alfvén waves, (iii) the Alfvén wave as a heating agent of the coronal loop structure observed in soft X-rays which allows us to envisage the magnetic field pattern in the corona, and (iv) the possible role of Alfvén waves in providing additional momenta to the solar wind, in order to explain the observation that the solar wind fast stream originates from the coronal hole. We attempt herein to review briefly these different aspects.

The cooling of sunspots by Alfvén waves

A series of papers was devoted to the problem of whether the Alfvén wave can play the role of the cooling agency in sunspots when it is excited through the overstability and propagates out of the region. This idea of Alfvén wave cooling was first proposed by Parker to replace the mechanism of inhibition of convective energy transport by the magnetic field as proposed many years ago by Biermann. According to Parker, the excitation and escape of Alfvén waves caused by the thermal overstable convection in the magnetized atmosphere serve as a cooling device, and the gas pressure lowered by the cooling may in turn facilitate the concentration of magnetic fluxtubes. The formation of sunspots itself may eventually be caused in its nonlinear regime. This concept has received support from some researchers (19.072.003; 19.072.085; 17.072.010) but Cowling (18.072.034) claims that Parker's mechanism is not acceptable both from thermodynamical and hydrodynamagnetic points of view. There would also be the difficulty with the eventual deposition of the missing flux. Beckers (19.072.012; 19.074.077) tried unsuccessfully to find observationally whether any appreciable part of this missing flux was found escaping into the corona. This result, or the smallness of the amount of escaping Alfvén wave flux into the corona, had been predicted theoretically by Uchida and his collaborators (11.074.052; 13.072.058) and confirmed by Thomas (Astrophys.J.225,275) in the context of coronal loop heating and of the resonant trapping of Alfvén waves in sunspots in the form of the umbral oscillation (we come back to these problems later). Parker's hypothesis may still be able to rely upon the energy leaking downwards into the deep layer, or possibly also upon the observed penumbral wave (08.072.053), but even in such cases the problem of the eventual disposal of the missing flux without any observable trace seems to pose serious difficulty (18.072.034). Spruit (19.080.011), however, has claimed that the problem of the missing flux can be resolved in a different way. According to him, all that is necessary is reduced heat conduction near the solar surface.

If this is true the bright ring expected both in Biermann's picture and in Parker's down-propagating case may become rather diffuse. Therefore, it may be said that the debate itself has also become rather "diffuse" without a definite conclusion being reached.

The oscillation in sunspot umbra, in magnetic knots, and the running penumbral waves

The problem of the oscillation of sunspot umbrae of non-300 sec period found by Giovanelli, Beckers and Schultz, and of magnetic knots (Giovanelli, Livingston and Harvey) has called for much discussion (Solar Phys.56,5). The mechanism of the excitation of the selected eigen-periods has been attributed to the overstability of the convection layer having a magnetic field with the resonant trappings of Alfvén waves in the region (13.072.058; 12.072.004), the observed oscillation being the secondarily induced compression. The running penumbral waves (08.072.053; 17.071.032) were considered to be waves leaking out of the overstable region (10.072.010) guided along the magnetic field of the penumbra (17.072.011). The observation that the umbral oscillation has a period different from the general 300 sec period oscillation may be taken to provide evidence against the hypothesis that 300 sec oscillation is a global one (Publ.Astron.Soc.Japan. 27,581; 14.071.042). Because if it is, the magnetic fluxtube of sunspots whose roots are imbedded in the denser interior should be affected by the oscillation there and may not be able to have other periods of oscillation.

The heating of the coronal loops

One of the problems related to Alfvén waves and provided with much new evidence by X-ray and EUV observations from space, is the problem of coronal heating. The Skylab observations showed that the hot corona around the active regions is directly related to the magnetic field, copying the field pattern of active regions. The background cooler corona is also related to other weaker background magnetic regions. The long-believed acoustic wave theory of coronal heating was considered inapplicable, at least as the predominant source. That the Alfvén wave was the possible agency for coronal heating was suggested by Uchida and Kaburaki (11.074.052) and by Piddington. In the above-mentioned search for evidence of sunspot cooling by Alfvén waves, Beckers (19.074.077) also examined the velocity amplitude of the loops above active regions, and found an unexpectedly low oscillatory velocity (19.074.045). Uchida and Kaburaki's proposal was to consider the non-linear wave-mode coupling which produces the decayable slow mode waves out of the energy-carrying Alfvén waves. Such waves are difficult to be dissipated in the coronal condition, and need a considerable wave amplitude (Beckers pointed out in this connection that the estimate they made for the wave-amplitude by using the WKB treatment left out the factor of area change, and accordingly, their estimate was an overestimate). Therefore, these observations of low oscillatory velocity indicated that Uchida-Kaburaki type mechanism might not operate too efficiently. Wentzel (18.074.140; 20.074.048) considered wave-mode coupling through the interaction in the chromosphere, producing short wavelength Alfvén waves with periods of 10 sec or less. He, also, pointed out that the Alfvén wave with a source of a finite (small) extent has the character of a MHD fast-mode wave rather than that of the usual Alfvén waves with an infinite wavefront extension. Wilson (Astrophys. J.225, 1058) also noted that the waves show a considerably different behaviour in a confined domain, and considered surface waves on the boundary between two media or in the skin of the loop structure. Ionson has proposed that the resonantly absorbed energy of such a surface wave may be deposited in the loop-skin and may heat up the coronal loops. Wentzel has also advanced this idea and considers that it might be the way out of the dilemma that the loops are intensely heated while they have rather low oscillatory velocity. No definite statement, however, can be made at the present stage that the Alfvén wave heating hypothesis is better than other hypotheses such as heating by the dissipation of electric currents which may be flowing along the loop (Astrophys. J.222, 317). The hypothesis of the heating of the loops by intermittent plasma injection from the heated footpoints seems inadequate to explain the majority of the loops except for some special ones, because the soft X-ray observation indicates a continual heating of the loops (Solar Phys. 57,103).

Acceleration of solar wind stream in the coronal holes

The fourth aspect is the possible role of Alfvén waves as the agent of the momentum addition to the solar wind fast stream. Frustrating enough, however, the solar wind fast stream comes from the coronal hole (09.074.083) where we can see rather little of what is occurring in the corona, but indirect evidence suggests that it is at least not an exceptionally high temperature region. On the contrary, it is likely that not only density but also the temperature is lower in the coronal hole than in the surroundings (20.074.009). We should therefore consider some additional device of acceleration other than the standard mechanism of thermal expansion wind if we want to explain the fast stream which is difficult to explain as even the thermal expansion wind from the warmer normal corona. There are several ways proposed to solve the difficulty but we here confine ourselves to the possibility of the direct momentum addition by Alfvén waves in the context of our present theme. The idea of acceleration was also motivated by the detection of the Alfvén wave in the solar wind at 1 A.U. by Belcher and Davis, although there is a possibility that it is expected insitu by instabilities involving Alfvén waves (20.062.045; Solar Phys. 57,445). Hollweg, in a series of papers (07.073.024; Solar Phys. 56,305), pursued the possibility of finding the wave source on the sun, and suggested the motion of the supergranule boundary as the source. Both he and Jacques (20.074.003) have considered the effect of the Alfvén waves as a source of momentum to the solar wind flow. Rosner and Vaiana (20.074.008) claim that this type of acceleration should take place in the coronal hole beyond the temperature maximum. It should be remarked here that other mechanisms, such as the formation of the additional de Laval point, besides the Parker critical point due to the configurational effect (18.074.110), offer little hope in explaining the fast stream, though no final preference can be made at this stage.

6. ENERGY FLOW IN THE CHROMOSPHERE AND CORONA
(R. Grant Athay)

Overview

This brief review of energy flow in the chromosphere and corona is intended to summarize our current level of understanding and to mention the major highlights of the 1976-1978 period. Because of the imposed brevity of the report, many relevant contributions must be omitted. Also, emphasis is on the quiet sun, which bypasses work related to flares and active regions.

The steady-state energy balance in the solar atmosphere may be expressed in the form

$$\nabla \cdot F_{in} = \nabla \cdot F_{out}$$

where F_{in} and F_{out} represent incoming and outgoing energy fluxes within a volume element. The problem of determining flux gradients is complicated by our inability to observe them directly. Instead, we observe only the net fluxes of radiation and mass flow reaching the vicinity of earth, and from these restrictive observations we attempt to derive all the relevant energy gradients. Such derivations, of necessity, require accurate modelling of the temperature, density, velocity and magnetic structure of the solar atmosphere.

The past few years have brought several important conceptual changes, some of which are still just emerging. Two important changes in modelling concepts are: (1) that the magnetic geometry deviates radically from a radial geometry even on a small scale, and (2) that even in the quiet sun systematic flows are important. The first concept has radically modified our evaluation of heat conduction and solar wind flow: the second has made us more aware of the importance of enthalpy transport and of work done against gravity. A change of perhaps more consequence is reflected in a notable shift in favour of magnetic heating of the corona as opposed to shock wave heating.

Measured Energy Fluxes

The only cases for which we observe energy fluxes that can be ascribed unambiguously to the chromosphere and corona are for the solar wind flow and for the emission in the EUV, X-ray and radio regions of the spectrum. Near the sun, the solar wind energy is carried mainly by thermal conduction and arises almost entirely outside the temperature maximum. We know very little about temperature gradients in the outer corona, and, as a result, we know very little about radial gradients in the solar wind flux.

The source of the high speed solar wind streams has been identified with coronal Holes, including some high latitude holes of polar origin (Hundhausen, Coronal Holes and High Speed Wind Streams, ed. Zirker, 1977, 225). Thus we must now conceive of the solar wind as emerging mainly from "nozzles" covering but a fraction of the solar surface rather than as a general expansion of the entire corona. Within the coronal holes, the solar wind flux appears to be the primary energy loss from the corona (Withbroe, ibid. 145).

EUV and X-ray fluxes from the sun at wavelengths shorter than λ 1700 arise in the chromosphere and corona. Approximately half of the flux is in Lyman alpha, which is of chromospheric origin. The remaining half comes mainly from the transition region and inner corona below the temperature maximum. The average EUV and X-ray flux exceeds the average solar wind flux by an order of magnitude. Energy flux in the radio portions of the spectrum is of little consequence.

Approximately 40 percent of the solar surface, as seen at temperatures from 2×10^4 to 10^6K, is covered by bright network. The network accounts for 75 percent of the observed flux at temperatures near 3×10^5K and for 60 percent at temperatures near 5×10^4 and 10^6K (Reeves. 18.073.034). There is little apparent difference in the network underlying coronal holes from the average network.

Network regions are characterized by a general downflow of material, but they are also the seat of spicule eruptions. Also, the network outlines the regions where the magnetic fields are strongest and where the field lines are most nearly vertical.

X-ray observations of the corona have provided convincing evidence that the soft X-ray flux is unusually large in regions where the magnetic field lines are closed, as in active regions and newly emerging dipole regions. It seems clear that the soft X-rays are of thermal origin and that high X-ray flux is a direct measure of high thermal energy content. The high energy content of the corona in closed field regions and the relatively high radiation losses from the network imply a high rate of energy input to these regions. Thus, in the corona the energy input is closely coupled, to regions of closed magnetic field lines, and in the chromosphere it appears that the energy input is highest in regions where the magnetic fields are vertical and strong.

Studies of photospheric and chromospheric oscillations with periods of 3 to 5 minutes strongly support the concept that considerable wave energy is present in the lower atmosphere and that at least some fraction of it propagates into the chromosphere and corona.

These brief preceding statements represent about the limit of what we have extracted from observations concerning the energy flow in the solar atmosphere without imposing some form of restrictive modelling of the phenomena observed.

Inferred Energy Fluxes

Models serve the dual purpose of allowing us to convert integrated fluxes into flux gradients and of identifying energy fluxes that are intrinsic to the models. By modelling the oscillations, for example we can infer the energy flux in the waves. By modelling the temperature and magnetic field structure of the corona and transition region we discover that thermal conduction plays a major role in the energy balance, and by modelling the chromosphere we discover that radiation losses from the low chromosphere exceed the combined losses from the high chromosphere, transition region and corona. It must be emphasized, however, that all such conclusions are model dependent.

Since the energy balance depends upon differentials of energy fluxes the problems related to energy balance place severe demands upon models, i.e., even a modest accuracy in the differential requires high accuracy in the quantity being differentiated. Nevertheless, models have provided considerable insight to identifying mechanisms of possible importance.

Radiation losses from the low and middle chromosphere involve separating out the chromospheric contributions to the strongest Fraunhofer lines. Such a separation requires an accurate theory of line formation and an accurate evaluation of collisonal excitation rates within the chromosphere. The latter depend sensitively on the chromospheric temperature. Current models indicate that the emission cores of the MgII and CaII resonance doublets and most of the central intensity in Balmer alpha and the CaII infrared triplet are due to electron collisional excitations the chromosphere and therefore represent a true loss of energy. The sum total of this radiation loss exceeds the chromospheric and coronal EUV radiation losses by about a factor of four.

Near the temperature minimum H^- opacity is significant and in addition, numerous Fraunhofer lines contribute to the "effective" mean opacity. Even small departures from radiative equilibrium require important energy inputs relative to the chromosphere and corona. Neither the radiative equilibrium model nor the true model is known with sufficient accuracy to provide a reliable estimate of the nonradiative energy term. It could easily be that the upper photosphere and temperature minimum regions require more non-radiative heating than either the chromosphere or the corona (Athay 17.003.016).

Deubner's (18.071.059) estimate of the very substantial energy flux in high frequency photospheric oscillations depends upon the assumption that the high frequency waves are pressure waves propagating vertically at the sound speed and, in addition, that the true velocity amplitude is smeared to a fraction of its value by radiative transfer effects. Each of these assumptions is plausible. They are reinforced by the work of Ando and Osaki (20.080.006) and Rhodes, Ulrich and Simon (Astrophys. J.218, 901), which clearly establishes the 5-minute oscillations as high order, nonradial p-modes of the global sun, and which, in addition, predict that the higher frequency modes should be present and should propagate. The energy inferred by Deubner appears to be sufficient to heat the low chromosphere.

Several attempts to obtain energy fluxes in waves in the chromosphere and transition region have been made with conflicting results. The most recent and most extensive results have come from the OSO-8 satellite. However, no one has claimed to find energy fluxes in excess of a few times 10^4 erg cm^{-2} sec^{-1}, which is at least an order of magnitude below the heating requirements of the upper chromosphere and corona (Athay and White, Astrophys.J.,226,1135). This failure plus the observed association between X-ray flux and closed magnetic field structure have undermined confidence in the theory of wave heating for the upper chromosphere and corona.

The estimate of the conductive energy flux in the transition region by Gabriel (18.073.076), which is the most reliable so far, is dependent directly upon the magnetic field geometry. Unfortunately, the geometry is based only on plausible assumptions rather than observations. Similarly, Pneuman and Kopp's (Solar Phys., 57, 49) estimate of the importance of enthalpy flux is dependent upon the assumption that the downflow velocities are typical of the most of the network rather than of just isolated segments of network. Observations have not yet provided an answer to that question.

In the low and middle chromosphere the main energy loss is radiation. Energy input by dissipation of sound waves still appears to be promising. Within the transition region and bordering regions major radiative losses occur. However, heat conduction, enthalpy transport and changes in gravitational energy associated with vertical motions are all of major importance. Whether heat input from waves or magnetic fields is important is still an open question. In the corona, heat input is again clearly required. Thermal conduction, both inward and outward and along closed field lines is of major importance. Radiation losses are of diminishing consequence, but still of importance. Enthalpy transport and gravitational energy changes are important in the outer corona and probably, as well in the inner corona. However, the latter is largely conjecture since we know little about motions in the inner corona.

Theoretical Models

Theoretical models in which energy mechanisms are specified and which predict solar wind flux and/or XUV fluxes are useful for identifying specific deficiencies in the assumed mechanisms and for helping to eliminate unimportant mechanisms. Unfortunately, one cannot be certain, at the present time, that such models contain all of the essential physics. The heat input is usually treated as an extrinsic source i.e., independent of the resulting atmospheric structure, whereas other energy flows, such as radiation and thermal conduction are completely intrinsic. A major problem occurs in the case of mass flows since these clearly may be a result of both extrinsic and intrinsic properties. The heat input may be accompanied by momentum input, as in the case of waves. Also, the resulting temperature and density configuration may be statically unstable. Most theoretical models ignore such problems, except in the case of the outer corona where the solar wind becomes important.

Gonczi, Mangeney and Souffrin (19.064.013) have discussed in some detail the basic assumptions and limitations that normally accompany treatments of heating by shock waves, and they propose improved methods for treating this problem. Flower and Pineau de Forets (18.074.035) have considered rather carefully a model in which shock heating, conduction, radiation and solar wind expansion are included. Kopp and Orrall (18.074.102) adopt a different approach in which the heat input is described by a boundary value and an exponential scale height. The heat input is balanced against radiation, conduction and solar wind flow through a magnetic geometry that diverges faster than r^2. They then select the two parameters of the heat input that give best agreement with observations.

Each of the above approaches, as well as others of similar type, has considerable merit and more work of this type is certainly to be encouraged. The models clearly represent many of the basic elements of the observations. However, until the magnetic field geometry is properly included and until the full effects of mass flow in the upper chromosphere,transition region and inner corona are properly included we cannot trust any detailed inferences regarding the heat input.

The modelling approach by Kopp and Orrall appears to be particularly promising in view of the recent pessimism about wave heating in the transition region and corona. On the other hand, the alternative to wave energy is magnetic field dissipation, and this makes it doubly important that the magnetic field geometry be compatible with the assumed heating.

Magnetic Field Dissipation

The supposition that the solar atmosphere is heated by dissipation of magnetic field energy is gaining popularity. Past theories have focused attention on magnetic reconnection in neutral sheets between merging fields of opposite sign. Such proposals have had the general disadvantage that energy is dissipated only at unique, and therefore presumably widely scattered, field locations. The heat input, however, is a widespread phenomenon.

Parker (14.076.039) proposes a means of avoiding this difficulty by noting that strongly twisted flux tubes would develop kink instabilities, and that in an aggregate of such twisted tubes magnetic reconnection between kinks in adjacent tubes would be widespread. In a much more radical departure from the normal picture,Rosner, Golub, Coppi and Vaiana (Astrophys.J.222, 317) propose that magnetic energy is dissipated by anomalous resistivity acting on force free currents. Since force free currents can arise from virtually any deformation of the magnetic field from its potential configuration, the existence of force free currents is expected to be widespread. Thus, the expected heating by dissipation of such currents would be widespread. Also, the dissipation mechanism is self contained in that it does not rely upon interactions between adjacent flux tubes. As with all other magnetic heating models, however, there are many unanswered questions. In particular, the required twisting and contorting of the magnetic field is left to a Genie still hiding in a magic lamp beneath the photosphere.

7. CORONAL HOLES
(J. B. Zirker)

Research on coronal holes and their associated high speed wind streams has progressed rapidly during the period 1976 through 1978. Just prior to this period, holes were observed as regions of low density and temperature in the corona, were identified tentatively as the source of high speed wind streams, and were recognized to rotate rigidly. Firm correlations between holes, streams and geomagnetic disturbances were beginning to appear. A nine-month Skylab Workshop on Coronal Holes (October 1975 - August 1976) stimulated many new advances (Coronal Holes and High Speed Wind Streams, ed. J. B. Zirker, Colorado, 1977; abbreviated henceforth as CHM).

This review considers only a few highlights of coronal hole research during 1976-1978. More complete details can be found in such reviews as Bohlin ("Phys. of Solar Planetary Environments," Am. Geophys.Union, 1,114, 1976), Harvey and Sheeley (Space Sci. Rev.1978), Hollweg (Rev. Geo.Sp.Sci., 1978), Orrall (OSO-8 Workshop Proc., 1978) and Zirker CHM; 20.074.009).

Observational Characteristics

Holes lie in unipolar regions of photospheric magnetic fields. The photospheric field in coronal holes differs very little from that outside them. Only when the photospheric fields are extrapolated to the corona, using the method of potential continuation (Altschuler et.al.,19-074.053), do holes stand out as essentially open field regions.

The field lines within a hole diverge non-radially. According to Levine et al. (Ap.J.,215,636, 1977), the cross-section of Skylab coronal holes expanded 10 to 100 times faster than radially; however, the north polar hole expanded in cross-section only seven times faster than radially (Munro and Jackson, 19-074.038).

Evidence for the behaviour of holes throughout the sunspot cycle is still fragmentary. According to Broussard et al.(Solar Phys., 56, 161, 1978), the area of polar holes shrank before the maximum of cycle 20. Just at maximum, holes were located between the polar caps and the sunspot belts, or near the equator. These holes were small and lasted less than two rotations. Coronal holes are largest and most stable (i.e. 3-10 rotations) during the declining phase of the sunspot cycle. During the current rising phase of the cycle, holes are small, transient and appear at high latitudes.

The Association of Holes and Streams

The holes and wind streams that occurred during Skylab (Hundhausen, CHM) showed a strong association. Of 69 central meridian passages of coronal holes ($\pm 30^0$ latitude), 75% are associated with streams at 1 A.U., 9% are possibly associated and 16% are not associated. All the large holes observed during Skylab have been positively identified with streams. The polar holes are also sources of high speed wind, as Rickett and Sime (J. Geophys.Res., 1978) have shown.

Prior to 1976, several authors reported associations between coronal holes, high speed wind streams, and recurrent geomagnetic disturbances. Recent work by Rickett et al. (18.074.019), Bell and Noci (18.074.021) and Sheeley et al. (18.074.128) confirms this connection.

Physical Models

One of the objectives of coronal hole research is to construct a comprehensive model of the run of physical parameters that will predict all of the observed radiation and account for the expansion of the corona. To date, this grand objective has only been partially achieved and serious discrepancies have turned up.

Preliminary analysis of some of the data has yielded important constraints on comprehensive models. Calculations of the potential magnetic field within a hole show that the field lines diverge nonradially in the middle corona. Kopp and Holzer (18.074.110) found that a diverging field geometry tends to lower the height of the critical point and to reduce the density at low altitudes in a coronal hole. Munro and Jackson (19.074.038) used Skylab observations of the white light corona to derive an empirical model of the density between 2 and 5 $R_\odot$. They showed that the cross-section of this hole expanded 7 times more rapidly than that of a radial cone, and that, for a reasonable particle flux at 1 A.U., the critical point lay between 2.2 and 3 $R_\odot$. Thus, the primary acceleration of the wind seems to occur below $R_\odot$

The earliest studies of coronal holes (e.g. Munro and Withbroe (08.074.022) showed that the temperature gradient in the transition zone, and hence the downward heat flux, was an order of magnitude smaller than in the quiet corona. These results have been confirmed with Skylab observations by Mariska (Astrophys.J. 225, 252, 1978) and by Withbroe.

The energy flux carried in high speed streams is about twice as high (i.e. 2 erg cm^2 sec^{-1}) at 1 A.U. as in the quiet wind, and if the observed expansion factors are applied, an energy flux of 5×10^5 erg cm^{-2} s^{-1} is needed at the base of a hole. This flux is an order of magnitude larger than required for the quiet wind and sets a severe constraint on solar wind theory. The energy carried in a wind stream is the predominant loss from a hole, far exceeding the downward heat conduction.

Models of the transition zone within a hole, based on Skylab EUV observations, have been very useful in setting boundary conditions for coronal expansion models. Withbroe and Mariska derive $N_e = 1.7 \times 10^8 cm^{-3}$ and $T_e = 7.6 \times 10^5$ at 1.03 $R_\odot$ within the north polar hole. Assuming a typical particle flux for this hole, Mariska derives a flow velocity of 15 km/sec at 1.08 $R_\odot$. This value agrees with the Doppler velocity measured in another hole by Cushman and Rense (18.074.003).

Unfortunately, no measurements exist of the maximum temperature or its location in a coronal hole. X-ray, EUV and white light observations imply temperatures no higher than about 2 million degrees, but considerable interpretation is necessary for this conclusion. Independent evidence consists of the ionization equilibrium of the oxygen, nitrogen, iron and silicon ions that are "frozen" in the wind (Bame, 11.074.041). These data imply that the maximum temperature in the wind stream is probably no higher than 2.1 million degrees at about 1.5 $R_\odot$. With such a low maximum temperature, the wind stream must derive energy from some other source than heat conduction to attain the observed energy flux at 1 A.U. Extended heating by waves and radiation pressure of waves (especially Alfvén waves, as proposed by Barnes and Hollweg), have been generally accepted as plausible working hypotheses.

Recent coronal hole models fall into three categories; semi-empirical (Dulk, 20.074.050), Munro and Mariska, OSO-8 Workshop Proc., 1978), energy balance models (McWhirter and Kopp, MNRAS, 1978; Rosner and Vaiana, 20.074.009) and MHD models (Suess, 20.074.015).

White light and EUV observations give a consistent density model of the north polar coronal hole for the period June-August 1973. The model Munro and Mariska extends from 1.05 $R_\odot$ to 5 $R_\odot$.

Dulk (20.074.050) attempted to reconcile Skylab EUV observations and high resolution radio maps of a coronal hole. No one model satisfied all the observations. Models that agree with the EUV data gave radio brightness by a factor of 2 to 3 too high, while models that agree with radio data gave EUV intensities a factor of 10 too low. Dulk (OSO.8 Workshop Proc.1978) suggests that EUV and radio observations could be reconciled if thermal diffusion, mass flow and ionization non-equilibrium are taken into account, but only if the overall density of the transition zone is an order of magnitude lower than usually believed. This seems unlikely, and the problem of reconciling radio and EUV measurements remain unsolved.

McWhirter and Kopp have solved the flow equations in a diverging geometry, taking into account the gradient of the pressure of an assumed Alfvén wave flux but neglecting wave dissipation. Alfvén wave pressure is an adequate mechanism for the acceleration of the wind above 2.2 solar radii. Unfortunately, the gradient of wave pressure is so steep below this radius that no stable solution could be found. This deficiency should affect all models for the whole atmosphere and stands out as one of the major difficulties with the wave acceleration mechanism.

Rosner and Vaiana investigated the effects of flow, field geometry, wave pressure and extended heating. They were able to satisfy constraints from X-ray, EUV and radio observations, but their model is still subsonic at the coronal temperature maximum, at 1.1 $R_\odot$. Thus, they are forced to conclude most of the wind acceleration occurs beyond the temperature maximum, possibly by wave pressure gradients, in conflict with the results of Munro and Jackson. Suess et al. extended the work of Kopp and Holzer. They considered polytropic one-fluid flow in the Munro-Jackson geometry and showed that it is possible to fit the observations if the magnetic field strength decreases away from the axis of a hole, and if the base temperature is about 2×10^6 K.

Thus, in the region where observations are most plentiful, i.e. between 1 and 5 $R_\odot$, semi-empirical and energy balance models are still inadequate. If Alfvén waves heat and accelerate the wind, they also seem to disrupt the lower atmosphere.

The Origin of Coronal Holes and its Relation to the Solar Interior

Bohlin and Sheeley (Solar Phys.56,125, 1978) have proposed a phenomenological model for the origin of coronal holes. The model depends upon the appearance, within the same or opposite hemispheres, of bi-polar magnetic regions with unbalanced fluxes. If these regions are sufficiently close, field line reconnection can take place to form an extended unipolar magnetic region, which is an equatorial hole.

McIntosh (18.074.045) and Hundhausen (CHM) have stressed the systematic eastward drift of the locations of new coronal holes throughout the Skylab era. Such order suggests a sub-photospheric mechanism that operates on a global scale, such as a solar dynamo.

Gilman (CHM) has analyzed photospheric magnetic fields and differential rotation measured during the Skylab period by R. Howard in an attempt to search for several kinds of dynamo effects on the holes. His results do not support the idea that the solar dynamo ultimately controls the origins and rotation of coronal holes.

Stix (19.074.079) also analysed Mt. Wilson photospheric magnetic fields. He suggests that two symmetric modes control the position of the three most conspicuous coronal holes observed during the Skylab period and that the rigid rotation of the modes accounts for the rotation of the holes.

Stix (19.074.079) also analysed Mt. Wilson photospheric magnetic fields. He suggests that two symmetric modes control the position of the three most conspicuous coronal holes observed during the Skylab period and that the rigid rotation of the modes accounts for the rotation of the holes.

Problems for the Immediate Future

To make further progress in understanding coronal holes and solar wind streams, we need additional critical observations, particularly in the region of wind acceleration, and further development of some physics.

Measurements of the temperature, density and velocity profiles in the critical regions between 2 and 20 solar radii are needed as additional constraints for models of coronal holes and wind streams. Recently, Woo (Astrophys.J.219,727, 1978) has derived a velocity profile of the mean solar wind between 2 and 20$R_\odot$ from phase-scintillation measurements of satellite telemetry signals. This is an extremely promising technique and hopefully will be pursued further in the future. The profile of hydrogen Lyman α can also yield information on coronal temperatures (Beckers and Chipman, 11.074.011). Any observations of the hypothesized Alfvén wave fluxes near the sun would help greatly. Synoptic observations of holes and photospheric fields, such as those being made at Kitt Peak National Observatory, are needed for a full cycle, for evolutionary studies. Developments are needed in the theory of electron heat transport in collisionless plasmas, and for the generation, dissipation, and conversion between modes of MHD waves. How is it possible for a mechanical wave flux of 10^6 ergs cm^{-2} s^{-1} to pass through the chromosphere and transition zone without totally disrupting the atmosphere ? Transition zone and coronal models have generally assumed local ionization equilibrium by ignoring the effects of flow and diffusion. These effects should now be incorporated in more refined models. One would expect that the evolution of coronal holes relates somehow to the solar dynamo. To understand this relationship we need long term measurements of the large scale circulation in the photosphere and of the large scale magnetic field migration over the sunspot cycle. In combination with continued observations of coronal holes from the ground, these dynamical observations may stimulate further development of dynamo theory.

8. THE FINE STRUCTURE OF THE CORONA AND THE ORIGIN OF THE SOLAR WIND
(G. E. Brueckner)

Several major reviews treating different aspects of the topic have appeared during the time period covered by this report. These are: "Coronal Holes and High Speed Wind Streams", 1977, J. B. Zirker, ed., Colorado Associated University Press (R1); J. V. Hollweg, "Some Physical Processes in the Solar Wind", 1978, Reviews of Geophysics and Space Physics 16, 689 (paper 8R0531) (R2), and: W. C. Feldman, J. R. Asbridge, S. J. Bame, J. T. Gosling "Plasma and Magnetic Fields from the Sun", in "The Solar Output and Its Variation", 1977, O. R. White, ed., Colorado Associated University Press (R3). Wherever reference is made to these review papers, the numbers associated to them will be used.

The Solar Wind and Coronal Holes (CHs)

It is customary to classify the corona into two different structures: Loops, that have closed magnetic fields and the remaining areas that have "open" magnetic fields. It is also generally thought that the solar wind originates from open magnetic structures. For the purpose of this report we will discuss only open magnetic field areas. Nolte et al. (18.074.025) found, that all high velocity solar wind streams observed during the Skylab period (1973 May -1974 Feb., declining solar activity) could be associated with large, nearequatorial CHs. Sheeley et al. (18.074.128) presented a similar correlation for the period 1973-76 for CHs with $\ell < 40^0$ helio-centric latitude. Cuperman and Dryer (Astrophys.J.223,601, 1978) found a weak anticorrelation between green line intensities and high speed streams (1962 -71). The magnitude of the anticorrelation coefficient increases with helio-centric latitude to a value of 0.6 at about 30^0. During Carrington rotations 1595 - 1604 (late 1972 - 1973) all high speed streams seem to originate from low latitude extension of polar CHs (Hundhausen R1). Nolte et al. (20.074.036) could not find a convincing correlation between recurring high speed wind streams and low latitude CHs at solar minimum, based on x-ray photographs of the sun (1976, Nov.). Schwenn et al. (J. Geophys. Res. 83, 1011, 1978) found a good correlation between high speed wind streams and equatorial extensions of polar CHs (March 1975) comparing the measurements at 0.31 A.U. (Helios I) and 1 A.U. (Imp 7/8). They find sharp boundaries (10° in latitude) between high speed stream and the surrounding "quiet" solar wind. This result contradicts earlier estimates of the amount of meridional flows by Munro and Jackson (19.074.038). It also is not in agreement with earlier estimates of the geometric extensions of CHs into the interplanetary space (Hundhausen, R1).

While there is no doubt about the origin of high speed streams in CHs, the results do not allow a final conclusion on which solar latitudes contribute most to the high speed streams.

During solar minimum the solar magnetic field is bent more toward the ecliptic plane and high speed streams become more visible in the ecliptic plane which is in agreement with the results by Bame et al. (18.074.077) and Gosling et al. (18.074.068) who found that high speed wind streams with velocities in excess of 700 kms^{-1} are more abundant during declining and minimum solar activity than near solar maximum. Bohlin (19.074.054) reported that during the Skylab period 20% of the solar surface was covered by CHs; the lower latitude CHs ($\ell < 60^0$) provided only 2% to 5% of total hole area. Broussard et al. (Solar Phys. 56,161, 1978) show that polar CHs as seen in x-ray and XUV photographs are absent at solar maximum, but very pronounced at minimum. However, the magnitude of the interplanetary field (Hedgecock 14.106.012) and the solar wind sector structure (Svalgaard and Wilcox 14.076.056) does not exhibit changes with the solar cycle.

Wagner (17.074.041) found a correlation between the magnetic field polarity in the "quiet" solar wind and that of high latitude CHs (1972 May- 1973 Oct.). He suggests that the "quiet" solar wind originates also from polar cap CHs. Because the average "quiet" solar wind does not show solar cycle variations while the polar caps do, the possibility, that the "quiet" solar wind originates in small or "mini" CHs scattered over the solar surface, most be left open.

Coronal Holes (CH) Models

The EUV, x-ray and radio observations may be explained by either hydrostatic or flow models of homogeneous CHs as has been demonstrated by Rosner and Vaiana (20.074.008). Munroe and Jackson (19.074.038) introduced into their model an a priori outflow which is deduced from high speed wind stream observations at 1 A.U. In this way they obtain a trans-sonic point of a homogeneous solar wind between 2.2 and 3 $R_\odot$. Energy must be deposited beyond this point for wind acceleration. Dulk et al. (20.074.050) found a discrepancy of electron density values which are derived from radio and XUV observations. Inhomogeneities within a CH could remove these differences. However, Trott and Lantos (Astron.Astrophys.70, 245, 1978)and Chambe (Astron.Astrophys.70,255, 1978) pointed out that because of uncertainties in the measurements and inversions homogeneous models can satisfy the observations. While homogeneous models of CHs cannot distinguish between hydrostatic equilibrium and flow pattern, this situation changes when inhomogeneities are taken into account.

Polar plumes are fine structures of CHs. Ahmed and Webb (Solar Phys.1978) - derived models of polar plumes from x-ray data. They showed that hydrostatic models require unacceptable low temperatures. Introducing an outward flow in polar plumes they found a convective energy flux of $2 \times 10^6 erg\ cm^{-2}s^{-1}$ which is in agreement with energy fluxes at the solar surface estimated by Hundhausen (R1) from high speed wind streams. This result shows the importance of inhomogeneities in all modelling efforts. Using medium resolution K-corona measurements Koutchmy et al. (20.074.001) found density fluctuations of a factor of 5 in a polar cap CH.

Small Scale Fine Structure of the Corona

Very small coronal structures ($\sim$ 1") called "coronal spikes" have been observed in the quiet sun during eclipses. Reliable physical parameters of coronal spikes have been determined for the first time by Koutchmy and Stellmacher (18.073.162). They found: (i) the coronal spikes have a diameter of approximately 1" at an altitude of 20,000 km above the solar surface; (ii) because of the absence of chromospheric emission lines the spikes are not of chromospheric origin, their spectrum is characteristic of Thompson scattering; (iii) a density of $Ne = 10^{10} cm^{-3}$at 20,000 km could be derived. This density exceeds that of the homogeneous background corona by two orders of magnitude. The observations suggest, that they are open field structures which must eventually blend into a more or less homogeneous solar wind. Such structures must have very large pressure gradients which would generate high outward velocities very close to the solar surface.

Coronal Heating and the Energy Input into the Solar Wind

The plasma outflow in high speed wind streams constitutes an energy sink of $3\text{-}9 \times 10^5$ erg cm^{-2} s^{-1} at the base of the corona (Hundhausen, R1). Radiation losses of the "quiet" corona may be of the same magnitude (Kopp and Orrall, 18.074.102). A few times 10^6 erg cm^{-2} s^{-1} are required to maintain the underlying chromosphere above the temperature minimum (Withbroe, R1). Therefore, the mechanical energy transmitted through the transition zone must be similar in magnitude. We have no direct coronal observation of the nature of the coronal heating mechanism. However, a large amount of material has been accumulated bearing on the question of the transmission of mechanical energy through the transition zone. Athay and White (Astron. Phys.J.226, 1135) found only low amplitude semi-periodic fluctuations in the transition zone. They estimate the combined energy flux of periodic and aperiodic sound waves to be 1×10^4 erg cm^{-2} s^{-1}. However, these estimates are based on observations which have a spatial resolution of 2" x 20" only, CIV line profiles of the same spatial resolution were analysed by Bruner and McWhirter and compared with profiles (Byerley et al. 1978, J. Phys.B.11, 613) calculated with the presence of acoustic waves. Bruner and Mcwhirter concluded that there is not enough flux to heat the corona by means of acoustic waves. However, from the observed line width (ξ turb. > ξ thermal) they leave the possibility open that fast MHD waves are travelling through the transition zone. From observations of Si IV line profiles (same spatial resolution 2" x 20") Billings, Roussel-Dupré and Francis (20.076.016) ruled out sinusoidal acoustic waves, sinusoidal acoustic shocks and normally propagated MHD disturbances. Vernazza et al. (14.074.005) found non-periodic, fast (< 1 min) intensity fluctuations (up to 50%) in chromospheric, transition zone and coronal lines. They interpret these intensity fluctuations as being caused by the passage of shock waves. Their observations were made with a spatial resolution of 5" x 5". In very high resolution (1"x 0.5")spectra, Brueckner et al. found superthermal material motions in the transition zone. The events are very small in size (∿ 1") but show blue shifted Doppler motions up to 400 km s^{-1}. They occur in otherwise normal regions of the quiet Sun. They can be interpreted as strong shock waves (M = 8) and can account for the heating of the corona and the mass flow of the solar wind provided that the estimate of their abundance, which is based on a large extrapolation, is correct. While the emission of these events amounts only to about a few percent of the total transition zone emission, their kinetic energy flux is approximately a factor of 4 larger than the mechanical energy flux of the whole transition zone deduced from excess line broadening. This finding underlines once more the importance of high spatial resolution observations.

Direct Observations of Mass Flows in the Transition Zone and Corona

Francis and Roussel-Dupré (20.074.062) could not find any net mass outflow in the transition zone of CHs greater than 4.3 km s^{-1}, observed with a spatial resolution of 2" x 20". The only observations of coronal mass motions are those of Cushman and Rense (18.074.003, 19.074.901) and Brueckner et al. (1977 OSO-8 workshop, Boulder, Col.). Cushman and Rense obtain net-outflow velocities of 16 km s^{-1} in a CH by comparing Doppler velocities of coronal emission lines in a CH with those in the quiet sun. Brueckner et al. found net coronal outflows of 5-15 kms^{-1} in the quiet sun by measuring Doppler velocities of coronal lines in a framework of low chromospheric emission lines which were assumed to be at their rest wave length's. However, because in neither case a full analysis of the optically thin line profile could be performed, it is impossible to derive mass flow from the line center Doppler shifts which are small when compared with the line width.

Boundary conditions of Solar Wind Theories Near the Solar Surface

As pointed out by Hollweg (R2) solar wind theories assume a hot, homogeneous corona as boundary conditions near the solar surface. As a consequence (Hundhausen, 08.003.077) electron heat conduction is the main energy source of solar wind acceleration. Also, mass conservation of a homogeneous corona close to the solar surface results in low initial outflow velocities. All models predict solar wind values at 1 A.U. which are at variance with the observations. Specifically, the predicted proton temperature at 1 A.U. is 1 order of magnitude lower than the observed one in the quiet solar wind and the electron heat conduction flux is predicted to be 1.5 orders-of-magnitude larger than the observed values. (For details see Hollweg R2). The standard models cannot explain the high speed streams, that are assumed to originate in the cooler coronal holes, where the heat conduction is even less. Nerney and Barnes (20.074.012) alter the boundary conditions and find that coronal proton temperatures of $4\text{-}5 \times 10^6$ K and moderate electron temperatures (1.5×10^6 K) can produce a solar wind at 1 A.U. which is in reasonable agreement with the observations, except for the electron temperatures. They require low densities at the solar surface (10^6 cm^{-3}). It is obvious that these changes of boundary conditions are not in agreement with the observations of a homogeneous corona. Moreover, for high speed streams very unreasonable conditions would have to be postulated. However, the introduction of fine structures such as the coronal spikes would alter the boundary conditions significantly. The superthermal jets as observed by Brueckner et al.can inject a large amount of kinetic energy at the solar surface into the solar wind. This would overcome another basic discrepancy of contemporary wind theories, which predict a substantial amount of electron heat conduction even at 1 A.U., while the observations show that 97 % of the solar wind energy is kinetic energy (Feldman et al. R3). Hollweg (C3) concluded that an additional energy source is needed to overcome the difficulties presented by contemporary solar wind theories. He proposed solar generated MHD waves to be the source of the needed additional energy. The superthermal jets (Brueckner et al. 1978) can be interpreted as very strong shock waves ($M = 8$) in the transition zone. They could provide the needed energy input. However, solar wind theories would have to incorporate boundary conditions at the solar surface which <u>unite</u> the processes of coronal heating and solar wind acceleration in the same heights regime near the solar surface. From an observational point of view, our lack of understanding of the boundary conditions for the solar wind is a direct result of our incomplete knowledge of the very small scale fine structure of the low corona.

9. CONCLUSION

Space telescopes and balloon-borne experiments have undoubtedly opened up new vistas of observation and interpretation of details of the solar atmosphere. Ground based instruments have also played a vital role in this progress, by the use of a new generation of solar telescopes, improved detectors and the new procedures of data recording, and image evaluation. As we come to the end of a summary of selected recent highlights of attainments in solar physics, we look forward with much optimism and enthusiasm to what we can learn of the solar atmosphere in the next triennium and the years thereafter.

It gives me much pleasure to place on record the splendid cooperation I have received from the Vice-President and the members of the Organizing Committee.

M.K.V. BAPPU
President of the Commission

24. PHOTOGRAPHIC ASTROMETRY (ASTROMETRIE PHOTOGRAPHIQUE)

PRESIDENT: C.A. Murray

VICE PRESIDENT: H.K. Eichhorn

ORGANIZING COMMITTEE: Ch. de Vegt, L.W. Fredrick, W. Gliese, R.S. Harrington, P. Lacroute, H.I. Potter, W.F. van Altena.

Although our Commission bears the title "Photographic Astrometry", we cannot ignore the development of modern techniques of observation and image detection. The long term prospects for astrometry are dominated by the plans for space astrometry now being considered by NASA and ESA.

A team consisting of van Altena (leader), Franz and Fredrick have completed their astrometric instrument definition study for the NASA Space Telescope. Preliminary designs have been developed for two instruments which are capable of measuring relative image positions with an accuracy of $\pm$ 0".002 for m <17. It is expected that parallaxes should be measurable to $\pm$ 0".001. A proposal for astrometric use of the ST, with W.H. Jefferys (Texas) as principal investigator, has been submitted to NASA. In addition to scientific programmes, which include parallaxes, binaries and faint solar system astrometry, the responsibilities of the team include a basic service of guide star selection and acquisition for general use with ST.

A Phase A study of the ESA proposal for a dedicated astrometric satellite, to be known as Hipparcos, was completed early in 1978; the science team for this study included Lacroute and Murray from this Commission. The study demonstrated the feasibility of Lacroute's original proposal to measure large angular distances by means of a two-field telescope. A possible programme would be to observe some 10^5 stars during a 2-3 year mission; relative positions, annual proper motions and absolute parallaxes would be obtained with accuracy $\pm$ 0".002 for m < 11. The impact of both projects on astrophysics was the subject of a conference on European Satellite Astrometry which was held in Padova in June 1978.

IAU Colloquium No. 48 on "Modern Astrometry" was held in Vienna, in September 1978. Formally this was a colloquium of our Commission, but Commission 8 was also closely involved. A resolution was passed, urging the Presidents of both Commissions to set up a Joint Working Group to coordinate work on improving the stellar reference frame. A working party of radio and optical astrometrists was also set up with K.J. Johnston (NRL, Washington, DC) as chairman for coordinating the identification and precise position measurement of radio sources and their optical counterparts for improvement of the relationship between optical and radio reference frames.

In these fields of activity, the interests of our Commission coincide with those of Commission 8. It is important to optimise the available effort in setting up and maintaining the stellar reference frame, and maybe the time is ripe for a review of the responsibilities and even the titles, of our two commissions.

The study of nearby stars has always been a major interest for our Commission.

Plans are reported by van Altena at Yale, for a revision of the General Catalogue; these formed the subject of one formal session at Vienna, and also much discussion. Several papers have appeared recently, on systematic parallax errors and the use of parallaxes for absolute magnitude calibration (Turon Lacarrieu and Crézé, Buchholz, Nørgaard-Nielsen; 19-111-001, 005, 008). Lutz has extended his procedure for estimating bias in absolute magnitude calibration from parallaxes, to include the effect of a Gaussian luminosity function and apparent magnitude cut off. Hanson (19-111-006) has used observed proper motion distributions to improve the Lutz-Kelker corrections, and has also found evidence for magnitude dependence in Allegheny parallaxes. Upgren, with his associates, has studied parallax errors (19-111-002) and has published an extensive review (20-111-001). Gliese reports plans for revision of his Catalogue of Nearby Stars. From data published since 1968 he has compiled three lists: (1) 293 stars with $\pi > 0''.045$ which are not yet in the Catalogue; of these 181 have trigonometric parallax measured. (2) 163 suspects. (3) 374 new measurements of stars in the Catalogue, 68 of which now have $\pi < 0''.045$

Other activities reported by Commission members are summarized in the following sections.

Trigonometric Parallaxes - Nearby Stars

Harrington reports that the US Naval Observatory is continuing its programme of measuring parallaxes of intrinsically faint stars, with the 61-inch astrometric reflector at Flagstaff. At present there are about 200 stars on the programme. Fourth and fifth lists containing 93 and 95 stars respectively, have been published (Publ. US Nav. Obs. XXIV, pt. 111 (1976) and pt. IV (1978)).

Upgren reports the completion, during 1976-8, of twelve lists containing parallaxes of a total of about 250 stars determined at van Vleck Observatory; most of these are K and M dwarfs found by Vyssotsky.

In collaboration with J. Stock and D.J. MacConnell, Upgren has begun a parallax programme using the 65-cm refractor of CIDA, the Venezuelan National Observatory. At a latitude of 8° 48' N, this telescope is well suited to the study of the systematic differences between northern and southern observatories. Stars for observation have been selected primarily from those already observed at Allegheny, McCormick, Yale or Cape, but other stars of astrophysical interest are being included. Coordination with programmes at van Vleck and Mt Stromlo (Ianna) is also planned.

At Sproul, parallaxes have been published for 98 stars from long-interval series; these are contained in 14 papers by van de Kamp, Lippincott, Heintz, Hershey and others. Long term behaviour of the Sproul refractor is being studied from data obtained since 1950. Experiments with larger numbers of reference stars and plate constants have shown minimal gain in accuracy.

Klemola has completed measurements for about 20 stars on the programme of the Lick 93-cm refractor; plates are now available for the remaining 80 stars of the programme, which was initiated by Vasilevskis.

The Yale parallax series has been resumed with the publication of results of 13 previously unmeasured series taken with the 26-inch refractor at Johannesburg and at Mt Stromlo (Astron. J. 83, p. 640, 1978).

van Altena and L.H. Auer have completed an analysis of Yerkes parallax plates, measured on PDS microdensitometers and reduced with digital image centring methods. Average accuracy from 25 single exposure plates is ± $''\!.004$ (s.e.)

van Altena (Bull. Am. Astron. Soc. 9, p. 599, 1978) reports results of a pilot programme using the prime focus of the Kitt Peak 4 m telescope. Observations to B $\sim$ 23 have been completed in three mid-latitude fields. Preliminary measurements of one field give a parallax accuracy of ± $''\!.0024$ (s.e.) from 22 single exposure plates. About 4000 stars in each field will be measured.

At Pulkovo, Kanaev, A.A. Kiselev, V.A. Sokolova and others are continuing the measurement of parallaxes of bright stars, double stars, carbon stars, red dwarfs and some circumpolar stars, with the 26-inch refractor (18-111-006).

A second list of 26 stars measured at Herstmonceux has been published (B.G.F. Scales: Mon. Not. Roy. Astron. Soc. 184, p. 101) and results for a third list of 19 stars have been obtained; these are mostly K and M dwarfs discovered by Vyssotsky.

Also at Herstmonceux, measurements have been made, with GALAXY, of about 15,000 images on each of more than 50 plates taken with the UK Schmidt (1.2 m) telescope, covering an area of $4\frac{1}{2}^{\circ} \times 4\frac{1}{2}^{\circ}$ near the South Galactic Cap. So far parallaxes have been obtained for 880 stars with B < 14, with an external s.e. of ± $''\!.018$. (Murray and Corben, Mon. Not. Roy. Astron. Soc. in press).

K. Kamper (David Dunlap Observatory) and Wesselink (Yale) have completed their study of the α - Proxima Centauri System from plates taken with the Yale and Cape refractors, and the double astrograph at El Leoncito. They find that the space motion of AB and Proxima are equal in size and direction, but point out that the orbital motion of Proxima around AB should be detectable within a century.

Deutsch has found perturbations with 6 and 12 year periods in the motion of 61 Cygni from Pulkovo and Sproul data, and estimates the mass causing the 6 year perturbation to be 4J. (19-118-001; Pis'ma v. Astron. Zh. 4, p. 95, 1978).

Proper Motions and Kinematics

Luyten has published a new catalogue of some 3600 stars with annual proper motion exceeding $0''\!.5$, and estimates that the total number of such stars with m_{pg} < 21 in the whole sky is about 4350. Unpublished machine-readable data for more than 2×10^5 stars with annual proper motion between $0''\!.18$ and $0''\!.09$, found with the automated-computerized blink machine, have been deposited with the NASA Data Center at Greenbelt, Maryland. A new catalogue (NLLT) which is expected to contain data for about 60,000 stars with annual proper motion exceeding $0''\!.18$ will be in press before the next General Assembly.

An analysis of the reduced proper motions and colours of 115,000 stars was presented by Luyten at the Russell Symposium (1977); this shows clearly that near-by M-dwarfs with small proper motion are not as abundant as has been postulated recently.

Klemola reports that by the end of 1978, second epoch plate material will have been obtained for nearly 900 of the 1246 Lick fields. Measurements have been completed for about 450 fields with $-3^{\circ} < \delta < 43^{\circ}$ and outside the zone of avoidance;

final reductions are in progress. Faint stars are used for plate constants, and galaxies give the reduction to absolute proper motion. Representative samples of stars from McCormick fields and the KSZ programme are also being measured.

The Yale Southern Observatory, in collaboration with Felix Aguilar Observatory of the University of San Juan is extending the region of overlap with the Lick programme; a partial overlap for $-23^{\circ} < \delta < 28^{\circ}$ will be obtained.

In USSR work continues at Pulkovo, Kiev and the Sternberg Astronomical Institute on the KSZ programme of measuring proper motions relative to galaxies. At Tashkent, A.G. Rakhimov has published 5 catalogues of proper motions in areas of the KSZ programme (20-112-011, 012, 013, 014, 016) and at Kiev, L.P. Panteleeva has studied the quality of galaxy images obtained with various astrographs (Soobsch. Gos. Astr. Inst. Sternberg, No. 193, p. 14, 1977). Investigations of magnitude differences between various proper motion catalogues by Rakhimov (20-112-010; Tsirk. Astr. Inst. Tashkent, No. 60, p. 9, 1976) and S.P. Rybka (19-112-007; Astrometrija i Astrofizika 35, p. 3, 1978) indicate a possible cause for divergence in the measurement of precession.

Quijano reports that at San Fernando (Cadiz) second epoch photography of 48 KSZ areas will start in 1979.

Work on star clusters at Yale, and at Hamburg where absolute proper motions of 40 clusters are being measured, has been reported to Commission 37. Also at Hamburg, U.K. Gehlich is studying Perseus OB2. While at Yale, Hanson redetermined the Hyades distance modulus (3.30 ± 0.06) from Lick proper motions. At Pulkovo, V.N. Frolov (20-153-035) has studied four clusters in Cassiopeia, and at Tashkent, Latypov has published proper motions in NGC 6709 and 6940 (20-112-015, 017). A.I. Yatsenko has measured motions of trapezium type systems, (Astrometriji i Astrofizika 55, p. 9). At Sydney, Robertson has published proper motions in NGC 6025 (20-112-019), and D. King in NGC 3532 (Jl. Roy. Soc. NSW 111, p. 1); King is now studying NGC 2516 and 4103.

Researches on kinematics include an analysis of AGK3 and SAO (South) proper motions by Dieckvoss for solar motion and galactic rotation (Astron. Astrophys. 62, p. 445) and velocity ellipsoids; investigation of space motions is continuing. B.F. Jones and Klemola have determined proper motions for 210 Sanduleak stars near the North Galactic Pole, and have measured a sample of faint blue stars in a field at $8^{h}\ 48^{m}$, $+ 18^{\circ}$ on Palomar Schmidt plates at the suggestion of Sandage. Jones continues his work on T Tau and other stars in the Taurus dark cloud. At Tokyo, T. Mikami has determined statistical parallaxes of F, G, K, M type stars using AGK3 proper motions and various sources for radial velocities (Pub. Astron. Soc. Japan, 30, pp. 191 and 209) and has also compiled data for 1490 M-type and 321 carbon stars (Ann. Tokyo. Astron. Obs. 17, No. 1).

Photographic Positions and Catalogues

Optical positions of radio sources are being determined at US Naval Observatory, including 15-cm transit circle and 20-cm twin astrograph observations.

Positions of 48 sources have been measured jointly at Cambridge (Schmidt) and Herstmonceux (Refractors, Isaac Newton Telescope); the results will be published in the proceedings of the Vienna Colloquium.

Klemola has compiled a catalogue of 5000 reference stars for JPL to be used for alignment of cameras on the Voyager Spacecraft, which will pass Jupiter in 1979.

Quijano reports that the Carte-du-Ciel astrograph at San Fernando (Cadiz) is engaged in observing 20 selected minor planets for the Leningrad ITA programme.

Lacroute and Velbousquet have completed the re-reduction of the Carte-du-Ciel plates between $-2^{o} < \delta < 31^{o}$, using the Strasbourg AGK2/3 as reference data (20-002-009).

Eichhorn and Hoffleit have continued work on the Yale -60° to -70° zone.

Nicholson reports that measurement of the Cape overlap survey plates has begun on the GALAXY machine at Herstmonceux; so far about 100 plates ($\delta = -44^{o}$) have been measured.

Work with the Sydney astrograph has been reported to Commission 8.

In the USSR, a fourfold overlap survey of the Northern Hemisphere by several observatories is planned.

Instrumentation and Measurement

The El Leoncito double astrograph was re-collimated by R. Platzeck (La Plata) and van Altena; new tailpiece assemblies, plateholders and an image dissector auto-guider are being installed.

A new 400/2000 Zeiss (Jena) astrograph has been installed at Goloseevo, (I.G. Kolchinskij, C.P. Major; Vestnik Akademii Nauk Ukr. SSR, 1978, 4).

A 400 mm astrograph has been constructed at Engelhardt Observatory, for observations of planets and the Moon (Sh. T. Khabibulin; N.G. Rizvanov; Izv. Engelhardt Astron. Obs. 1976. No. 41 - 42, p. 104).

Performance of the 400/2000 astrograph of the Zvenigorod Observatory has been studied by Muhammed Omar Sulejman. (Nauchn. Inf. 1978. No. 39).

Pilowski has proposed a new method of measuring the relative positions of galaxy and AGK3 star images on Schmidt plates, using narrow band filters for magnitude and colour compensation, with the coated layer removed from the central regions over the galaxy image.

P. Connes has put forward a comprehensive proposal for improving the accuracy of ground based parallax measurements, including a special helium filled telescope (ESO Conference. Optical Telescopes of the future).

A new Zeiss ASCORECORD measuring machine has been installed at San Fernando (Cadiz).

Reduction Methods

A.A. Kiselev and S.M. Chantariya have developed an algorithm for determining proper motions on the AGK3 system, by combining Carte-du-Ciel and modern photographic positions, which they have applied to the region of α Per (19-112-011).

A geometrical method of plate reduction has been developed by Kiselev, O.V. Kiyaeva and O.A. Kalinichenko, and applied in the investigation of distortion in wide-angle cameras (19-031-231).

Yu. F Zinchenko and Podobed have described a modification of the overlapping plate method to be used for the polar catalogue (20-041-055).

Firneis reports on a statistical investigation of astrometric positions by means of the parameter and error covariance matrices; theoretical formulae have been tested by Monte Carlo methods.

Dommanget has adapted his programmes for the reduction of comet and asteroid positions, for use with star catalogues stored on disc. He has also developed his new method of reduction of astrometric plates, which will be applied in fields containing open clusters.

Acknowledgements

I am very grateful to all who responded to my request for information on which this report is based.

C. A. MURRAY
President of the Commission

25. STELLAR PHOTOMETRY AND POLARIMETRY

(PHOTOMÉTRIE ET POLARIMÉTRIE STELLAIRES)

PRESIDENT: M. F. McCarthy
VICE-PRESIDENT: J. A. Graham
ORGANIZING COMMITTEE: A. Behr, G. V. Coyne, A. Feinstein, M. Golay, P. W. Hill, A. R. Hyland, T. Markkanen, K. Osawa, F. Rufener, W. Schoneich, V. Straizys, J. Tinbergen

1. INTRODUCTION

Three items of special importance should be noted at the outset for they necessarily involve changes in the format of this text as compared with reports of previous years: 1) With the improvement in techniques for automatic data collection, conservation and reduction plus the successful extension of photometric studies to new domains of the electromagnetic spectrum through space studies and advances in infrared research, the results achieved in photometric and polarimetric studies have undergone a very large expansion; 2) Advice and suggestions from the members of Commission 25 in the course of the past three years point out to the Organizing Committee the importance of limiting the activities of Commission 25 to those service facilities and techniques connected with the observation, recording and processing of photometric and polarimetric data. This means that we must leave to other Commissions of the Union the discussion and analysis of data obtained; so, for example, multicolour data obtained by the techniques proper to Comm. 25 will be discussed and intercompared rather by colleagues of Commission 45; similarly we must leave the exciting discussion of conclusions from photometric and polarimetric studies to other Commissions and specifically to those concerned with Galactic Structure and Dynamics (Comm. 33), with Variable Stars (Comm. 27) and with Interstellar Matter (Comm. 34). 3) The final item is a practical one but must, with regrets, be mentioned here: the general deterioration in communication of world postal services has made interchanges with members of the Commission much more difficult during the past triennium; on the other hand we have witnessed the improvements in the abstracting services of our colleagues in Heidelberg and the growth and splendid development of data centers in many parts of the world: Strasbourg, Moscow and Washington.

In Section 6 we bring to the attention of the Commission some of the new techniques developed during the triennium, plus an adequate sample (which cannot be complete) of references to the various kinds of methods being employed in photometric and polarimetric studies of stars. Even here, for reasons of space, a cryptic reference system but one easily understood by members will be adopted. Thus in the listed references we shall list first the number (Volume, Section and Running Serial Number) of entries in Astronomy and Astrophysics Abstracts, then the name of the first author listed, with an asterisk attached in the case of multiple authors, and then a sometimes shortened version of the title. Given the purpose of this reference system, which is to point out to the members important references which might otherwise be missed, it is hoped that the abbreviations will be understood by our colleagues. Normally no references will be made to individual astronomical objects; details concerning Variable Stars, galactic and globular clusters, peculiar stars, stars embedded in emission nebulosity, etc. will be found in other sources. An exception will sometimes be made here when in the opinion of the author some specific aspect or item of importance to the methodology of photometric research is discussed. The references given here will refer normally to articles and not to abstracts. With these preludes we come to the substance of the triennial report.

2. ADVANCES IN TECHNIQUES OF STELLAR PHOTOMETRY

During the triennium since the Grenoble Assembly the changes in stellar photometry which had been foreseen in previous reports have become a reality; this report will reflect certain of these and will try to outline the salient features of some of them. Many of the improvements have been a growth careful and planned following the advances of the classic photometries of the 1950's, while others are associated directly with the tremendous breakthroughs in recording and detection techniques which characterized the later years of the decade of the 1970's. We summarize some of these as follows:

1. The influence of techniques developed for other purposes which incidentally but substantially forward the progress and quality of stellar photometric studies. Among these we include the recent improvements in the sensitivity in receiver design involving a most reliable and stable discernment and measurement of the S/N ratio of radiation from the stars.

2. The marked extension of the range of photometric sensitivity into the regions of the space ultraviolet and into the optical and radio regions of the infrared. The extension into the far UV has been achieved largely through the successful launching of orbiting space observatories whereas the work in the near IR is to be attributed rather to the improved photographic emulsions and their sensitization by a variety of new techniques and in the far IR to the development of detectors of increased sensitivity and stability such as the doped crystals of germanium which have lead to the Germanium-arsenide receivers.

3. A parallel development of software for reducing the numerous observations secured with exactness and ease. In many cases these advances allow observations of stars on nights of less than perfect photometric quality provided only that observations of star and of star plus sky can be made simultaneously. This development of the supplementary software refers both to improvements in counting techniques and in automatic data recording and reduction; one thinks of the programs of recent introduction in many observatories such as the popular photometric program (PPP) devised by W. Kunkel and associates at Cerro Tololo; this program involves a kind of running dialogue between the master programmer and the astronomer at the telescope; it places at the disposal of the observer the "instant aids" of a "first look" at his photometric results by providing a rough solution for a color observation; this can after examination be consigned to its place in the recorded data bank for later elaboration and refinement or as necessity may demand can induce the observer to repeat, while there is time those observations which later might have to be abandoned and discarded.

4. Perhaps the major advancement in the past three years is due to the developments of the panoramic detectors in ever larger arrays. Developments here bid fair in the 1980's to revolutionize photometry by giving it a sound observational basis for the detection and measurement of very faint sources. This with the limited resources available earlier was hardly possible.

5. Parallel to these developments in space and at ground stations in the far UV and in the near and far IR has been the growth and availability of electronic technologies which allow for reliable interfacing with photometers and with photon counters. We mention here at the same time another by no means minor contribution to the photometric 'risorgimento' we enjoy today; this is the development in the theory and practice of the photographic process, which has provided us with emulsions of much increased sensitivity and with a wealth of hypersensitizing and filtering techniques so indispensable for photometric success especially with extremely faint sources.

6. Certainly an outstanding feature of the advances in the growth of present day photometry has been the increased availability of so many new light gathering facilities. One thinks at once of the numerous new large reflectors which have become available in the past triennium; many are equipped with Ritchey-Cretien optics and field correctors which permit deep penetration of space plus a wide field of observation available for photometry. This has opened serious photometry

of faint sources to our colleagues throughout the world. Along with these improvements in large reflecting telescopes one notes the presence at high altitude observing stations of new or improved fast wide angle cameras of the Schmidt and Matsukov types with a variety of new recording devices for stellar photometry and polarimetry.

For these improvements on the ground and in space (one thinks here of the works in progress now in the preparation of photometers and polarimeters for the Space Telescope and for the ESA and other laboratories in space) we are most grateful and ready to accept the challenges presented for future photometric and polarimetric research.

3. BOOKS AND CONFERENCES ON STELLAR PHOTOMETRY

Photometric progress and problems have been covered extensively and well in several books published during the triennium, among which we list the following:

1. Multicolour Stellar Photometry by V. Straizys (Russian) Vilnius 1977. An English Version of this book is in preparation and the Russian original has a full Table of Contents and excellent summary in English. This book together with the Introduction to Astronomical Photometry by M. Golay 1974, D. Reidel, Dordrecht provides a firm background for one entering the field of stellar photometry or preparing to teach a course in this subject.
2. Light by R. W. Ditchburn 1976, Academic Press, London.
3. Image Science, by R. Shaw 1976, Academic Press, New York.
4. Detection and Spectrometry of Faint Light by J. Meaburn 1976, Reidel, Dordrecht.

Several printed books are the product of international and national conferences, symposia and colloquia, study weeks and summer courses. We list some of these which have treated topics of interest to members of Comm. 25.

Two IAU conferences have a direct connection with problems of stellar photometry. Since the introductions to the texts of each conference provide full organizational details and other structural items of these meetings we omit further elaboration here.

In Nov. 1977 IAU Symposium No. 80 was held at the U. S. National Academy of Sciences in Wash., D. C. in honor of the centennial of the birth of H. N. Russell. The title of the Symposium was The HR Diagram. The editors of the proceedings, published in 1978 by Reidel Co., Dordrecht are A. G. Davis Philip and D. S. Hayes. A study of the index will point up many of the photometric topics treated including a discussion of the combination of different photometric systems and specifically the Geneva and the Vilnius photometries through a considered selection of receiver-filter combinations.

In July 1978 IAU Colloquium No. 47 sponsored by Commissions 25 and 45 was held at the invitation of the Pontifical Academy of Sciences and of the Vatican Observatory in the Synodal Hall of the Aula Nervi in Vatican City State. The title of the Colloquium was Spectral Classification of the Future and it honored the centennial of the death of Angelo Secchi, S.J. Editors of the proceedings published by the Vatican Observatory in 1979 are M. F. McCarthy, S.J., A. G. Davis Philip, and G. V. Coyne, S.J. Here the interaction between stellar spectroscopy and stellar photometry were explored; the role of panoramic detectors and prospects for automatic data collection and reduction of both photometric and spectroscopic observations were discussed. It is becoming increasingly evident that with the arrival of new detectors, photometric systems may be established throughout the spectrum and that these may have a wide variety of different wavebands. The importance of communication and cooperation among researchers was stressed: standard systems must be preserved and extended, while new problems will require new experiments but should not lead to a proliferation of a host of new systems, photometric or spectroscopic. A specific point commended to Comm. 25 for discussion was the importance of designating a spectral type which is derived from multicolour photometry rather than from the examination of spectral images with a different nomenclature than the one used in the MK system.

Another conference which concerned problems of stellar photometry led to the publication in 1977 of The Distribution of Stars in the Direction of the Galactic Polar Caps. This was a Joint Discussion at the Grenoble Assembly which was edited by M. F. McCarthy, S.J. and A. G. Davis Philip and is published in Highlights of Astronomy Vol. 4, Part II; by D. Reidel Co., Dordrecht; Gen. Ed. is E. A. Muller. Of specific interest to stellar photometrists were the discussions of faint photometry made with large Schmidt cameras and the plate errors and colour corrections to be encountered together with the problems of transfer of magnitudes and photometric interpolation.

The Proceedings of a meeting of the Society of Photo-optical Instrumentation held at Reston, VA in Mar. 1976 has been published in 1977 by Palos Verdes Estates in California; Ed. C. Freedman (20.012.060). The title of the conference is Low Light Level Devices for Science and Technology.

Photo-electronic Image Devices is the title of the symposium held at Imperial College, London in Sept. 1974; Eds. B. L. Morgan, R. W. Airey, D. McMullen and published by Academic Press, NY in two parts, A and B.

Attention of photometrists is called to the publication of the Proceedings of the Working Group on Photographic Processes which met in May 1978 at ESO in Geneva under the Chairmanship of R. West of ESO. Discussions of problems of plate hypersensitization were excellent.

Another Union sponsored conference of interest to photometrists was held at Paris-Meudon Sept. 6-8, 1976 on Astronomical Applications of Image Detectors with Linear Response. The proceedings are published by the Obs. of Paris-Meudon in 1977; Eds. M. Duchense and G. Lelievre, as IAU Coll. No. 40.

Commission 25 in cooperation with Commission 30 on Radial Velocities and with Commission 45 on Spectral Classification and Multicolour Photometry requested the approval of the Executive Committee of the Union for a Symposium or Colloguium to be held in Halifax, Nova Scotia at the time of the Montreal General Assembly 1979. The title suggested was Automatic Collection and Reduction of Data from Photometric, Polarimetric, Spectroscopic and Radial Velocity Observations. The Presidents of the Commissions invited Dr. R. West of ESO to serve as Chairman of the Scientific Organizing Committee. The Executive Committee did not grant approval for this conference. The Commissions concerned express their gratitude to Dr. West for the labours he undertook in the preliminary planning stages.

A workshop on Problems of Calibration of Multicolor Photometric Systems will be held at Dudley Observatory in March 1979. A. G. Davis Philip is the Chairman of the Organizing Committee. The workshop will consider the problems of calibrating multicolor photometry with astrophysical parameters such as temperature, gravity and abundance. It is planned to publish the proceedings before the time of the General Assembly in Montreal in 1979.

Attention was called above to improvements in the bibliographical aids available for studies in stellar photometry and polarimetry. Advances here have been as exciting as those in the fields of instrumentation.

We cite with appreciation the bibliographic references and lists of catalogues available as contained in the publications of CDS (Stellar Data Centre at Strasbourg) and edited by C. Jaschek.

An extensive bibliography which presents a listing of 775 papers concerned with the Stromgren four colour photometry and the H Beta photometry from the years from 1950 to 1976 has been prepared by A. G. Davis Philip and C. L. Perry and published in Vistas in Astronomy, 1978 Pergamon Press, London, Ed. A. Beer. A figure in this paper well illustrates the growth in the number of articles published per year and the "publication explosion" mentioned at the beginning of our current report.

The full gamut of problems and prospects for future data reduction in astronomy was treated at IAU Colloquium No. 35 on the Compilation, Critical Evaluation and Distribution of Stellar Data held at Strasbourg, France in August 1976. Edited by G. Wilkins, it was published by D. Reidel Dordrecht in 1977.

4. REPORTS ON STELLAR PHOTOMETRY FROM MEMBERS

W. Buscombe (Northwestern Univ.)

The results of observations made at Cerro Tololo by H. J. Augensen to secure UBV photometry for selected high velocity stars are presented. Three objects: LTT 3987, LTT 5334 and HD 123598 have excess ultraviolet flux δ (U - B) greater than 0.15 mag. Of these, only LTT 5334 has a very unusual galactocentric orbit with an eccentricity of about 0.5. Also the space motion exceeds 150 km/sec for LTT 2575, LTT 2744, LTT 3709, LTT 3834, LTT 4822, LTT 6061. Of these LTT 2744 has the largest space motion ($\backsim$ 300 km/sec relative to the sun) and a highly elliptical orbit (e = 0.9).

H. Eelsalu (Astrophysical Observatory of Tartu)

An Observational Check on a Theoretical Constraint Imposed on the Photometric Curve for Faint Photographic Stellar Images (Tartu Astr. Obs. Publ. 46, 1978) by H. Eelsalu, et. al. A rigorous photometric and statistical processing of faint out-of-focus stellar images on a non-sensitized photographic plate has been carried out. The darkening law confirms that predicted by V. Riives and tested by us earlier.

A. Feinstein (Observatorio Astronomico de La Plata)

J. C. Muzzio (1978, submitted to the Astron. J.) found small differences between the main wavelengths of the wide and narrow filters of the Hβ filter sets give rise to important color effects on the β index. A difference of 10 A changes the β index about $0\overset{m}{.}01$ for every 1 mag. of (B - V).

H. Marraco has been obtaining BV polarimetry of southern stars in R Associations, OB stars in HII regions and field Be stars.

A. Feinstein and H. Marraco reported that through the measures with interference filters of the Hα, Hβ, and Hγ lines in Be stars, (the emission indices e_α and e_β) were obtained. Both indices describe the amount of the emission and are very well correlated with spectroscopic data. The absolute magnitude of the Be stars can be obtained through the β index corrected for emission (subtracting e_β).

J. A. Graham (Cerro Tololo Inter-American Obs.)

Graham is engaged in a long-term program for setting up standard stars in the Harvard E-regions for UBVRI photometry on the Kron-Cousins system. The standards will reach 16th magnitude and are based on Cousins' photometry of brighter stars in the field. Special attention is being paid to covering a large magnitude range within a small area of the sky so that the standard stars can be used to check the calibration of area detectors. A supplementary list of stars is planned to cover a large color range at magnitudes between 10 and 12.

B. Hauck
(Institut d'Astronomie de l'Universite de Lausanne et Observatoire de Geneve)

The group of our Institute has continued its efforts to collect all photometric data. The list of catalogues can be found in the Information Bulletin of the CDS. However, we would mention in particular the catalogue of Mermilliod and Nicolet (Astron. Astrophys. Suppl. 1977, 29, 259) giving all measurements in the UBV system and that of Nicolet giving (for the UBV) a mean value for 53,000 stars (in press). A new version of the uvbyβ catalogue ($\backsim$ 16,000 stars) is now in preparation and will be ready at the time of the General Assembly.

Mermilliod is working on comparative studies of young open clusters and has found (18.153.029) a new gap in the main sequence between spectral types B7 and B8. He is now working on a better definition of the observational isochrones in various photometric diagrams, using for this purpose the new edition of his catalogue on UBV and MK types for star clusters (18.153.023).

Hauck has continued his study of the Ap stars and a photometric parameter of peculiarity has been established for the Ap stars measured in the Geneva system (18.113.015, and Astron. Astrophys. in press).

North has made a preliminary study with a view to using a system (following a proposal by Straizys) employing the U, B1, B2 and V filters of the Geneva system and P, Z and S of the Vilnius system. With these filters many properties of both systems are conserved and it also seems to be very good for the study of the cool stars.

A. U. Landolt (Louisiana State Univ.)

Landolt is observing secondary UBVRI photoelectric standard stars around the celestial equator. On the order of 400 stars in the range $7 < V < 17$ and $-0.2 < (B - V) < +2.0$ between $+10^{\circ} < \delta < -10^{\circ}$ are being observed at Cerro Tololo. The photomultiplier is a RCA 31034. This is an expansion of work published (Astron. J., 78, 959, 1973).

E. Rybka (Krakow)

Among activities in stellar photometry in the past three years are the following:

1. Catalogue of magnitudes of HR stars in the uniform P_{44} and V systems. Uniwersytet Jagiellonski, Krakow 1977, p. 74. Magnitudes of 9110 HR stars have been reduced to the system of E. C. Pickering (Harvard Annals Vol. 44) and then expressed in Johnson's V system. The average mean error of one entry was diminished from $\pm 0\overset{m}{.}10$ to $\pm 0\overset{m}{.}07$.

2. A review article in Polish concerning the UBV photometric system: Problem standardow w szerokopasmowej fotometrii UBV. Postepy Astronomii, 26, No. 1, pp. 19-30 Warszawa 1978.

3. The final catalogue of 229 photometric standards in the UBV system near the Selected Areas 1 - 115. This work is in press in Acta Astronomica Warsawa. It contains the completion of the author's investigations concerning the determination of photometric standards distributed uniformly over the sky in the northern hemisphere.

5. STELLAR POLARIMETRY
(G. V. Coyne)

Summary

Significant advances have been made in technique for high resolution polarimetry. Moderate resolution ($\sim$ 2 A- 20 A) has been achieved with the technique of tilt-scanning with narrow band interference filters. The highest resolutions (less than 0.5 A) have been obtained by coupling an echelle spectrograph to various types of panoramic detectors.

These techniques have been used to study intrinsic polarization in various types of stars including: 1) rapidly rotating early type stars with extended circumstellar envelopes and emission-line spectra (Be stars) and Wolf-Rayet stars; 2) close dynamical systems in which stars are distorted by their close proximity or are losing mass into gas streams or some common envelope; included here are dwarf novae and X-ray binaries; 3) cool red stars with very extended atmospheres or those with circumstellar dust shells, including the regular pulsating variables of the Mira type, carbon stars, R CrB stars and other cool giant and supergiant stars.

Techniques

In recent years considerable effort has been made to increase the spectral resolution with which the intrinsic polarization from various types of stars is measured. At each increase in spectral resolution further significant detail has been revealed in the polarization. These observations have contributed to the study of the stratification in the extended atmospheres about early and late type stars.

A few spectropolarimeters have been designed by using the tilt scanning properties of narrow band interference filters (the process is similar to tilting a Fabry-Perot etalon) to enable limited spectral regions ($\sim$ 60 A) to be scanned and polarimetry to be performed sequentially at many wavelengths in the scan range (Clarke and McLean 1975). More recently wavelength scanning instruments have been used in which the polarizing optics have been coupled with the standard grating monochromators and multi-channel Cassegrain scanners. One such instrument is in use by the Royal Observatory Edinburgh. Early fears of severe instrumental polarization effects when using gratings have been overcome by always locating the polarimeter ahead of the spectrometer in the optical train. Using some birefringent device as an optical retarder (sometimes variable) followed by a fixed, perfect polarizer, ensures that the spectrometer is fed with fully polarized light having a constant direction of vibration, only the intensity of the light is modulated by the operation of the retarder. The most commonly used "fixed" retarders are achromatic wave-plates (made for example from quartz and magnesium fluoride) which have a great advantage, especially for spectropolarimetry, of giving a wavelength independent modulation as they are rotated. Pockels cells (electo-optic crystals) and photo-elastic modulators (piezo-optic crystals) are the best known variable retarders. Although they are not achromatic they have the advantage of having no moving parts. They also have higher modulation frequencies. A Pockels cell is used with a spectrograph and Reticon panoramic detector at Steward Obs., Univ. of Arizona. In addition, the photoelastic modulator has a wide field of view. However, because of readout limitations, these advantages are minimized for the detector arrays discussed below.

Undoubtedly, the most significant innovation in astronomical spectropolarimetry has been the recent development of multi-channel instrumentation, made possible by the application of linear and two dimensional photon-counting detector arrays. One such device, employing a Digicon image tube has been constructed at the University of Arizona according to a design by K. Serkowski (McLean, et. al. 1979). In the Digicon, photoelectrons, emitted by an S-11 photocathode on which the optical spectrum is imaged, are magnetically focussed onto a linear array of 106 silicon diodes. Magnetic stepping over half diode intervals gives effectively 214 channels. A rotating superachromatic half-wave plate and a fixed prism in front of the spectrometer are used for measuring the linear polarization Stokes parameters simultaneously in all channels. This instrument may be used in a low or high resolution mode. The low resolution mode uses a grism giving a resolution of 30 A between 3000 A and 7000 A; the high resolution mode uses an echelle which gives a resolution of $\sim$ 0.3 A around Hβ A CID will also be employed as a detector with this system. It is expected that all new spectropolarimeters will be designed to use two-dimensional photon counting image detectors with high quantum efficiency, such as Vidicons, Reticons, CID and CCD arrays.

Observational Results

A general review of new developments in the polarimetric study of Be stars has been given by Coyne (1976). More recently, variations have been found in the linear polarization at a resolution of $\sim$0.5 A across the Hβ line of the Be shell stars γ Cas, ϕ Per, ψ Per and ζ Tau (McLean, et. al. 1979). At the line center the polarization increases and there are secondary minima in the polarization in the extreme line wings. The flux profile, obtained simultaneously with the polarimetry, have double emission peaks separated by a central absorption. The variations in polarization are due to differential effects of Doppler broadening of spectral lines on the polarized flux produced by scattering in an oblate circumstellar envelope which is expanding and/or rotating and is not viewed exactly edge-on. These observations have assisted in the modelling of stratification effects in the circumstellar envelopes (Poeckert and Marlborough 1978, McLean 1978).

The latest review of polarization in late type stars was given by Shawl (1974). Landstreat and Angel (1977) have measured a number of late type variables at a resolution of about 40 A. The polarization varies across all of the strongest absorption features, principally the TiO and ZrO molecular bands. Coyne and Magalhaes (1977) have detected changes in the polarization of the semi-regular variable V C Vn in the H β spectral region. The maximum polarization shifts to shorter wavelengths as the star approaches minimum light. A detailed study of Mira has recently been made by McLean and Coyne (1978). The spectrum from 3500 A to 7000 A has been studied with a spectral resolution of 50 A. The polarization decreases across each of the TiO bands. Changes in the polarization have been studied at a spectral resolution of 0.5 A across the H β line in emission near maximum light. The emission flux is polarized about four times more than the continuum. Models employing fluorescent radiation at the moving shock front to explain these remarkable changes are being explored.

References

Clarke, D. and McLean, I. S.: 1975, Mon. Not. Royal Astron. Soc. 172, 545.

Coyne, G. V.; 1976, in Be and Shell Stars, I. A. U. Symposium No. 70, Ed. A. Slettebak, Reidel Co., Dordrecht. p. 233.

Coyne, G. V. and Magalhaes, A. M.: 1977, Astron. J. 82, 908.

Landstreat, J. D. and Angel, J. R. P.: 1977, Astrophys. J., 211, 825.

McLean, I. S.: 1978, Mon. Not. Royal Astron. Soc. (March 31).

McLean, I. S. and Coyne, G. V.: 1978, Astrophys. J. Letters, (December 1978).

McLean, I. S., Coyne, G. V., Fuecker, J. and Serkowski, K.: 1979, Astrophys. J. 226.

Poeckert, R. and Marlborough, J. M.: 1978, Astrophys. J., 220, 940.

Shawl, S. J.: 1974, Planets, Stars and Nebulae Studied with Photopolarimetry, Ed. T. Gehrels, Univ. of Arizona Press: Tucson, p. 821.

6. SELECTED REFERENCES TO STELLAR PHOTOMETRY

As indicated earlier we present here a list of abbreviated references to research in stellar photometry. References have been grouped under the following main topics: I. Photometers and II. Photometric Reduction Techniques. Papers are identified by the index number from Astronomy and Astrophysics Abstracts published during the past two years; this Volume Number, followed by the author's name and then a sometimes very shortened version of the title. Use of the following symbols allow this abbreviated presentation to be made:

PM (Photometer or Photometry)
PEP (Photoelectric Photometry)
* (et. al., the names of other authors are not listed here)
... (the full formal title has been abbreviated for lack of space)

I. PHOTOMETERS

17.032.012	E. Spiller*	Coatings for the Far UV Region
17.034.004	A. Gur'yonov	Investigation of PMs...
17.034.016	R. Schielicke	On a Semi-automatic Iris-type PM
17.034.018	V. Ivanov*	Electro PM - Computer System...
17.034.035	B. Ozak*	A Device for Automatic Sensitivity Regulation...
17.034.036	N. Lebedev*	...Photomultiplier for Recording Weak Light Signals
17.034.043	M. Torr*	PM Calibration Error...
17.034.048	M. Abbas*	Infrared Upconversion for Astronomical Applications
17.034.052	J. Davidson*	A Sky Compensating Filter PM
17.034.063	T. Moss	Infrared Detectors
17.034.068	A. Avotins*	Two Channel Electrophotometer
17.034.070	S. Krawczyk*	The Three Channel Microphotometer
17.034.078	C. Cosmovici*	A High Resolution Fabry-Perot Tilting Filter...
18.034.006	W. Furtig	A New Universal Stellar PM
18.034.014	M. Pehk	An Amplifier for an Infrared PM
18.034.016	E. Burke*	A New PM Design...
18.034.020	S. Babichenko*	High Sensitivity Wide-Band Stellar PM
18.034.023	O. Maksumov	Control Unit of a PM...
18.034.041	G. Paturel	New Direct Sensitometer for Photographic PM
18.034.054	J. Kizla	A Two Channel Infrared PM
18.034.060	G. Lygna*	Calibrations
18.034.061	D. Dawson*	Testing a PEP
18.034.063	J. Oliver	Fast Electrometer Amplifier...
18.034.080	C. Coleman*	Electronographic PM
18.034.081	R. Altenhof	Design of a Large Aperture Infrared Optical System
18.034.085	J. Horak*	Automatized PEP...
19.032.556	M. Laget*	Ultraviolet 3 Channel PM in S183 Experiment
19.034.056	E. Olsen	On the Stability of the PM System...
19.034.058	A. Bukach*	Intensity Micro PM with Step Drive
19.034.064	E. de Lara*	A PM Using Silicon Diodes...
19.034.082	W. Feibleman	Ultraviolet Response of In Ga As P Photocathodes
19.034.084	J. Dobrowolski*	Colored Filter Glasses...
19.034.085	K. Fritze*	A Scanning Micro PM with an On-Line Data Reduction...
20.031.272	B. Barlow	Ground Based Low Light Level Astronomy
20.031.300	J. Lorre*	Recent Developments at JPL in the Application of Digital Image Processing...
20.034.016	P. Boyce	Low Light Level Detectors...
20.034.024	G. Holah*	Far Infrared Interference Filters
20.034.025	S. Babichenko*	Use of Absorption Filters in UV Spectro PM...
20.034.028	K. Shivanandan*	Recent Advances in Far Infrared Detectors...
20.034.029	F. Vescelus*	A Large Imaging Array CCD Program
20.034.030	R. Aikens*	Astronomical Applications of Charge Injection
20.034.032	G. Carruthers*	Low Light Level Imaging Devices...
20.034.041	D. McMullan*	The Electronographic Camera
20.034.042	G. Taylor*	...Large Area CCD Imaging Systems
20.034.048	E. Kellogg*	The Photicon
20.034.051	C. Coleman	High DQE Detectors
20.034.058	P. Rybski*	The McDonald Obs. Digital Area PM
20.034.059	B. Campbell	The DDO Diode Array Spectrometer
20.034.064	S. Furlani*	The Multifunction PEPM of the Torino Obs.
20.034.065	A. Kubicela*	A D. C. PEPM
20.034.066	R. Tull	Making Every Photon Count...
20.034.067	M. Res*	Filter Glasses with Bypass Characteristics...
20.034.072	C. Baddiley	A Wide Field Infrared PM
20.034.074	M. Kimmitt	Recent Development of Infrared Detectors
20.034.077	D. Lemke	Infrared Instrumentation at Calar Alto
20.034.084	H. Tsunemi*	A High Speed PEPM...

I. PHOTOMETERS (continued)

20.034.085	W. Weiss	A Project Study for a Continuum PM
19.034.001	N. Maseyk	A PIN Diode PM
19.034.002	A. Corney	Silicon PIN Diodes in Astronomical PM
19.034.005	R. Gehrz*	A Low Resolution Infrared Array Spectrometer...
19.034.016	I. Aslanov*	...Application of Electronographic Image Converter...
19.034.017	A. Lallemand*	Developments of Electronographic Lallemand Cameras...
19.034.018	G. Kron	The Kron Electronographic Camera
19.034.019	P. Felenbok	...Electrostatic and Electromagnetic Lallemand Cameras
19.034.020	J. McGee	Present Status of the Spectracon
19.034.021	D. McMullan*	The RGO Electronographic Cameras
19.034.022	G. Carruthers	Electronographic Cameras for Ultraviolet Space Astron.
19.034.025	G. Wlerick*	PM by Electronography...
19.034.026	A. Boksenberg	...Image Photon Counting System
19.034.027	L. Robinson	Image Tube Scanners
19.034.028	B. Fort*	...Detective Quantum Efficiency of an Image Photon Counting System
19.034.029	V. Cenalmor	...A TU System for Counting Photons
19.034.031	J. Lowrance	SEC Tube Development...
19.034.032	J. Westphal*	Use of Silicon Target (S-T) and Intensified Silicon Target (SIT) Camera Tubes for PM...
19.034.033	T. Ingerson*	...Integrating Vidicon Detectors at Cerro Tololo
19.034.035	W. Livingston	Diode Rays - A Review
19.034.036	R. Tull	Self Scanned Digicon...
19.034.037	G. Walker*	The U.B.C. Diode Array Systems
19.034.038	R. Aikens*	The Kitt Peak CID Systems
19.034.043	D. Currie	A Photon Counting Array PM Using an Intensified Charge Coupled Device
19.034.044	J. Schumann	PM Performance of Silicon Photodiodes
19.034.045	M. Lampton	Microchannel Plates and Their Applications...
19.034.046	J. Timothy	Performance..of Photon Counting Detector Arrays...
19.034.047	A. Broadfoot*	Channel Plate Electron Multiplier...
19.034.048	M. Walker	The Use of Electronographic Image Devices...
19.034.050	P. Rybski*	...Intensified Dissector Area PM
19.034.053	D. McMullan*	Developments in Electronography...

II. PHOTOMETRIC TECHNIQUES

17.031.214	M. Osipenko*	...Fluctuational Character of the Amplitude of Stellar Image Vibrations
17.031.233	J. Lorre*	Recent Development at JPL in the Application of Digital Image Processing Techniques to Astron. Images
17.031.235	J. Sherman	...Restoration of Atmospherically Degraded Images...
17.031.244	S. Ghedini	General Principles of PEP
17.031.273	F. Roddier	...Astronomical Images Degraded by Atmospheric Fluctuations
17.031.403	G. Auner	Computer Program for the Determination of Photographic Magnitudes
17.031.405	D. Fischel	Oversampling of Digitized Images
17.031.408	U. Numanova	Computer Process for Transforming the Data of an Iris PM into Magnitudes
17.031.409	E. Hog	Automatic Measurement of Photographic Plates...
17.061.043	L. Nordh*	Infrared PM at Stockholm Obs.
17.082.001	K. Birkle*	Seeing Measurements in Greece, Spain, Africa and Chile
17.082.043	W. Traub*	Theoretical Atmospheric Transmission in the Mid and Far Infrared at Four Altitudes
17.082.050	A. Hoag	City Sky Glow Monitoring at Kitt Peak
17.082.056	J. Shapiro	Diffraction Limited Atmospheric Imagery

II. PHOTOMETRIC TECHNIQUES (continued)

17.082.057	C. Roddier	Measurements of the Atmospheric Attenuation...
17.082.063	R. Crane	Extinction by Condensed Water
17.082.072	W. Tam*	Infrared Spectral Radiance of the Sky
17.082.082	R. Pike	A Sample Computer Model for...Light Pollution
17.082.086	A. Takechi*	Sky Brightness at Dodaira
17.113.001	W. Warren	Rotational Velocity Effects in uvby, β Systems...
17.113.003	P. Warren*	Galactic Center Sequences in B and V
17.113.004	A. Cousins*	Numerical Simulation of Natural PM Systems
17.113.005	A. Cousins	VRI Standards in the E Regions
17.113.006	M. Penston*	Emission Free Photographic PM of Stars...
17.113.013	E. Oblak*	ubvy β PM and MK Spectral Classification
17.113.016	A. Heck	Applications of Multivariate Analysis...
17.113.024	D. Turner	...Random Scatter in Variable Extinction Analyses
17.113.026	R. McClure	Standard Stars for DDO PM
17.113.033	V. Nikonov	...Problems of Fundamental Stellar PEP
17.113.034	V. Burnashev*	Transformation of Similar Photoelectric Systems...
17.113.041	G. Alekseev*	PM Observations Aboard Salyut 4 and Soyuz-Apollo...
17.113.048	L. Houziaux	Comparison Between Celescope Magnitudes and Ultraviolet Spectrometric Measures
17.113.053	B. Hauck*	Sensitivity Functions of PM Systems
17.113.054	P. Jennens	Development of the UBViyz PM System...
18.021.027	J. Dooley	Two Dimensional Interpolation of Irregularly Spaced Data Using Polynomial Splines
18.031.210	G. Belvedere*	...PEP of Non Periodic Fast Phenomena...
18.031.214	P. Craven*	Photographic PM from Skylab
18.031.221	H. Landis	Some Thoughts on the How and Why of PEP
18.031.246	K. Janes*	UBV PM from Low Dispersion Slitless Spectra
18.031.328	G. Marette	Absolute Calibration Methods for Ultraviolet Flux
18.031.423	J. Burke	Estimating Objects from...Blurred and Grainy Images
18.082.098	N. Mandwewala	Analysis of Rufener's Method for the Atmospheric Extinction Reduction
18.082.102	R. Barletti*	...Verticle Profile of Atmospheric Turbulence...
18.113.005	V. Straizys*	Band Width Effects...and Black Body Colors
18.113.011	J. Forte	UBV Photometry...
18.113.012	A. Bernard	Photoelectric Standard Sequences...
18.113.013	P. Veron*	UBV Sequences...
18.113.014	K. Topaktas	RGU Photometry...
18.113.023	D. Crawford*	Relation Between Blanketing Parameter and Abundances...
18.113.025	T. Thuan*	A New Four Color Intermediate Band PM System
18.113.026	M. Bessell	UBVRI PM with a Ga-As Photomultiplier
18.113.028	A. Zvereva*	...PM of Stars on Board...Cosmos 215
18.113.029	G. Florsch	Introduction to Astronomical PM...
18.113.031	B. Nicolet	...Magnetic Tape for the UBV System
18.113.038	A. Cousins	Standard Stars for VRI PM...
18.113.039	B. Gronbech*	Standard Stars for PEP...
18.113.051	M. Breger	Evaluation of Stellar Spectro PM
18.113.053	A. Heck	Calibration of Luminosity Criteria...
18.113.058	B. McGimsey*	Photoelectric UBV Sequences...
18.113.060	T. Simon	Broad Band 20 μ PM...
18.113.063	D. Crawford*	Abundance Effects on uvby PM
18.113.066	E. Mendoza	Narrow Band PM of A and F Stars
18.113.067	D. Whittet*	Infrared PM Extinction Curves and R Values...
18.113.068	J. Dean	BVI PM of LMC Supergiants
18.113.076	K. Ogura*	UBV PM of Stars in Fields of Emission Nebulae...
18.113.077	S. Wramdemark	B and V Magnitudes and Spectro PM Quantities...
18.113.078	D. Dawson	...PM of Faint Standard Stars
18.113.080	J. Sudzius	Effects of Variations of the Interstellar Reddening...

II. PHOTOMETRIC TECHNIQUES (continued)

18.113.084	O. Eggen	Intermediate Band PM of Late Type Stars
18.113.085	G. Ponomareva	...the 3 Colour PM System of the 50 cm Matsukov...
18.113.086	E. Mendoza	Multicolour PM of Metallic Line Stars...
18.113.087	E. Mendoza	...$H\alpha$ OI PM of Young Objects
18.113.088	A. Antalova	UBV Photographic PM...
18.114.113	R. Bell	Derivation of Abundances through PM and Spectroscopic Methods
19.002.047	F. Rufener	Catalogue of PM Parameters for Stars Measured in the Geneva System
19.004.010	D. Herrmann	N. G. Pogson and the Definition of the Astro PM Scale
19.031.232	J. Pilkington	Data Reduction Techniques for Electronography
19.031.233	S. Worswick	Electronographic PM
19.031.237	A. Penny	Electronographic Stellar PM
19.031.238	A. Blecha*	Stellar PM by Electronography
19.031.239	L. Panaiotou	Posphorescent Sources for PM Calibration...
19.031.241	H. Dyck*	Calibration of...33 μ PM
19.031.244	K. Nandy	Ultraviolet PM from the S2/68 Observations...
19.031.245	B. Nicolet*	Critical Evaluation of PM Data
19.031.253	A. Lucchese	...Methods for Reducing Visual Observations...
19.031.267	W. Wenzel	Infrared Star PM
19.031.271	R. Bryant	Experiments with a Visual PM
19.031.272	E. Avetisov*	Reduction of Astronegatives...
19.031.274	V. Pop*	Reduction of PE Observations to UBV...
19.031.294	A. Herzog*	Direct Plate Automated Reduction Techniques
19.031.297	B. de Batz*	..Technique for Astronomical Imagery in the Infrared
19.031.406	P. Kelton*	Optimization of Multi Element Detector...
19.036.001	A. Millikan	...Plate and Film Experiments for Electronography
19.036.003	A. Kaye	Detective Quantum Efficiency Gains...
19.036.004	G. Allbright	Methods of Hypersensitization...
19.036.005	J. Heudier	The Renovation of Astronomical Photography...
19.113.002	C. Butler	Photoelectric Sequences near South Galactic Pole...
19.113.005	R. Gibb	Southern Photoelectric Standard Stars...
19.113.006	W. Couch	Determination of Stellar Surface Conditions from Multicolour PM
19.113.007	A. Heck	uvby β PM...
19.113.010	B. Gronbech*	Photoelectric $H\beta$ PM...
19.113.011	D. Crawford*	Interstellar Reddening Relations in the UBV, uvby and Geneva Systems
19.113.012	J. Lub*	Properties of the Walraven VBLUW PM System
19.113.015	R. Hilditch	New P. E. Sequences...
19.113.025	A. Ardeberg*	...PM Study of Stars and Interstellar Medium...
19.113.026	C. Garmany*	...Stromgren PM...in 27 McCormick...Fields
19.113.029	R. Hering*	Photoelectric UBV and RI Sequences in SMC
19.113.030	G. Schnur	Two UV Sequences...
19.113.031	L. Palmer	uvby, β PM System and Stellar Temperatures
19.113.033	M. Kazanasmos*	...PM Standards of Wirtanen and Vyssotsky
19.113.038	R. Woolley*	The Harvard Magnitude Scale
19.113.042	I. King*	Magnitude-Diameter Relations for Star Images on the Palomar Sky Survey Prints
19.113.045	P. Harmanec*	Photoelectric PM at the Hvor Obs...
19.113.048	M. Bessell*	Southern Standards...
19.113.057	J. Alexander	On the Accuracy of (R - I) As a Temperature Indicator
19.113.058	A. Cousins	U_cBV Magnitudes and Colours of South Circumpolar Stars
19.113.066	V. Piirola	PM Systems and Extinction Conditions...
20.002.025	M. Collins	Astronomical Catalogues: 1951-1975
20.002.038	M. Golay	Catalogue of PM "Star Boxes" in the UBVB B V G System
20.031.060	J. Eggert*	...Visual and Photoelectric...Imaging Quality...

II. PHOTOMETRIC TECHNIQUES (continued)

20.031.216	C. Christian*	Multivariate Analysis of Spectro PM
20.031.272	B. Barlow	Ground Based Low Light Astronomy
20.031.300	J. Lorre*	Recent Developments At JPL in the Application of Digital Image Processing Techniques to Astron. Images
20.031.325	H. Landis	...the Human Eye and PE Visual PM
20.031.329	L. Rusconi	Estimate of Optimum Measuring Time in Synchronous PM...
20.031.413	V. Kosykh	Automation of...PM Measurements
20.036.002	G. Lengauer	Nomogram for the Determination of the Limiting Mag...
20.113.002	F. Rufener	Note on Catalogues of the PM Parameters of Stars Measured in the Geneva Obs. System
20.113.006	N. Kiselev	Investigation of the PM System of the AZT-8 Telescope
20.113.007	V. Zhilin	...Accuracy of Fundamental Quasi-Monochromatic Stellar PEP
20.113.009	A. Moffat*	...the Balmer Discontinuity in UBV Reductions
20.113.010	A. Landolt	A Photographic Magnitude Sequence...
20.113.017	J. Drilling	UBV PM of Stars in a Region Near the SGP
20.113.020	P. Kulkarni*	Near Infrared PM of Some Stars
20.113.021	G. Kuriliene*	Intrinsic Color Indices of Supergiants...
20.113.022	G. Kuriliene*	Intrinsic Colours...in the Vilnius System
20.113.023	G. Kuriliene*	Color Excess Ratios in...uvby System
20.113.036	O. Eggen	Intermediate Band PM...
20.113.040	U. Dzervitis	Middle Band Photometric System for Classification...
20.113.042	N. Galinskij*	Isocon TV PM of Stars
20.113.046	A. Kharitonov	Comparison of Results of Absolute Spectro PM and UBV PM
19.034.039	J. McNall*	An Intensified Self Scanned Array Detector System...
19.034.040	J. Choisser	A Review of Image Tubes...
19.034.042	J. Geary*	An Intensified Solid State Photon Counting System

7. CONCLUDING NOTE

In this Report Sections 1, 2, 3 and 6 were prepared by M. F. McCarthy; Section 4 contains the Communications received from Commission members; Section 5 gives the Summary on Stellar Polarimetry prepared by G. V. Coyne

ADDENDUM

The following Report on Photometry and Polarimetry has been received from J. Tinbergen (Leiden Obs.).

The Leiden Light Collector and five colour photometer is being moved from Hartebeestpoortdam in S. Africa to La Silla (ESO) in Chile; it will be known as the Dutch National Telescope.

Leiden Observatory is putting into service a 12 channel photometer-polarimeter, optimised for the study of stellar absorption lines, but capable of most other types of photometry or linear polarimetry.

J. Lub and J. W. Pel have given a detailed discussion of the properties of the Walraven 5 colour photometry, which is a very useful companion to that of Stromgren (Astron. and Astrophys. 54, 137, 1977). They have applied their results to large scale studies of classical Cepheids (Pel) and RR Lyrae variables (Lub) in the Galaxy. Van Genderen has used the five colour system to study Cepheids in the Magellanic Clouds and X-ray binaries in the Galaxy.

Tinbergen has nearly completed a polarimetric study of stars in the solar neighborhood (35 pcs.). The local dust content is even lower than earlier studies indicated. A region of regular magnetic field can be seen but it does not extend over the whole volume studied. Byproducts of this investigation are a list of about

180 low-polarization standards (Astron. and Astrophys., in press) by Piirola and Tinbergen and a suspicion that stars of spectral type F0 and later show intrinsic polarization that could be related to magnetically active regions on these stars (Zwann and Tinbergen, in press).

Bastiaansen (Leiden) has started a programme of coordinated measurements of extinction, linear and circular polarization all as a function of wavelength; the aim is to put stricter limits on models of the interstellar dust. The programme comprises 13 southern stars with proven circular polarization.

V. Straizys communicates the following items concerning activities in photometry in observatories of the U.S.S.R. in the past report period.

In the Special Astrophysical Observatory the instrumental complex MANIA (Multichannel Analysis of Nanosecond Intensity Alterations) was constructed. The experiment is carried out for detecting relativistic objects (neutron stars, black holes) due to superfast variability of their optical brightness and polarization (Shvartsman et al., Commun. of SAO Nos. 19 and 20, 1976). The complex was used to observe X-ray stars, fast-variable radio sources, objects with continuous optical spectra etc.

In the Crimean Astrophysical Observatory the photometry of the stars with TV-technique were continued (Petrov, 17.122.044; Prokofjeva et al., 19.142.120). Photoelectric photometry here was concerned mainly on variable stars. The satellite "Kosmos-215" was used for ultraviolet photometry of 36 stars (Dimov et al., 18.032.508; 18.113.028).

Zdanavicius et al. (Bull, Vilnius Obs., 45, 3, 1978) constructed and tested an eight channel stellar photometer with constant rapid rotation of the filter wheel. The photometer ensures quasi-simultaneous photometry in all eight colors and this permits one to make measurements of color indices during "non-photometric" nights. Kalytis, Ralys et al. (Bull. Vilnius Obs., 48, and 49, 1978) suggested a convenient criterion for selection of optimum photomultipliers and other astronomical receivers and investigated different components of photometric equipment of the Vilnius observatory.

The plans to join the Vilnius and Geneva photometric systems are being discussed between both observatories. The preliminary variant of the joint system (VILGEN system) was realised by the Vilnius astronomers (Straizys, in "Multicolor Stellar Photometry", 1977, p. 209). The joint system was investigated theoretically by Straizys, North and Hauck.

The work of the Vilnius Astronomical Observatory was concentrated mainly on the further photometry using intermediate band seven color Vilnius photometric system. The system is described most completely in the monograph by V. Straizys "Multicolor Stellar Photometry" (Mokslas Publishers, Vilnius, 1977).

S.D. Sinvhal of the Uttar Pradesh State Observatory, Naini Tal, India, reports on work done there on infrared photometry; J and H band magnitudes of 50 stars brighter than visual magnitude zero and mainly of spectral types K and M have been reported. (Kulkarni, P.V., Ananth, A.G., Sinvhal, S.D., and Joshi, S.C., 1977, Bull. Astr. Soc. India, 5, 52.

M. F. McCARTHY, S. J.
President of the Commission

26. ETOILES DOUBLES (DOUBLE STARS)

PRESIDENT : P.Muller
VICE-PRESIDENT : O.G.Franz
COMITE D'ORGANISATION : A.H.Batten, M.Fracastoro, R.S.Harrington, S.L.Lippincott, C.E.Worley.

I. Observations courantes, Calculs d'orbites et de trajectoires

A - MESURES VISUELLES CLASSIQUES

En trois ans, environ 16 000 mesures ont été obtenues par un certain nombre d'observateurs dont les plus actifs ont été W.D.Heintz (6 000), C.E.Worley (3 300), F.Holden (2 050), P.Couteau (1 650), P.Muller (1 550), le groupe de Belgrade (1 200), R.L. Walker (500). Durant cette période, 10 000 mesures seulement ont été publiées, 12 000 sont en instance.

L'activité des observateurs a augmenté par rapport aux trois années précédentes, mais elle reste le fait d'un petit nombre. Un effort se dessine vers les couples négligés, souvent depuis longtemps et parfois à tort. La qualité moyenne des résultats se maintient, mais les instruments en service sont en général anciens, rarement équipés de micromètres enregistreurs (Washington, Nice) qui assurent un meilleur rendement pour une moindre fatigue de l'observateur. Le ciel austral bénéficie au moins d'un instrument plus moderne (Cerro Tololo). A Moscou (lunette de 30 cm, G.A.Starikova) et à Copenhague (lunette de 30 cm, N.P.Wieth-Knudsen) les mesures se font avec le micromètre à double image. Il faut noter avec satisfaction les débuts, à la fin de 1976, d'un programme au nouveau 65 cm de Mérida (Venezuela), à un peu moins de 9° N (D.J.MacConnell, A.J.83, p.436, 1978).

Des couples nouveaux ont été découverts par Couteau (402), et au nombre de quelques dizaines chacun par Heintz, Muller et Popovic. W.J.Luyten poursuit sa recherche systématique d'étoiles naines et ajoute 2 000 couples relativement écartés à grand mouvement propre commun.

Les mesures et les découvertes dues aux techniques nouvelles sont signalées au chapitre réservé à celles-ci.

B - MESURES PHOTOGRAPHIQUES

L'Observatoire Sproul travaille au rythme de 5 500 plaques par an sur des couples astrométriques à fortes parallaxe. Quatre autres établissements ont appliqué cette méthode : U.S. Naval Observatory (lunette de 66 cm), Turin (télescope astrométrique de 1 m), Lembang (lunette de 60 cm) et Lick (lunette de 91 cm) ; de ce dernier viennent de paraître des résultats obtenus entre 1948 et 1960 (H.M. Jeffers et S.Vasilevskis, AJ. 83, p.411, 1978).

C - CALCULS D'ORBITES ET DE TRAJECTOIRES

Depuis novembre 1975, un peu plus de 125 déterminations d'orbites elliptiques ont été publiées, la plupart d'abord par la voie des Circulaires d'Information. On y relève 37 orbites premières, pour moitié préliminaires. Plusieurs concernent des géantes ou des variables, avec pour but une approche des masses. On observe que les calculateurs sont parfois trop pressés, ou ne se sont pas suffisamment informés ; c'est ainsi que 10 orbites premières reposent sur une dernière observation qui remonte à plus de 10 ans, et jusqu'à un maximum de 19 ans. Treize orbites se rapportent à des couples connus depuis moins de 40 ans, et 9 à d'autres connus depuis moins de 20 ans. Le nombre des orbites très bien déterminées n'a pas changé ; il est tributaire essentiellement des mesures photographiques.

En trois ans 18 trajectoires rectilignes ont été déterminées, dont 17 par P.Muller (A. and A., Suppl. 32, 165, 1978).

II. Basic Documentation

A- DOUBLE STAR DATA BANK (USNO)

Between 1974 Oct. 1 and 1978 Sept. 20, 350 new pairs have been included (Index Catalogue), and 8 300 new observations from 20 lists have been added (Observation Catalogue). A small backlog of unpunched data now exists, but C.E. Worley has good expectation that it will be removed by the time of the General Assembly. The project to complete the Observation Catalogue by addition of the older (pre-1927) mesures continues. Below is a tabulation of the progress in this project :

	Grenoble	Sept. 1978
Means added to the catalogue	47 000	75 000
Means punched but not yet added	28 000	54 000
Total	75 000	129 000

This data has been extracted from 543 separate lists. C.E. Worley anticipates that the full 129 000 will be physically present in the card file by the time of the Montreal meeting. This data represents virtually all of the 20th century observations, and a great proportion of the earlier material. Although it is not too esay to make a good estimate, C.E.Worley believes that there are probably not more than 40 000 unpunched means ; he projects that this endeavor will be virtually complete within 3 years.

B - COPIES

La copie de Nice est tenue à jour sous la responsabilité de P.Couteau à l'aide des envois de cartes de C.E.Worley. Avec l'accord de ce dernier et du Royal Greenwich Observatory, la copie de Herstmonceux a été remise, pratiquement dans l'état où elle avait été reçue, à l'Observatoire Royal de Belgique où J.Dommanget l'a prise en charge. Jusqu'à nouvel ordre, ce transfert ne s'accompagne pas des obligations extérieures qui étaient au départ attachées à ces copies.

A Nice est tenu à jour l'atlas d'orbites créé en juillet 1973 par P.Couteau et P.Morel. Des compléments sont envoyés tous les six mois à ceux qui en détiennent un exemplaire.

La Circulaire d'Information de la commission a été éditée régulièrement, avec l'aide de l'Observatoire de Paris, par les soins de P.Muller (numéros 68 à 76) qui par ailleurs, en concertation avec Nice et Uccle, prépare un 4e Catalogue d'Ephémérides de couples orbitaux, à paraître avant l'Assemblée Générale.

G.N.Sulukvadze a publié un catalogue de 412 systèmes du type Trapèze (Abastumansk Obs. Bull. 49, 1978).

III. Etudes d'ordre statistique, théorique et astrophysique

A - SPECTROSCOPIE ET VITESSES RADIALES

D'importants progrès ont été enregistrés ces dernières années par suite de l'intérêt croissant des astrophysiciens pour les couples visuels. H.A.Abt a obtenu les spectrogrammes nécessaires à la classification spectrale en cours de quelque 3 000 composantes de binaires visuelles. T.W.Edwards a publié (17.118.014) les types spectraux et les classes de luminosité observés pour les composantes de 208 couples orbitaux dont seuls les spectres composites étaient connus T.E.Lutz et J.H. Lutz en ont fait de même (19.118.009) pour 91 couples brillants dont ils ont également obtenu les magnitudes UBV.

Dans le domaine des vitesses radiales des composantes de binaires, citons les travaux de R.F. Griffin (très nombreuses observations systématiques de composantes de couples orbitaux), H.A.Abt et S.G.Levy (étude de plusieurs centaines de spectres de couples orbitaux sur une quinzaine d'années), E. van Dessel (réduction de quelque 90 spectrogrammes de 30 couples orbitaux obtenus à l'ESO, à l'OHP et à Ondrejov), A.H.Batten, J.M.Fletcher, C.L.Morbey et C.D.Scarfe (programme portant à la fois sur des couples visuels et spectroscopiques et des systèmes multiples). A tous ces observateurs sont venus s'ajouter récemment Fr.Fekel (Austin) et H.A.mcAlister (Atlanta).

Un certain nombre d'objets particuliers, intéressants à divers égards, on fait l'objet d'études qu'il serait trop long de détailler ici. Il s'agit souvent de systèmes composites où l'on retrouve des composantes spectroscopiques ou photométriques, ou le soupçon de compagnons invisibles.

J.Dommanget et O.Nys achèvent la préparation d'une seconde édition de leur catalogue des Vr relatives dans les couples visuels orbitaux ; E.van Dessel en a déjà tiré trois listes de couples intéressants dans l'immédiat à divers égards dont une publiée (20.118.018), une autre diffusée à Grenoble, la dernière en cours de publication. Ces trois auteurs ont établi un catalogue de toutes les déterminations individuelles de Vr portant sur des couples orbitaux.

B - SYSTEMES MULTIPLES, COMPAGNONS PLANETAIRES ET AMAS OUVERTS

La grande fréquence des binaires dans les amas ouverts s'est vue confirmée par C.Bettis (19.153.016) d'une part, J.P. Dabrovski et W.D.Beardsley (19.153.016) d'autre part P.Baize donne une nouvelle détermination de la parallaxe des Hyades à partir des parallaxes dynamiques de 10 binaires visuelles de l'amas. (L'ASTRONOMIE,1978). L'étude de mouvements perturbés a porté sur divers couples connus (P.Baize, 20.118.015 et L'ASTRONOMIE 92,1978,211), A.N.Deutsch et O.N.Orlova, 19.118.001), tandis que S.Lippincott exploite l'importante collection de clichés de Sproul pour rechercher d'autres cas de compagnons invisibles. La difficulté de conclure dans certains cas attire des objections de W.D.Heintz (17.117.018)faisant valoir, sur la base de la distribution de certains éléments orbitaux, le rôle que pourraient jouer des effets systématiques observationnels. P. van de Kamp dans un travail de synthèse, procède à une étude statistique de la densité spatiale et de la densité massique de 19 compagnons obscurs. (19.117.042 et Yearbook of Astronomy 1977).

En ce qui concerne plus spécialement les systèmes multiples, signalons que H.A.Abt et Mrs E.S.Biggs ont commencé la constitution d'un catalogue des données spectroscopiques les plus complètes de leurs composantes, dont l'élaboration devrait prendre plusieurs années.

Les problèmes relatifs à la dynamique et la stabilité des petits systèmes multiples ont continué à faire l'objet des recherches de R.S.Harrington, le conduisant à montrer que les orbites planétaires stables peuvent exister dans de tels systèmes (20.117.014 et Mercury, 7,1978,p.34). V.Szebehely et K.Zare (19.117.034) ont plus particulièrement porté leur attention sur les systèmes triples et les différents types hiérarchiques.

Les travaux du colloque consacré par la commission aux systèmes multiples, signalé dans le rapport précédent, ont été publiés par O.G.Franz et PARIS PISMIS (Revista Mexicana de Astronomia y Astrofisica, N° spécial 3, Mexico mai 1977).

C - STATISTIQUE

En statistique proprement dite, H.A.Starikova (Astron. Zh. sous presse) a montré que la fréquence des composantes des couples stellaires vis-à-vis de celle des étoiles simples est maximale (0,91) parmi les types spectraux A7V et minimale (0,31) parmi ceux de type G5V et que les composantes des couples de la série principale ont sans doute généralement mêmes spectres et mêmes magnitudes mais que la dispersion de leurs écarts croît des types jeunes aux types avancés. R.S.Harrington et M.Miranian (20.118.005) estiment que les effets de sélection observationnels suffisent à expliquer la distribution observée des excentricités orbitales. L'orientation des plans orbitaux fait l'objet d'une recherche par J.Pensado et J.F.Lahulla (17.118.019). J.A.Fernandez (19.118.010) a repris son étude sur la possible corrélation entre les vitesses de rotation des composantes des binaires visuelles ; il trouve une telle corrélation pour les couples à composantes de classe IV ou V et, de plus, de types spectraux voisins.

J.Dommanget a poursuivi ses recherches statistiques sur les distances de séparation et des différences de magnitudes des couples de l'Index Catalogue et sur la fréquence relative des étoiles multiples et des amas ouverts.

D - ASTROPHYSIQUE

Dans ce domaine, le problème essentiel semble rester celui de l'origine et de l'évolution des binaires.

D'après H.Abt et S.G.Levy, en conclusion de leur étude

sur la multiplicité parmi les étoiles de type solaire (signalée plus haut : 17.117.014), les couples visuels se répartiraient en deux grandes catégories : ceux de période inférieure à 100 ans et les autres. Les premiers résulteraient de la fission d'une protoétoile commune, les autres de deux protoétoiles proches en voie de contraction. Par contre, l'étude des quatre amas : α Per, les Pléïades, Coma Ber et Praesepe conduit C.Jaschek à admettre un processus unique de formation de tous les couples stellaires.

Les implications astrophysiques de nos recherches sur les binaires se rencontrent sous des aspects très divers. Ici encore, nous ne pouvons citer tous les travaux sur des cas individuels. Notons cependant les tentatives d'interprétation de l'émission X observée dans le système de Sirius (R.Mewe, 18.118.013 et K. Breche, 18.126.037), la détermination des masses dans cinq couples dont la primaire est une variable à longue période par P.Baize (sous presse), celle des masses et des magnitudes individuelles (UBV) de 110 couples orbitaux de faible Δm par P. Laques et R.Despiau (17.115 009), lesquels constatent un bon accord global avec les modèles théoriques, mais aussi la non-unicité de la relation masse-luminosité pour les étoiles de masses intermédiaires. Notons enfin l'étude de S.Lippincott (20.117.022) sur les couples astrométriques observables visuellement dans certaines circonstances et dont nous manquent les caractéristiques physiques ; il s'agit d'objets à fort Δm et qui sont des classes III, IV et VI ou V avancés.

IV - New techniques ; Programs and Results

Photoelectric measures of magnitude differences on the UBV - and the Strömgren systems of 215 pairs with separations as small as 0".6 have been obtained by K.D.Rakos, Vienna Observatory, and his collaborators. The observations, carried out with area-scanning photometers at the Figl Observatory, Austria, the Mauna Kea Observatory, Hawaii, and the European Southern Observatory, Chile, also produced measures of position angle and separation for 140 of these objects.

Using an area-scanning photometer on the 1,8 m Perkins telescope at Lowell Observatory, O.G.Franz continued his program of UBV measures of double stars with particular emphasis on pairs containing at least one variable component.

From the Hoher List Observatory, E.H.Geyer reports that a programm of measuring magnitude differences in nearby binaries of $\rho > 2''$ is in progress with the use of an area scanner.

Exploratory use of imaging detectors for photometric and astrometric observations of double stars has continued. A.V. Hewitt, U.S. Naval Observatory, and A.J.Penny, Royal Greenwich Observatory, used electronographic cameras and PDS microphotometers in separate investigations of this type (IAU Coll. N° 40, Paris-Meudon, 1976).

At the Royal Observatory of Belgium, J.Dommanget has continued work on a television system intended for astrometric observations of visual pairs. Occultations of stars by the moon are playing an increasingly important role in the discovery and measurement of close double stars. Particularly active programs of occultation observations are being carried out by W.I.Beavers at Iowa State University, by D.S. Evans and this associates at the University of Texas, by C. de Vegt at the Hamburg Observatory, and by N.M. White at the Lowell Observatory. Important contributions also come from amateur astronomers. A list of newly discovered pairs is regularly compiled by D.W. Dunham and published in the Occultation Newsletter.

Speckle interferometry has become a productive tool of binary star research. H.A.McAlister, Georgia State University, has used the 4 m and the 2,1 m telescopes of Kitt Peak National Observatory since the fall of 1975 to obtain more than 4000 speckle-interferometric observations of known and suspected binary stars. Four lists of observations are in print or in press (Aph. J.), and lists of newly resolved binaries have appeared in IAU Comm. 26 Circ. d'Inf. N°s 70 and 73. Measured separations are in the range of 0$''$,031 to 3$''$,6 ; the limiting accuracy of measurement is ± 0°,2 in position angle and ±0,6% in angular separation.

G.P.Weigelt, University of Erlangen, has used the 1,06 m telescope at the Hoher List Observatory to carry out speckle interferometry and-holography of visual binaries to a limiting resolution of 0$''$,15. While these observations reached a limiting magnitude of about 9,5, measurement of stars of magnitude 12 may be possible. A program of re-observing nearby binaries with separations less than 1" is in preparation.

V - Divers

Les travaux des deux derniers colloques de la commission mentionnés à la fin du rapport de 1976 sont parus, le premier avec un retard dû à des difficultés locales et grâce au dévouement de J. Dommanget et de M. Fracastoro (20.012.019), le second édité par O.G.Franz et Paris Pismis sous forme d'un numéro spécial (N° 3, mai 1977) de la Revista Mexicana de Astronomia y Astrofisica.

Je remercie les auteurs qui ont collaboré au présent rapport, dans l'ordre de ses parties P.Couteau, C.E.Worley, J.Dommanget et O.G. Franz. On constate que les limites de compétence de la commission s'étendent sans cesse, ce qu'il nous faut faire apparaître dans le choix de nos nouveaux membres ; sans doute devrons-nous créer des groupes de travail, peut-être inter-commissions, dans certains domaines connexes.

Le déclin des observations classiques, malgré le dévouement de quelques-uns, se poursuivra avec le détournement inévitable des jeunes vers les techniques nouvelles, ceci bien que le résultat final ne puisse qu'être négatif avec l'abandon de la plupart des couples orbitaux, et notamment des milliers de découverte récente, et à terme un déficit évident quant aux masses. Il sera trop tard alors pour que les astrophysiciens viennent nous supplier d'observer davantage ; à eux de le comprendre dès maintenant et de mieux soutenir nos efforts.

P.MULLER
Président de la Commission

27. VARIABLE STARS (ÉTOILES VARIABLES)

PRESIDENT: J. Smak.
VICE-PRESIDENT: J.D. Fernie.
ORGANIZING COMMITTEE: N. Baker, M. Breger, A.N. Cox, M.W. Feast, H. Gursky, W.E. Kunkel, W. Wenzel.
EDITOR OF INFORMATION BULLETIN: B. Szeidl.

1. INTRODUCTION (J. Smak)

The contents and format of this Report was a subject of extensive consultations with all Commission Members and, in particular, with the Members of its Organizing Committee. Following the opinion expressed by the majority of Members, it has been decided that the Report should consist of a series of reviews covering all organizational problems and selected research areas. The Authors had freedom of choice between:

A. conventional presentation, with an emphasis on completeness of the survey and references, as it was the custom until now, and

B. critical review, or an essay, with an emphasis on the most important and prospective aspects and trends.

An effort was also made to avoid unnecessary duplication of work and, as a result, certain topics have been left entirely for other Commissions.

The complete list of Sections and Authors was published in I.B.V.S. No. 1448 to enable all active workers in the field to send their reprints, preprints, and other relevant information to the respective Authors.

During the period under review the following meetings took place on topics which are entirely, or at least partly, within the area of Commission 27:

IAU Symposium No. 75, Star Formation, Geneva (Switzerland), 6-10 September 1976.

IAU Symposium No. 80, The Hertzsprung-Russell Diagram, Washington, D.C. (U.S.A.), 2-5 November 1977.

IAU Symposium No. 83, Mass-Loss and Evolution of O-type Stars, Qualicum Beach (Canada), 5-9 June 1978.

IAU Colloquium No. 42, The Interaction of Variable Stars with their Environment, Bamberg (F.R.G.), 6-9 September 1977.

IAU Colloquium No. 46, Changing Trends in Variable Stars Research, Hamilton (New Zealand), 27 November - 1 December 1978.

IAU Colloquium No. 47, Spectral Classification of the Future, Vatican City State, 11-15 July 1978.

Conference on Novae and Related Stars, Paris (France), 7-9 September 1976.

Several excellent review articles covering the subject of variable stars have been published in "Annual Review of Astronomy and Astrophysics" and other similar publications. The extremaly useful role of "Astronomy and Astrophysics Abstracts" could hardly be overestimated. Our Commission's "Information Bulletin on Variable Stars" continues its important service.

2. ARCHIVES OF UNPUBLISHED OBSERVATIONS (M. Breger)

The Archives of Unpublished Photoelectric Observations of Variable Stars was created to provide two permanent archives, one in Great Britain and another one in the Soviet Union for Socialist countries. The Archives can replace lengthy and expensive tables in scientific publications by a single reference to the archival file number. Furthermore, many observations are never used for scientific publications, and the Archives could make such observations available to other astronomers at a time when they might become very important.

In 1977, Dr. W.S. Fitch resigned as Coordinator of the Archives after many years of active service and was succeeded by Dr. M. Breger. At present, the Archives contain 59 assigned file numbers, of which the first 43 were previously described.

Astronomers who wish to obtain unpublished photoelectric measurements on variable stars may do so by requesting at, cost, whole files (not partial files) from either

Mrs. E. Lake, Librarian
Royal Astronomical Society
Burlington House
London, W1V ONL, Great Britain

or

Dr. E. Makarenko
Odessa Astronomical Observatory
Shevchenko Park
Odessa 270014 U.S.S.R.

Astronomers who wish to submit unpublished photoelectric observations of variable stars to the Archives, should submit two copies as well as a brief descriptive cover sheet to the Coordinator (address listed below). Alternatively, one of the two copies should be sent directly to Dr. Makarenko (U.S.S.R.). The Coordinator will assign file numbers and forward the observations to London (and Odessa, if necessary). New files must be printed or hand-written in black ink. The printed part should be no larger than 8.5 by 11 inches (21.6 by 28 cm). If a new file number is required for scientific publications (in place of extensive tables of measurements), the file number can be assigned by the Coordinator before receipt of the actual measurements.
The address is:

Dr. M. Breger
Department of Astronomy
University of Texas
Austin, TX 78712 U.S.A.

The new acquisitions are files IAU (27). RAS-

44. uvby measurements of WW Aur by Etzel.
45. UBV measurements of RU UMi by Hogg.
46. BV measurements of V1500 Cyg (Nova Cyg 1975) by Armbruster, Blitzstein, Hull and Koch.
47. UBV measurements of HX Ara by Duerbeck and Walter.
48. uvbyRI, Hα measurements of U Cep by Olson.
49. UBV measurements of TU Hor by Duerbeck.
50. BV measurements of EE Peg and S Equ by Catalano and Rodono.

51. BV measurements of KO Aql by Blanco and Cristaldi.
52. ubvy measurements of HR 3413 = HD 73340 made by Heck and Manfroid and submitted by Renson.
53. Photometric observations of TT Ari by Duerbeck (still in preparation).
54. Photometric observations of AH Tau by Duerbeck (still in preparation).
55. ubvy magnitude differences between HR 3413 and HR 3440 by Renson and Sterken.
56. Photoelectric magnitude differences between HD 148898 (Omega Oph) and HD 148211 by Renson and Maitzen.
57. Photoelectric observations of HD 125248, HD 134793 and HD 184905 by Blanco, Catalano and Strazzulla.
58. B measurements of 14 Aur relative to 18 Aur made by Fitch and Wiśniewski.
59. Photoelectric photometry of 7 Ap stars: HD 59256, 66255, 66605, 83368, 83625, 94660 and 96616 made by Manfroid.

3. GENERAL CATALOGUE OF VARIABLE STARS (M.S. Frolov and P.N. Kholopov)

The Moscow team headed by Prof. B.V. Kukarkin and Dr. P.N. Kholopov continued accumulating all the data on variable stars. The Third Supplement to the third edition of the General Catalogue of Variable Stars, containing information on 699 variable stars designated in 1974-1975 and improved information on 3923 previously designated variable stars was published. The Second Special Supplement to the third edition of the General Catalogue of Variable Stars, containing the list of 3186 stars arranged in the order of right ascensions for the equinox 1950.0 was prepared and was published in the Special supplement to the bulletin "Variable Stars", Moscow, 1976.

The 61st, 62nd, and 63rd Name-Lists of Variable Stars were prepared and were published in IBVS No. 1068, 1248, and 1414 correspondingly. These Name-Lists contained the data on 200, 1045, and 313 objects, respectively.

Now the work on preparing for print the New Catalogue of suspected variable stars (one of the volumes of the Fourth edition of GCVS) is nearly finished.

4. CEPHEIDS (R.S. Stobie)

A. Introduction

This essay will attempt to outline the progress and major areas of development in the field of cepheid variable stars over the last three years. The list of references is not comprehensive and for a more complete list the reader is referred to Astronomy and Astrophysics Abstracts.

During the relevant period two reviews have been published on the observational aspects of cepheids, one by Pel (Conference on Current Problems in Stellar Pulsation Instabilities) and the second by Stobie (IAU Coll. 29), the latter having special reference to multiperiodic phenomena in cepheids.

B. Multicolour photometry of galactic population I cepheids

In what must rank as one of the finest contributions to the observational study of cepheids, Pel (AA Sup 24, 413) has published Walraven five-colour photometry for about 150 cepheids in the Southern Milky Way. The quantity and quality of this data is outstanding and must represent the most extensive, uniform, high accuracy body of data ever obtained on cepheids. Lub and Pel (AA 54, 137) have discussed the properties of the Walraven system and the data have been analysed by Pel (AA 62, 75) and by Pel and Lub (IAU Symp 80, 229). It is shown that there exists a narrow intrinsic cepheid locus in the (V-B), (B-L) diagram. Using this locus, colour excesses are determined for 156 cepheids. For a number of cepheids no consistency can be obtained for the unreddened colour in all three two-colour diagrams. Pel considers that these discrepant cases are probably caused by the presence of photometric companions and concludes that at least 25 per cent of all cepheids are members of binary systems. The (B-L), (L-U) diagram is particularly sensitive to differences in line blanketing and Pel shows that there is a remarkably small range in heavy element abundance for short period population I cepheids, whereas a low metal content population II cepheid is clearly separated on this diagram. Fitting the five-colour data to model atmosphere spectra, temperatures, gravities, bolometric light curves, radius variations and the equilibrium values of these quantities are derived from the intrinsic colours of 98 cepheids with $P < 11$ days. The observational systematics of population I cepheids are studied in the instability strip. In the HR diagram an excellent fit of the data to the theoretical fundamental blue edge is obtained for $Y = 0.28$ and $Z = 0.02$. The present cepheid temperatures are significantly cooler than most of the cepheid temperature scales currently in use. As a result the pulsation and evolutionary masses are now in better agreement and the remaining mass discrepancy is not significant relative to the errors in observation and theory.

The problem of determining the intrinsic colours of individual cepheids continues to be of key importance in cepheid research. Dean et al. (MN 183, 569) demonstrate the usefulness of the BVI photometric system in determining the reddenings of galactic cepheids. The reddening free locus in the (B-V), (V-I) colours is constructed for cepheids by piecing together the loci for individual cepheids and then fixing the zero point from stars in clusters and associations whose reddenings are known independently. This locus, used in conjunction with reddening lines is employed to obtain colour excesses for about 60 cepheids. Extensive BVI photoelectric photometry has been obtained for galactic and extragalactic cepheids (Dean MNASSA 36, 3; Dean et al. Mem RAS 83, 69; Martin and Warren, unpublished). Comparison of colour excesses obtained with different photometric systems shows in general reasonable agreement although there are systematic differences with earlier work and, in particular, a strong systematic effect with spectroscopic reddenings. Photometry of neighbouring field stars has been used by Feltz and McNamara (PASP 88, 699) to determine the colour excesses of eleven galactic cepheids. A subgroup of cepheids with small and well-established reddening values is recommended to serve as standards for the calibration of reddening-free indices with intrinsic colour indices. Nikolov and Ivanov (Per. Zv. 20, 63) have compiled a new list of colour excesses of 161 cepheids based on previously published work. Canavaggia et al. (AA 43, 275) suggest that, in establishing an effective temperature scale for long-period cepheids, the discrepancies encountered by Schmidt can be resolved by deriving colour excesses from six-colour photometry. Schmidt (ApJ 203, 466) has investigated the colours of cepheids and yellow supergiants in open clusters. He finds that colour-colour relations for cepheids differ significantly from star to star and additionally that the non-variables may obey a different colour-colour relation from the cepheids. Szabados (Mitt. Sternw. Budapest - Szabadsaghegy No. 70) has presented new UBV photoelectric observations on 38 northern cepheids of period less than 5 days. Extensive UBVRI photometry for galactic and extragalactic cepheids has been published and discussed by Eggen (ApJ Sup 34, 1; ApJ Sup 34, 33). Ivanov and Nikolov (Proc. of Third European Astr. Meeting, p405) using previously published photometry have considered classical cepheids and their spatial distribution in our Galaxy.

C. Cepheids in open clusters and associations

Strenuous efforts have been made to increase the number of cepheids known in open clusters and associations in order to improve the calibration (especially the zero-point) of the PLC relationship for galactic cepheids. The available data on cepheids in clusters and associations has been summarised by van den Bergh (IAU Coll 37). One may identify a subset of "classical" calibrators which consists of the seven cepheids in clusters which are common to the lists of van den Bergh and of Sandage and Tammann. In attempting to define the zero point of the PLC relationship (with other coefficients adopted from the relationship for LMC cepheids) it is revealing that the "classical" calibrators show a very much smaller scatter than that obtained by inclusion of all calibrating cepheids (Martin, Warren and Feast, preprint).

Harris and van den Bergh (ApJ 209, 130) have shown that the 5.9 day cepheid CS Vel is a probable member of the cluster Ruprecht 79. Van den Bergh and Harris (ApJ 208, 765) have presented new UBV photometry for the highly reddened cluster Lynga 6 and the 10.8 day cepheid TW Nor. The reality of the cluster and the membership of the cepheid is supported also by Walraven photometry (Thé, AA 60, 423). Van den Bergh et al. (ApJ 208, 770) have shown that the cepheid SV Cru is not a member of the sparse cluster Ruprecht 97. From photometric and spectroscopic data, Turner (AJ 82, 163) has concluded that the 18.9 day cepheid VY Car is a probable member of the association Car OB1. Eggen (ApJ Sup 34, 1) has concluded that the stellar associations, of which the cepheids RS Pup and 1 Car have previously been thought to be members, do not exist. Despite its very different reddening, Warren (MN 178, 21P) has shown that the cepheid GU Nor is at the same distance as the galactic cluster NGC 6067. The membership of V367 Sct to the cluster NGC 6649 has been questioned by Flower (ApJ 224, 948). Eichendorf and Reipurth (IBVS 1450) have noted that the variable star HR 4511 is a possible very long period, low amplitude classical cepheid in the cluster Stock 14.

D. Extragalactic cepheids and the PLC relationship

A major BV photographic study of cepheids has been completed by Butler for 72 variables in the SMC (AA Sup 24, 299) and 98 variables in the LMC (AA Sup 32, 83). For cepheids in common with other observers comparison of the photometry shows good agreement with photoelectric data but systematic trends are revealed with previous photographic surveys. Martin, Warren and Feast (preprint) have completed an analysis of photoelectric BVI observations of 78 LMC cepheids. Again a comparison of the photoelectric observations of cepheids in common between different observers shows encouragingly small residuals. Both these studies have stressed the importance of selection effects in accounting for differences between observers in the coefficients obtained in the PL and PLC relationships. It is apparent now that in order to compare the properties of cepheids between one galactic system and another, and even to derive relationships within one system, a large statistically unbiassed sample of stars is required.

Previous arguments for the inclusion of a C-term in the PL relationship have rested primarily on theoretical grounds. Indeed, because of the difficulty in determining the value this coefficient observationally, the canonical value of the C-coefficient has been adopted as 2.52 on semi-theoretical, semi-empirical arguments. Butler and Martin et al.'s work shows that observationally the existence of a C-term considerably reduces the scatter in the PL relationship, thus enabling one to derive a more accurate distance modulus. The value of the C-coefficient however turns out to be close to the value expected if differential reddening from cepheid to cepheid were the primary effect. It is thus important to establish whether the observational value of the C-coefficient is determined by differential reddening (Clube and Dawe, preprint) or is an intrinsic property of the cepheids. The BVI photometry of Martin et al. establishes conclusively that the C-term in the LMC is caused primarily by intrinsic differences between cepheids and is not a consequence of differential reddening. However, Madore (RO Bull 182, 151) has found significant differential reddenings for cepheids in the LMC and the SMC from UBV photometry. He has also recalibrated the PL relationship by substituting the reddening free parameter W = V - 3.1 (B-V) for the visual luminosity. Madore (MN 177, 215) has derived colour excesses for 36 long-period cepheids in the Magellanic Clouds from UBV photometry. Excellent agreement is found between the photometric reddenings and independently calibrated spectroscopic reddenings, thus indicating that no gross intrinsic colour differences exist between long-period ($P \gtrsim 20$ days) cepheids in the Galaxy and the Magellanic Clouds. Eggen (ApJ Sup 34, 33) has obtained UBVRI observations of 21 cepheids in the LMC and SMC with periods greater than 35 days. He concludes (ApJ Sup 34, 1) that the LMC and galactic cepheids are very similar but that the long-period SMC cepheids are hotter on average at a given period. Van Genderen (AA 54, 737) has intercompared Walraven five-colour observations of 25 cepheids in the SMC with galactic cepheids. When corrected for absorption the position of the cepheids in the two-colour diagram shows that the SMC cepheids are bluer in comparison to their galactic counterparts.

On the question of amplitude as a third parameter in the PL relationship, Butler (RO Bull 182, 159) has found that for cepheids in the period range $0.4 < \log P < 0.9$ the data for the SMC tend to support Sandage and Tammann's conclusion (ie. the largest amplitudes occur at the blue side of the strip) whereas the data for the LMC show no clearly defined trend. On the other hand Madore (RO Bull 182, 151) has presented evidence, for cepheids in the LMC, SMC and the Galaxy, that the largest amplitudes occur at the red side of the strip. Also Yakimova (Soobsch. Gos. Astron. Inst. Shternberga, 177, 19) has analysed van Genderen's photometry of SMC cepheids and finds largest amplitudes at the red side of the strip. The clearest statement on the question of amplitude as a third parameter comes from Pel's work (Pel and Lub, IAU Symp 80, 229) where, at least for galactic cepheids, it is apparent that amplitude is not a good parameter to measure the position of a cepheid inside the instability strip as it varies too little with temperature, and for the low amplitude variables we cannot distinguish the red-edge cepheids from the blue--edge first overtone pulsators by amplitude alone.

Eichendorf and Reinhardt (AA 61, 827) have derived a statistically sound method to construct an envelope to a point diagram. This procedure has been applied to the period-amplitude diagrams of cepheids in the Galaxy and M31 with the conclusion that the upper envelopes are not significantly different. Van Genderen (AA 65, 147) has investigated the envelope line in the period-amplitude diagram for population I cepheids. He finds that the envelope line of the SMC and LMC differ significantly from each other and that the LMC line is intermediate between that of the SMC and the Galaxy) M31. A correlation between this progression and the average metal abundance in a stellar system is indicated.

Becker et al. (ApJ 218, 633) have constructed frequency-period distributions of cepheids in seven galaxies in the Local Group. These distributions have been compared with theoretical predictions. It is shown that evolutionary tracks of stellar models of intermediate mass are highly sensitive to changes in chemical composition. As either Z or Y is increased the entire frequency-period distribution is displaced to larger values of the period.

Cogan (AA 49, 17) has derived the moss-luminosity relationship of cepheids in the SMC and finds observational evidence for a change in slope at $M \sim 10\ M_\odot$ in agreement with theoretical predictions.

E. Mass (or radius) anomalies for population I cepheids

Considerable progress has been made on the resolution of the cepheid mass anomalies although further work requires to be done before some of the resolutions can be regarded as definitive. As pointed out by A.N. Cox (preprint and Proc., Conf. on Stellar Pulsation Instabilities) and J.P. Cox (Proc., Conf. on Stellar Pulsation Instabilities) there are six indirect methods of deriving the mass of a cepheid, which we shall call M_{evol}, M_{theor}, M_{puls}, M_{Wess}, M_{bump}, M_{beat}. In general M_{theor} agrees with M_{evol} but the other masses do not and they have been listed roughly in order of increasing discrepancy, culminating in M_{beat} which is typically a factor of 2 to 3 lower than M_{evol}. The problem is to obtain agreement between these different mass determinations.

Cogan (ApJ 221, 635) has preferred to express these anomalies in terms of radius. He has assembled radius and temperature determinations of classical cepheids from the literature and compares these with the theoretical relationships. Different methods of radius determination give significantly different results. In the log R, log P diagram it is pointed out that it is very difficult, for the double--mode and bump cepheids, to avoid a slope of unity which strongly conflicts with other evidence on population I cepheids. These radius discrepancies are similar to the mass discrepancies because any mass discrepancy can be translated into a radius discrepancy (of a factor 3 lower) by means of the PRM relationship.

The mass discrepancies have been discussed by many authors (Schmidt, ApJ 203, 466; Cogan, ApJ 211, 890; Cox et al., ApJ 212, 451, 214, L127, 220, 996, 224, 607; Takeuti, Sci. Rep. Tohoku Univ. 59, 67; Petersen, AA 62, 205; Pel and Lub, IAU Symp 80, 229; Faulkner, Proc. astron. Soc. Australia 3, 124). The new distance scale for the Hyades and the cooler effective temperatures obtained by Pel have largely removed the discrepancy between M_{puls} and M_{evol}, and the remaining difference is not considered significant. A.N. Cox (preprint) has discussed M_{wess}, derived from Wesselink radii, and concludes that, despite showing a large scatter, M_{wess} is systematically low compared to M_{evol}. However, considering the uncertainties and possible systematic errors in the Wesselink techniques, this discrepancy does not appear very significant.

The two remaining mass discrepancies, which appear to be really significant, are M_{bump} and M_{beat}. Discussion on the M_{beat} mass anomaly will be deferred to the section on double-mode cepheids, but we would note at this point that the only theory so far which can account for both the M_{bump} and M_{beat} anomalies is that of A.N. Cox and his collaborators. This theory involves a major alteration to the envelope structure by requiring a helium rich zone to produce a chemically inhomogeneous envelope. However, one theory which can explain the M_{bump} anomaly without recourse to a chemically inhomogeneous envelope is that developed by Carson and Stothers (ApJ 204, 461). The new Carson opacities considerably alter the structure of the envelope and detailed non-linear pulsation calculations (Vemury and Stothers, BAAS 9, 360 and preprint) have shown that the phase of the bump is changed in such a way that the mass anomaly virtually disappears. It will be very interesting to see precisely what effect these opacities have on the masses of double-mode cepheids.

F. Double-mode cepheids

Double-mode (or beat) cepheids have attracted considerable attention in recent cepheid research mainly because the mass anomaly exhibited by these stars is the greatest of all the mass anomalies (assuming that they are normal population I cepheids). The existence of the double periodicity means that these stars form a particularly sensitive probe of the structure of the stellar envelope and the theory of pulsation. The properties and importance of eleven, confirmed double-mode cepheids have been reviewed by Stobie (MN 180, 631). In the range 2-4 days these stars form a considerable fraction of the total number of cepheids of all population types. Their spatial distribution and radial velocity dispersion is similar to that of population I cepheids as is their chemical composition (Pel and Lub, IAU Symp 80, 229). Their position in the HR diagram shows no concentration to the red side of the strip but, if anything, they are associated more with the transition between fundamental and first overtone pulsation (Pel and Lub; Cogan ApJ 221, 635).

Faulkner (ApJ 216, 49; ApJ 218, 209; Proc. astron. Soc. Australia 3, 124) has established by Fourier techniques that strong mode interaction (equal in energy to the energy contained in the fundamental mode) is present in the double-mode cepheids TU Cas and U TrA (see also Kozar, IAU Coll 29, 183). A significant third period of low amplitude has been discovered in TU Cas, although the reality of this third period has been disputed by Cox et al. (ApJ 224, 607) and Hodson, Stellingwerf and Cox (preprint).

If double-mode cepheids are mode switching then observable effects should already be detectable since, according to Stellingwerf, the e-folding timescale of the mode switch is about 80 years. We should note at this point that, if only a single mode switch occurs in the evolution of a star across the instability strip, this timescale is quite inconsistent with the observed number of double-mode cepheids.

Shobbrook and Faulkner (preprint) have detected an amplitude change in U TrA over a timebase of 20 years where, although the total energy of the pulsation has remained constant, the first overtone energy has grown at the expense of the fundamental. Hodson et al. have examined the observations of TU Cas spanning 67 years but were unable to conclude whether or not the variation in amplitude found was real. However, Gieseking and Radeke (AAp Sup 33, 207) indicate that new observations of TU Cas do not agree with Faulkner's Fourier fit and they suspect a change in the mode interaction. Illés and Szabados (IAU Coll 29, 165) are continuing to obtain photoelectric observations of TU Cas. Madore et al. (MN 183, 13) found no evidence for a change in amplitude in the case of V367 Sct over a period of 50 years.

The mode identification of double-mode cepheids has generally been adopted as the fundamental and first overtone modes of radial pulsation. However Takeuti (IAU Coll 29, 171), on the basis of the excitation of different modes, considers that the question of pulsation mode is still open. Petersen (preprint) has examined the period ratios of the fundamental and first three overtones of radial pulsation and concludes that alternative mode identifications do not solve the problem.

One crucial member of the double-mode cepheid class is V367 Sct which is a probable member of the open cluster NGC 6649 (Efremov and Kholopov, Per. Zv. 20, 133; Kobayashi et al., Sci. Rep. Tohoku Univ., 59, 67; Madore et al. MN, 183, 13). However, Flower (ApJ 224, 948) has cast doubts on the cluster membership of V367 Sct and indicates that either the cluster or the cepheid is anomalous in some way. It is important to establish beyond all doubt whether or not this cepheid is a cluster member.

A remarkable spectroscopic effect has been found by Barrell (preprint) in that the frequency and strength of Hα emission in double-mode cepheids far exceeds that observed in single-mode cepheids of similar period. The strength of the emission appears to be correlated with the phase of the observation.

The evolutionary status of double-mode cepheids is a vital question. As Stobie (MN 180, 631) has pointed out their existence requires a substantial modification to our current understanding of pulsation theory and/or stellar evolution theory. Although it has been considered that the masses and radii of these stars are correct and hence that they occur in some unexplained evolutionary stage (Petersen, AA 62, 205; Cox et al. ApJ 212, 451; King et al. ApJ 195, 467) the present weight of evidence is heavily in favour of these stars being normal population I cepheids which happen to be pulsating simultaneously in two modes (Efremov and Kholopov, Per. Zv. 20, 133; Faulkner, Proc. astr. Soc., Australia, 3, 124; J.P. Cox, Proc., Conf. on Current Problems in Stellar Pulsation Instabilities; Pel and Lub, IAU Symp 80, 229). The problem then is to explain why the double-mode cepheids have such low masses and radii compared to population I cepheids of similar period.

To this end considerable effort has been expended on determining the sensitivity of these masses and radii to various assumptions. Petersen (AA 62, 205) has investigated the effect of varying theoretical input parameters on the periods of stellar envelope models and concludes that the masses and radii are accurate to 9 and 4 per cent respectively. Cogan (ApJ 211, 890) and Saio et al. (Sci. Rep. Tohoku Univ. 51, 144) have found that the period ratio of first overtone to fundamental modes of pulsation can be substantially reduced by deep envelope convection and greater values of the mixing length parameter.

It is difficult, however, to see how this can explain the M_{beat} discrepancy as double-mode cepheids are not concentrated to the red side of the instability strip. Cox et al. (ApJ 214, L127) have shown that the inclusion of rotation cannot change the period ratio sufficiently to resolve the mass discrepancy. Faulkner (ApJ 218, 209) has suggested that, because of the observed high degree of mode interaction in double-mode cepheids, period changes may occur due to the non--linear mode interaction. However, non-linear effects on the theoretical periods of a double-mode cepheid have been investigated by Cox et al. (ApJ 220, 996). They conclude that the first overtone to fundamental period ratio agrees with the linear theory value independent of the stage of mode-switching and hence independent of the non-linear coupling.

The only theory which has been able to account successfully (see, however, Cogan ApJ 225, L39) for both the M_{beat} and M_{bump} discrepancies is that due to Cox et al. (ApJ 214, L127; ApJ 222, 621). This theory requires the structure of the stellar envelope to be altered in a major way so that the periods and period ratios are changed in the right direction. The envelope structure required is that produced by a chemically inhomogeneous envelope with the inclusion of a helium rich layer ($Y = 0.75$) throughout the hydrogen and helium convection zones ($T < 70{,}000^{\circ}K$). The physical cause of this helium enrichment cannot be diffusion from below. Instead it is considered that a helium deficient stellar wind can maintain this chemical inhomogeneity.

G. Anomalous cepheids in dwarf elliptical galaxies

The anomalous cepheids are a group of cepheids typically in the period range 1 to 3 days which do not obey the PL relationship of cepheids in globular clusters being systematically overluminous for a given period. Such variables occur in dwarf elliptical galaxies, in the SMC (Landi-Dessy variables) and there is one case known in a galactic globular cluster, NGC 5466 (Zinn and Dahn ApJ 81, 527). Hodge and Wright (AJ 83, 228) have presented results on 23 variables in the Leo I dwarf galaxy and in particular show that 12 of these variables are anomalous cepheids. These variables have been discussed by Norris and Zinn (ApJ 202, 335) and Demarque and Hirshfeld (ApJ 202, 346) who both independently reached the same conclusion that these variables are young, extremely metal-deficient ($Z \lesssim 10^{-4}$) objects of mass approximately 2 $M_{\odot}$. Zinn and Searle (ApJ 209, 734) have obtained spectrophotometric scans of some of these variables and deduce that, provided these variables are pulsating in the fundamental mode, they must be 3 times as massive as RR Lyrae variables. Deupree and Hodson (ApJ 218, 654) have investigated the dependence on mass and chemical composition of the location of the instability strip and also conclude that the anomalous cepheids are more massive than the RR Lyrae variables. The existence of such stars may require us to revise drastically our precepts on star formation in dwarf elliptical galaxies as, if these stars are single systems, their mass implies that they are 10^9 years old. Norris and Zinn indeed conclude that star formation has been occurring over the last 1-8 x 10^9 years in dwarf elliptical galaxies. On the other hand Renzini et al. (AA 56, 369) have considered the possibility that the anomalous cepheids occur in binary systems where mass transfer has taken place, in which case their age could be 10^{10} years.

H. Miscellaneous

The Baade-Wesselink technique continues to be an important, independent method of calculating the radius of a pulsating star. The effects of observational error on the solution has been examined empirically by Evans (ApJ 209, 135).Evans (ApJ Sup 32, 399) has obtained new photometric and spectroscopic observations of cepheids in order to match accurately in phase the light and radial velocity curves. Wesselink radii are derived for 14 classical cepheids and the data used to intercompare period-radius relationships. Scarfe (ApJ 209, 141) has redetermined the Wesselink radius of Zeta Geminorum. Balona (MN 178, 231) describes how the principle of maximum likelihood may be applied to determine the radius of a pulsating star given the light, colour and radial velocity observations. This statistical method represents an improvement over the normal Baade-Wesselink approach and has been applied to 54 well-observed cepheids giving radius errors of no more than 10 per cent. Budding (Ap Sp Sc 48, 249) has applied optimal curve fitting procedures, based on the Baade-Wesselink methodology, to radial velocity data and six-colour photometry of the classical cepheids δ Cep and η Aql to obtain improved absolute magnitudes. Opolski (IAU Coll. 29, 175) has derived a radius for the double-mode cepheid TU Cas consistent with the low values predicted from period ratios.

A new method (angular diameter method) has been developed to derive the distance of a variable star (Barnes and Evans MN 174, 489); Barnes et al., MN 174, 503; MN 178, 661). The validity of this method requires that a variable star during its pulsation cycle should follow the surface brightness - colour relationship determined by non-variable stars and evidence has been presented to support this for the V-R colour (Barnes and Evans, MN 174, 489). The method requires V, R colour observations and a radial velocity curve and is considered capable of yielding distances accurate to 15 per cent. Fernie (MN 180, 339) has extended the method of Barnes et al. by obviating the need for a radial velocity curve. His method uses the observed close relationship between the total radial displacement and the product of the period and light amplitude. The radial displacement amplitude is equated to the angular amplitude derived by the surface brightness method in order to derive the distance to the cepheid. Evans (MN 181, 85P) has compared phase shifts observed between light and radial velocity curves with phase shifts determined by the angular diameter method of distance determination and generally finds good agreement.

Ultraviolet observations from OAO-2 have been obtained for eight cepheids (Hutchinson et al. AJ80, 1044; Hutchinson, Cepheid Modeling, p5; Hill, Cepheid Modeling, p31). For some of the cepheids the ultraviolet observations show excellent agreement with ground--based photometry in the wavelength region of overlap and are consistent with a simple extrapolation of the light curve properties from the visible region. This is not the case with β Doradus which in the ultraviolet shows two extra flux bumps in addition to the well known bump near maximum light. A model is constructed in which all three bumps are associated with the arrival of multiple shock waves at the stellar photosphere.

Photometric evidence for companions has been obtained by studying the colour-colour curves of cepheids (Janot-Pacheco, AA Sup 25, 169; Madore, RO Bull 182, 151 and MN 178, 505; Pel, AA 62, 75). The evidence is consistent in pointing to the fact that at least 25 per cent of cepheids have photometric companions (most of which are presumably in a binary system with the cepheids).

The question of distinguishing different pulsation modes in cepheids has never been as clear cut as in the case of the RR Lyrae variables. However, evidence is mounting that cepheids with low amplitude, sinusoidal shaped light curves and occurring at the blue side of the instability strip are probable first overtone pulsators (Gieren, AA 47, 211; Ivanov and Nikolov, ApL 17, 115; Connolly, BAAS 9, 360). The strongest evidence on this point for galactic cepheids has been presented by Pel and Lub (IAU Symp 80, 229) where, in a diagram plotting fractional radius amplitude versus distance in effective temperature from the fundamental blue edge, a group of low amplitude, blue pulsators is clearly distinguished from the remainder of the pulsators, very reminiscent of the difference between RR Lyrae type-c and type-ab variables.

Henriksson (AA 54, 309) has searched for low amplitude variability in the cepheid instability strip. Out of 121 stars selected from the Bright Star Catalogue to lie statistically in the cepheid instability strip, 15 stars have shown detectable light variations.

Period changes in cepheids have been studied by Hoffleit (IBVS 1131), Szabados (Mitt. Sternw. Budapest-Szabadsaghegy, No 70) and Petit (IBVS 1402). Hoffleit has shown that a remarkably strong correlation exists between the absolute value of the fractional period change and the period for variable stars in the instability strip. RR Lyraes, cepheids of both populations, semi-regulars and Miras all appear to fall on the same relationship. Petit has verified this in detail for cepheids of all types. Szabados has presented evidence showing that the period changes are abrupt in the overwhelming majority of cases and has noted examples where secular change in the light curve have occurred.

Broglia et al. (AAp Sup 33, 339) has studied the photometric behaviour of the exceptional population II variable, RU Cam, over a period of 10 years. After its abrupt change in light curve characteristics in 1964, RU Cam subsequently appears to have maintained a low amplitude, quasi-sinusoidal variation.

Spectrophotometric determination of the radii and effective temperatures of cepheids has been carried out by Fracassini et al. (IAU Coll 29, 127) and Rautela and Joshi (Ap Sp Sc 40, 455).

5. BETA CANIS MAJORIS STARS (J. Rountree Lesh)

The renewed interest in the β Canis Majoris problem that began in the late 1960's has continued through the period since the last General Assembly. New techniques, such as observations from spacecraft, polarization measurements, and high time-resolution spectroscopy, have been applied to these stars. A number of new variables have been found, and previously known variables have been more extensively studied. Both theoretical and observational work has been directed to finding the answers to such questions as the evolutionary state of the β Canis Majoris stars, their mode of pulsation, and their instability mechanism. No general agreement has been reached on any of these subjects, however, and much work remains to be done.

A. Reviews and General Studies

A review paper on the observational aspects of the β Canis Majoris stars was published by Lesh and Aizenman (ARAA 16, 215). Theoretical problems were reviewed by J.P. Cox at the Los Alamos Solar and Stellar Pulsation Conference, and by Osaki at IAU Colloquium No. 46. Aizenman and Lesh reviewed the problem of mode identification in β CMa stars at the Goddard-Los Alamos Conference on Stellar Pulsation Instabilities.

Shobbrook (MN 184 43; 185, 825; IAU Coll. 46) approached the problem of the evolutionary state of the β CMa variables by obtaining new, accurate β photometry for the bright stars in the region of the "instability strip". He concluded that the strip is so narrow that it is not resolved by the data, but that about 3/4 of the stars in the strip, to a distance modulus of 8.0, are variable. Shobbrook inferred that the variables must be very near the end of core hydrogen burning. On the other hand, Sareyan (Thesis, Univ. of Nice) argued that β CMa stars are found in a much larger region of the HR diagram than had previously been supposed, and that they cannot all be represented by a single evolutionary state. In another general study, Ferrari d'Occhieppo (IBVS 1243) found that there are at least 25 β CMa stars per 10^8 cubic parsecs, so that these variables are almost as frequent as, for example, Mira variables.

Meanwhile, M.A. Smith and his coworkers (Smith and Karp, Solar and Stellar Pulsation Conference; Smith, ApJ 215, 574, Conference on Stellar Pulsation Instabilities, and 4th Trieste Colloquium on Astrophysics; Smith and McCall, ApJ 223, 221) found short-timescale profile variations in a number of early B stars, some of which were β CMa variables and some of which were constant in light. They deduced that the profile variations were due to complex nonradial oscialla-tions, and hypothesized that the β CMa stars are a subset of a larger class of "line-profile variable B stars."

B. Searches for New Variables

Major survey problems in the southern hemisphere have been undertaken by Balona (Mem RAS 84, 101) and Jerzykiewicz and Sterken (Act Ast 27, 365; ESO Mess. 11, 5; IAU Coll. 46), in an effort to find new β CMa variables and to improve our knowledge of the statistics of these stars. Balona found 8 β CMa candidates, including 4 in the galactic cluster NGC 3293. Jerzykiewicz and Sterken observed 14-24 variable stars in the summer sky, but concluded that at most 4 of them were β CMa stars. Percy and his coworkers (Percy and Lane, AJ 82, 353; Jakate, AJ 83, 1179) surveyed samples of stars in both heispheres. While they did not report any β CMa candidates in the field, Jakate observed 3 candidates in the cluster NGC 4755. Meanwhile, 2 β CMa candidates were discovered in serendipitous fashion by Haug (ESO Mess. 9, 14) and Lesh and Wesselius (AA, in press).

C. Observations from Spacecraft

Spacecraft have been used with increasing frequency in the last few years to make photometric and spectroscopic observations of β CMa stars in the ultraviolet. Photometric investigations have been made with the OAO-2 (Lesh, ApJ 208, 135; 219, 947), TD-1A (Burger, Thesis, Free Univ. Brussels); Beeckmans and Burger, AA 61, 815) and ANS (Lesh and Wesselius) satellites. These studies have generally shown that the ultraviolet light curves are nearly in phase with the visual light curves, and that the amplitude of the variation increases with decreasing wavelength. The ultraviolet and visual light ranges can be used to deduce temperature changes and radii for some of these stars.

The spectrometers on board the Copernicus satellite were used by Hutchings and Hill (ApJ 213, 11) and by Lesh and Karp (IAU Coll. 42, 4th Trieste Colloquium on Astrophysics) to study variations in the line profiles of several β CMa stars. Hutchings and Hill confirmed the equivalent-width variation in the C IV lines of β Cep, while Lesh and Karp found a consistently blueward asymmetry in the Si III profiles of ν Eri and σ Sco. They attributed this to the superposition of a stellar wind on radial or nonradial pulsation. They also observed multiple components in the profiles of 12 and 16 Lac.

D. Polarization Measurements

Polarization measurements have been made for the first time on a number of β CMa stars. Rudy and Kemp (MN 183, 595), observed the circular polarization on the wings of the Balmer lines in 9 variables, in a search for the presence of magnetic fields. They obtained positive results on β Cep and probably on γ Peg, and suggested that fields as large as 1 kG may be typical of this class of stars as a whole. Odell (BAAS 10, 401) suggested that nonradial pulsation in β CMa stars should produce a net polarization, which would fluctuate with the pulsation period. Schafgans and Tinbergen (AA, in press) searched for such a time-dependent polarization in β Cep, with negative results.

E. Ground-Based Observations of Individual Stars

Traditional observing methods continue to be used on both well--known and recently discovered β CMa variables. Recent emphasis has been on spectroscropy at high resolution in time and wavelength. Here we briefly cite the observations star by star.

γ Peg: Spectrographic and photometric observations were used by Sareyan, Valtier, and Le Contel (AA 44, 215) to redetermine the period, which they found to be constant. A more detailed analysis (AA Sup 25, 129) by the same authors showed the existence of short--period light fluctuations, which were correlated with profile variations. Peters (ApJ Sup 30, 551) matched equivalent width data, profiles, and the continuous flux distribution for this star with an interpolated model atmosphere of T_{eff} = 21,500 K, log g = 3.7, and found no abnormalities. Smith and McCall (ApJ 221, 861) deduced from high spectral and time resolution observations that γ Peg is undergoing radial pulsation, with two separate rising and falling shells.

12 Lac: Heard et al (JRASC 70, 213) studied line width and asymmetry variations and radial velocity curves, and concluded that either radial or nonradial pulsation could explain the observed phenomena. They found the period to be decreasing by 0.34 s/century. Sato (ASS 48, 453) found a similar rate of decrease for the primary period, and studied the physical properties of the pulsation using Wesselink's method and the line profile variations. Allison et al (AJ 82, 283) showed that there is line doubling at certain phases in the pulsation cycle, and that the true radial velocity amplitude is therefore twice that obtained previously.

HR 6684: Bolton et al. (PASP 87, 595) found a blue light range of 0.033 mag and a radial velocity range of 18 km/s, while Pike (MN 184, 265) found 16.2 km/s for the velocity range. Rosenzweig (IBVS 1125) concluded that both the accepted period of 0.139889 d and its 1 c/d alias of 0.162723 d are possible periods for this star.

BW Vul: Cherewick and Young (PASP 87, 311) found no correlation of the Hβ index with phase. Goldberg et al. (AJ 81, 433) described the line profile variations in detail, and computed displacement and acceleration curves for the stellar atmosphere. Stamford and Watson (IAU Coll. 46) proposed a radial shock model for BW Vul. A current rate of period increase of 1.8 s/century was found by Valtier (AA 51, 465), as well as by several of the above authors. However, Goldberg et al. found that the secular period increase before 1955 was 3.7 s/century, and Tunca (IBVS 1386) assumed a discontinuous increase shortly after 1970-71.

Miscellaneous: Radial velocity measurements confirmed the membership of λ Sco and χ Sco in the class of β CMa variables (Lomb and Shobbrook, MN 173, 709), while β Cen was observed to show line profile changes with the same period as its radial velocity variations (Lomb, MN 172, 639). On the other hand, the β CMa variations in α Vir have almost died out (Lomb, MN, in press). Photoelectric observations were used to look for period changes in α Lup (van Hoof, MNASSA 34, 73), δ Cet (Tunca, IBVS 1259), and σ Sco (Sterken, AA 43, 321). Among the β CMa candidates observed were V986 Oph (Jerzykiewicz, Act Ast 25, 81), V600 Her (Hill et al., AA 51, 1), 53 Psc (Sareyan et al., AA, in press; Conference on Stellar Pulsation Instabilities), and 53 Ari (Lane, IBVS 1202).

F. Theory

Where β CMa stars are concerned, the primary problem of the theoretician remains the discovery of the instability mechanism. Different variations on the standard stellar models have recently been constructed and tested for stability. Special attention in recent years has been devoted to nonradial osciallations - their stability, and the observable properties that could be expected if they are present.

The possible effect of the adopted opacities on the stability of stellar models was investigated by Stothers (ApJ 210, 434) and Stellingwerf (AJ 83, 1184). Stothers computed theoretical models of 10.9 and 15 $M_\odot$ stars, using Carson's opacities. His ZAMS models had cooler effective temperatures than models based on earlier opacities, and intersected the β CMa instability strip only once, in the core-burning phase. But all the radial and nonradial modes tested ($0 \leq \ell \leq 5$) were stable. Stellingwerf studied the destabilizing effect of the He^+ ionization edge (which produces an opacity feature near $T = 1.5 \times 10^5$K) on 12 and 8 $M_\odot$ models. The models remained stable, but became unstable if the opacity "bump" was artificially enhanced. J.P. Cox (private communication) has pointed out that "envelope excitation mechanisms" like the above have the advantage of producing an "instability strip" with negative slope - i.e., parallel to the main sequence, as observed for β CMa stars.

Osaki and Shibahashi (PASJ 28, 105, 199, 533) continued to work on the stability of massive stellar models in the secondary contraction and hydrogen shell-burning phases. A 10 $M_\odot$ model was found to be stable against quadrupole ($l = 2$) oscillations in the contraction and shell-ignition stages, and trapped nonradial gravity modes were vibrationally stable in the early shell-burning stages of 5 and 11 $M_\odot$ models. For 20 and 40 $M_\odot$ models, some g-modes with $\ell \simeq 10$ were overstable in the early shell-burning stage, but this instability was short-lived. Similarly, some gravity modes trapped in the zone of increasing molecular weight were unstable to Kato's mechanism in stars of 15 and 30 $M_\odot$ with a semiconvective zone, but the value of ℓ was too high to explain β CMa-type oscillations.

Sreenivasan and Wilson (Conference on Stellar Pulsation Instabilities) also concluded that the β CMa phenomenon is unlikely to be connected with semiconvection, since mass loss from 15-30 $M_\odot$ stars would prevent them from exhibiting semiconvective zones, while these zones would be small in any case for 10-12 $M_\odot$ stars. Van der Borght (PASA 3 (1), 89), on the other hand, thought that the overstable oscillations produced by the onset of nuclear reactions in a hydrogen-burning shell might be responsible for the β CMa oscillations, since they result in a periodic variation of the luminosity.

Aizenman and Weigert (AA 56, 457) studied the nonradial quasi-adiabatic stability of some very unusual models for β CMa stars - pre-main-sequence models with complete mixing, and post-main-sequence models with mass loss. All the models turned out to be stable. Meanwhile, Aizenman et al. (AA 58, 41) found that rapid shifts of frequency (bumping or avoided crossings) of the linear, adiabatic, nonradial modes of oscillation occur in stars evolving away from the main sequence, and thought that these might be connected with the beat phenomenon in β CMa stars.

In the area of mode identification, Kubiak (Act Astr. 28, 153) calculated line profiles and radial velocity curves for nonradially oscillating stars with values of ℓ up to 4. He found that single--period variables with sinusoidal velocity curves could be identified with stationary (m = 0) oscillations. Stamford and Watson (MN 180, 551; PASA 3 (1), 75; IAU Coll. 46) computed light, color, and radial velocity variations for radially and nonradially oscillating models. A comparison of the color variations with observations suggested the presence of both radial and nonradial oscillations with $\ell = 2$. Dziembowski (Act Astr. 27, 204) showed that nonradial oscillations cannot be distinguished from radial ones on the basis of the amplitude ratio of the radial velocity and light variations. He suggested instead using a generalization of the Baade-Wesselink method.

6. DELTA SCUTI STARS/DWARF CEPHEIDS (M. Breger)

A. General

More than 200 of these short-period pulsators are now known. A new review paper is expected to be published in the February, 1979 issue of the PASP. Two new variability surveys have been published. McInally and Austin (MN, 184, 885) tested 29 field stars for variability and found 6 variables. They report an unusually high period of $0^{d}.361$ for one star (HD 185969), but more observations of this star would be useful. Slovak (Ap.J., 223, 192) found a relative incidence of variability of 25% in the Alpha Perseus cluster, in good agreement with that found by Breger in other clusters (IAU Symp. 67, 231). The rotational velocity of the variables in Alpha Perseus resembles that in the similar Pleiades cluster in that the main-sequence variables have a slightly lower average rotational velocity than the constant stars.

The inverse radial velocity curve lags the light curve in V by about 0.09 periods, according to Breger, Hutchins and Kuhi (Ap.J., 210, 163). They also find the velocity to visual light amplitude ratio to be 92 km $s^{-1}mag^{-1}$. Dziembowski (Acta Astr., 27, 203) shows from calculations of nonradial p modes that nonradial osciallations cannot be distinguished from radial pulsation on the basis of amplitude ratios of radial velocity and light variations. He suggests the use of the Baade-Wesselink Method instead.

Antonello, Arienti, Fracassini and Pasinetti (Astr. Ap., 66, 37) reported very large radial velocity and equivalent width changes in three early A stars, based on photographic work. However, Breger, Light and Scholtes (Astr. Ap., in press) repeated the observations of two stars with a photoelectric spectrum scanner and found no variations. A group at Allegheny Observatory (Beardsley et al., preprints) believes that several hot stars, including Vega, are short-period variables. Kurtz (IBVS, 1436) reports the variability of an Ap star inside the instability strip, HD 101065 (Przybylski's star) with an unusually short period of 12.1 minutes. Kurtz and Wegner (Ap.I., in press) raise the interesting speculation that the high magnetic field of 2200 gauss forces the star into a high nonradial pulsation modes.

Stellingwerf (Ap.J., in press) examines the pulsational drivin zones with linear stellar models. Apart from the usual He II and H ionization zones he finds an additional (third) driving zone caused by the coincidence of the maximum photon flux with the frequency of the He II ionization edge.

Valtier, Baglin and Auvergne (preprint) have examined the importance of the H ionization zone to drive pulsation. This is important since the pulsating δ Del stars may be deficient in He in the He II ionization zone due to the diffusion. Only a marginally positive result for the H ionization was found.

B. Periods

There exists considerable evidence that these short-period variables pulsate in radial as well as nonradial modes. However, relatively little agreement on the exact period structures seems to exist from observer to observer. The problems can generally be traced to insufficient data, and sometimes, to deficient methods of analysis. A good summary, which is only slightly dated, is given by Fitch (Proc. IAU Coll., 29, 167). Valtier and co-workers (Astr. Ap. Suppl., 15, 115 and 18, 235) report that no stable frequencies exist for HR 432, 515, 812, 8006 and 9039. Warman (preprint), however, disagrees and finds stable periods which fit Valtier's data as well as earlier observations. McAlary and Percy (PASP, 87, 789) report three radial periods for 4 CVn. Warman, Pena and Arellano-Ferro (AJ, in press) find a stable multiperiod structure for this star over the last 21 years. A series of preprints by Kurtz conclude generally complex (nonradial) pulsation modes. Wizinowich and Percy (PASP, in press) find a radial period ratio of 0.774 for 44 Tau. Groups at Uttar Pradesh (Sinvhal, Gupta, Joshi and Rautele) and Texas (Loumos) also report a large observational multiperiodicity programme.

The theoretical problem with the high P_1/P_0 ratio for the large-amplitude variables may be resolves. J. Otzen Petersen (Astr. Ap., preprint) shows that in the P_1/P_0 range occupied by the Delta Scuti stars, the period ratios are extremely sensitive to composition and cannot provide unique information on stellar masses. Most of the observed period ratios can, however, be fit with normal masses and compositions. Cox, King and Hodson (Ap.J., in press) as well as Stellingwerf (Ap.J., in press) reach very similar conclusions. The largest radial P_1/P_0 ratio known, viz. 0.778 for SX Phe, can be fit by a 1.1 solar mass star with Population II composition (in good agreement with observations). Simon (Ap.J., preprint) explains the observed period ratios through a different hypothesis involving a resonant interaction between two excited modes. He requires low masses.

C. Large-Amplitude Variables

Recent evidence suggests that amplitude of pulsation is not a good discriminator of evolutionary status, mass and composition. The subdivision between Delta Scuti stars and Dwarf Cepheids (=RRs=AI Vel=amplitudes larger than 0.3 mag) appears not to be justified. Metal-poor variables appear among both the low-amplitude and large - amplitude variables (as pointed out by Eggen and others). In a large number of papers, two groups of workers led by McNamara and Breger respectively, have analyzed individual large-amplitude variables. Good summaries of their work can be found in McNamara and Feltz (PASP, 90, 275) and Breger (Poc. Solar and Stellar Pulsation Conf., Los Alamos, 47). These papers show that most large-amplitudes variables share the evolutionary status, abundance and mass with the smaller-amplitude stars. Balona and Martin (MN, 184, 1) find a Wesselink radius of 2.9 solar radii for RS Gru and deduce a post-main sequence state of evolution. The Barnes-Evans method fails to give realistic radii for RS Gru, presumably due to the presence of a companion.

The large-amplitude variable, SX Phe, is no longer the only extremely metal-poor, high-velocity case known. In an important discovery, Berg and Duthie (Ap.J. Letters, 215, L25) found GD 428 to be variable. McNamara and Feltz (PASP, 90, 275) confirm the variability and the star's lack of metals. They also derive log g = 4.18.

D. Pulsation and Metallic-Line Characteristics

Since the original announcement on the constancy of classical Am stars, several investigations have been made to confirm or refute this hypothesis. One of the reported exceptions, 32 Vir, has been examined by Kurtz, Breger, Evans and Sandmann (Ap.J., 207, 181). They find that the Am star has a normal A-star companion which they believe is pulsating. Mitton (preprint), on the other hand, suggests that the primary metallic-line star does indeed pulsate but that it is not a classical Am star and should be classified as an evolved Am star (δ Del) instead. In a new variability survey, Stickland has confirmed the short-period constancy of classical Am stars (Observatory, 97, 11).

Pulsation seems to be excluded only for the classical Am stars. The δ Del stars (evolved Am) seem to have an affinity for pulsation. Kurtz (Ap. J. Suppl., 32, 651) finds the δ Del stars to be a spectroscopically inhomogeneous group. Many spectra are similar to those of evolved Am stars with fewer overabundances of rare earths. Two Marginal Am stars (Am:) stars were also found to be variable by Kurtz (preprint).

7. T TAURI STARS AND RELATED OBJECTS (G.F. Gahm)

New observational results and current ideas concerning the physical structure and evolution of nebular variables, like stars of the Orion population, have been discussed at the IAU Symposium No. 75 in Geneva, 1976; the IAU Colloquium No. 42 in Bamberg, 1977; the Colloquium on "Early stages of stellar evolution" in Kiev, 1975 (published by Naukova Dumka, Kiev, 1977) and the IAU Colloquium No. 52 in Tucson, 1978. The proceedings of these meetings, like the reviews by Strom et al. (Ann. Rev. Astron. Astrophys. 13, 187) and Cohen (Quart. J. Roy. Astron. Soc. 19, 177) cover a large body of recent work and the present report is intended only as a brief overview of certain observational data and some current problems.

The T Tauri stars and the nebular Ae and Be stars show emission lines in the optical spectral region. In general, the broad emission lines are seen superimposed on an absorption line spectrum which can be classified according to spectral type and luminosity class. In many objects the absorption line spectrum is more or less veiled by a blue excess continuum, which in extreme cases extends into the red spectral region. The spectral surveys by Rydgren et al. (Astrophys. J. Suppl. 30, 307) and Cohen and Kuhi (1978, in press) indicate that when the blue spectral region is totally veiled, the red spectral region usually indicates a spectral type of late K or M. This is a very important result since many of the strongly veiled stars previously were classified as types F or G. Earlier statements concerning the positioning of stars in the HR diagram and relations between various observed parameters (like A_v) and spectral type should thereby be modified. Some of these statements can be proven to be invalid.

A good criterion on luminosity of the stellar component of the late-type stars is probably the strength of the red CaH bands. These were used by Mould and Wallis (Mon. Not. Roy. Astron. Soc. 181, 625) to demonstrate that the stellar component fall above the main sequence in the subgiant region. It should always be remembered that the relation between spectral type/luminosity class and effective temperature/M_v does not necessarily follow a "normal" law.

Two special characteristics of the absorption line spectra can be pointed out. Firstly, the lines are generally broader than for field stars of the same MK class. This effect is usually attributed to larger than normal rotational velocities of the stars. If so, the positive relations found for stars in NGC 2264 by Cohen and Kuhi (IAU Symp. No. 75 p. 203) between v sin i on one hand and degree of Hα emission and luminosity on the other hand could provide us with most important information concerning loss of angular momentum of the stars with evolutionary phase. The question remains open, whether the rotational energy liberated by the spin down of the star is transported away in form of mass loss or whether a considerable fraction reappears as radiation energy from an emitting volume around the star. Secondly, the Li I doublet lines at λ 6707 are unusually strong. This property is now well established through the extensive spectral survey by Herbig (Astrophys. J. 214, 747). It seems now quite safe to state that the surface Li abundance of T Tauri stars simply reflects the abundance of the original proto cloud and that the strength of the Li lines is one of the most important indicators of youth of a star.

A considerable number of optical spectrograms of certain T Tauri stars has accumulated over the last 35 years or so. It appears, that even if there are statements in the literature about changes in the appearance of the absorption line spectrum with time, many stars are repeatedly found to be of the same spectral type. For instance, there are now some 100 spectrograms or spectral scans taken of DI Cephei (for references covering the last three years we quote Gahm et al., Variable Stars 20, 381; Cohen and Kuhi, in press and Mundt, in press). The spectral type of DI Cep stays close to K0. Therefore, there seems to be a strong indication for this and other stars that the light and colour variability is not caused primarily by dramatic changes in the effective temperature of the stellar component. Rather, we must seek the origin of the brightness variations in the emitting gas envelope and/or in circumstellar dust shells. Before entering a discussion on these non-stellar components we note that a most thorough spectral and polarimetric study of the Herbig Ae and Be stars has been undertaken by Garrison and Anderson (Astrophys. J. 218, 438; ibid. 221, 601) and Garrison (Astrophys. J. 224, 535) who found little if any intrinsic difference between the nebular and non-nebular emission line stars of these spectral types. We also note that in the case of stars belonging to the FU Orionis class, a most dramatic change in spectral appearance takes place. (Herbig: Astrophys. J. 217, 693; Elias: Astrophys. J. 223, 859).

The emitting gas envelope is manifested in the emission lines and the continuous excess emission that is seen in the ultraviolet, the blue (the blue veiling) and the infrared spectral regions. For the latter we can state that since there seems to be little doubt that hydrogen free-bound emission gives a large contribution to the UV excess emission it follows that at least part of the IR excess originates from gas emission. However, the IR excess can also be explained as emission from grains in a circumstellar dust envelope. Quite divergent opinions exist on which of these processes dominate. It may be that both effects are present with different relative importance on different stars (Rydgren et al.: Astrophys. J. Suppl. 30, 273; Kuan: Astrophys. J. 202, 425; Rydgren: Publ. Astron. Soc. Pacific 88, 111; Cohen and Schwartz: Mon. Not. Roy. Astron. Soc. 174, 137; Thompson and Reed: Astrophys. J. Lett. 205, L 159; Warner et al.: Astrophys. J. 213, 427; Simon and Dyck: Astron. J. 82, 725; Wenzel: Astron. Nachrichten 296, 183; Rydgren: Publ. Astron. Soc. Pacific 89, 823; Schwartz: Astron. J. 83 785; Wenzel and Brüchner: Mitt. Veränderliche Sterne 8, 35).

At radio continuum wavelengths only a few objects have been detected so far. For others upper limits have been set. (Altenhoff: Astron. Astrophys. 46, 11; Harris: Mon. Not. Roy. Astron. Soc. 174, 601; Brown et al.: Mon. Not. Roy. Astron. Soc. 175, 87P; Woodsworth and Hughes: Astron. Astrophys. 58, 105; Schwartz and Spencer: Mon. Not. Roy. Astron. Soc. 180, 297). Emission at mm and cm wavelengths clearly originates in circumstellar ionized gas envelopes.

The first broad band flux measurements of Orion population stars at wavelengths shortward of 3000 A were obtained by de Boer (Astron. Astrophys. 61, 605). More detailed spectra obtained with the IUE satellite and covering part or all of the spectral region $\lambda\lambda$ 1150--3200 A are now available for T Tau (Penston et al.: in preparation), RU Lup (Fredga et al.: in preparation), S Cr A (Appenzeller and Wolf: Astron. Astrophys., in press) and DI Cep (Gahm et al.: in preparation). The presence of emission lines from several times ionized elements like C IV and Si IV indicates that in addition to the low temperature shell component seen in the optical spectral region there are shell regions of very high temperature ($\sim$ 100 000 K) around the stars. These observations introduce new and interesting aspects relevant to the understanding of the nature of the objects.

One of the existing major theoretical problems is to explain why the stars are surrounded by such energetic and luminous emitting gas shells and to formulate models describing the physical structure of these shells. Quite different ideas on these matters are in circulation.

The discovery of inverse P Cygni profiles that occasionally are seen present in the higher Balmer lines of a number of objects has led to models that, at least for these stars, favour infall of surrounding interstellar material as the main source of energy for the emitting gas envelope. The profiles of the sometimes very broad emission lines are understood on the basis of the velocity field of the infalling gas (Ulrich: Astrophys. J. 210, 377; Appenzeller and Wolf: Astron. Astrophys. 54, 713; Bertout: Astron. Astrophys. 58, 153; Wolf et al.:

Astron. Astrophys. 58, 163; Bertout et al.: Astron. Astrophys. 61, 737; Appenzeller: Astron. Astrophys. 61, 21: Appenzeller et al.: Astron. Astrophys. 63, 289). As a main cause of the brightness variations of the stars the possibility of changes due to density fluctuations in the infalling material has been pointed out (Mundt: Astron. Astrophys.: in press). For the proto-type of these so-called YY Orionis stars, namely YY Ori itself, Walker (Astrophys. J. 224, 546) arrived at the conclusion that the brightness variations are mainly due to variable opacity in a circumstellar dust shell in the line-of--sight to the star.

Many of the references quoted above in connection to observations of energy distributions of Orion population stars also include discussions on different aspects of variations due to changes in circumstellar dust layers and/or in circumstellar ionized gas. Closely related to the question of variations in the ionized gas are the observations leading to statements about extremely rapid (on time--scales of minutes or hours) changes in the intensities or profiles of certain emission lines or in the ultraviolet continuum (Kolotilov and Zaitseva: Variable stars 20, 153; Kuan: Astrophys. J. 210, 129; Zaitseva and Luyty: Variable stars 20, 266; Mundt: in press). Undoubtedly variability in the properties of the ionized gas plays at least some role in the brightness variations. On the other hand, one can find objects which seem to vary exclusively because of changes in a circumstellar dust envelope. One exemple is HR 5999 discussed by Thé and Djie (Astron. Astrophys. 62, 439).

Ideas on magnetic fields extending from the surface of the stars and acting as a relay mechanism for transporting variable amounts of mechanical energy into the circumstellar ionized gas and lining up circumstellar dust grains to produce circumstellar polarisation has been developed by Petrov and Schcerbakov (Proc. 3rd European Meeting, Tbilisi p. 163) and Gehrsberg and Petrov (Sovj. Astron. Lett. 2, 449). This model seeks to explain the energy source of the emitting ionized gas, the cause of the changes in brightness as well as the mechanism behind the FU Orionis type phenomenon.

We know that the energetics of a star in formation is governed by the energy liberated from accretion of interstellar material and also that young stars eventually will develop surface phenomena of chromospheric nature, e.g. similar those on the sun. Consequently, the problem is to understand when and why the circumstellar gas emission switches over from one energy source to the other. And clearly, the solar chromosphere is not dependent on accretion of interstellar material for its existence.

A closely related question is how to derive bolometric luminosities for the stars. If the excess emission is due to infalling material, the energy flux liberated should not be included in the computation of the bolometric luminosities for comparisons with theoretical evolutionary tracks. If, on the other hand, the energy comes from a mechanical flux generated in the star, rotational braking or any other internal source, the excess emission should be included.

This review has concentrated on nebular variables with a distinct stellar component. During the last years a number of variable but non-stellar objects have been discussed as possible pre-main-sequence objects. These have not been discussed here.

We close this review by mentioning that a number of works have been directed towards a description of the molecular content in the circumstellar or interstellar surroundings of Orion population objects. The following molecules have been looked for with various success: OH Rudnitskii: Sovj. Astron. J. 53, 1225; Lang and Willson:

Astrophys. J. Lett.: in press; Gahm et al.: Astron. Astrophys.: in press. CH_3OH Buxton et al.: Astron. J. 82, 985; H_2 Beckwith et al.: Astrophys. J. Lett.: 223, L 41. A search for fluorescent diatomic modecules was done by Gahm et al. (Astron. Astrophys. Suppl. 27, 277).

8. RR LYRAE VARIABLES (B. Szeidl)

Two important catalogues of RR Lyrae stars were published. The graphical catalogue (Cacciari and Renzini AA Sup. 25. 303) gives the period amplitude diagram, the period-frequency distribution and the period cumulative diagram for RR Lyraes in the globular clusters, in the solar neighbourhood and in some samples situated near the galactic centre, in the galactic halo and in the nearest extragalactic systems. The bibliographical catalogue, available on magnetic tape and on microfiches, gives bibliographical references for 5855 field RR Lyrae stars with 6607 entries (Heck, Lakaye AA Sup 30. 397).

Bookmeyer, Fitch, Lee, Wiśniewski and Johnson have finished their UBV photometry on about 180 field RR Lyrae stars. The second list (Revista Mexicana de Astron. y Astrof. vol. 2. No. 3. 235) contains observations of 70 stars. The two lists give mean colours and magnitudes at light extrema for 174 RR Lyraes adequately observed during the course of this program.

Dean, Cousins, Bywater and Warren published BVI photometry on 9 southern RR Lyraes (Mem. RAS 83. 69). Lub obtained light and colour curves in the Walraven VBLUW system for 90 field RR Lyrae and short period variables (AA Sup 29. 345). Using these observations he discussed the reddening and blanketing corrections and physical properties of RR Lyrae stars (ESO Sci. Prepr. 13).

Considerable work has been devoted to studying the metal content and other physical characteristics of RR Lyrae stars. From uvbyβ photometry of SW And (McNamara, Feltz PASP 89. 699) and SS Psc (McNamara, Redcorn PASP 89. 61) the metal abundance, effective temperature and gravity of these stars were derived. UBV photometry of RR Lyrae stars in ω Cen and M4 (Sturch, AA 46. 133; PASP 89. 349 and 90. 264) made possible to discuss the problems of metallicity and cluster reddening. The observed UV excess in ω Cen confirms the variations in [Fe/H] found from recent spectroscopy.

[Fe/H] values have been spectroscopically determined and analysed for a large number of field RR Lyrae stars (Butler, Carbon, Kraft BAAS 8. 347; ApJ 210. 120 and Rodgers ApJ 212. 117) and cluster variables (Butler ApJ 200. 68; Freeman, Rodgers ApJ 201. L71; Butler, Dickens, Epps ApJ 225. 148; Manduca, Bell ApJ 225. 908). Observations of RR Lyrae stars in ω Cen have revealed a diversity in the composition of these stars.

Kukarkin reviewed the various metallicity criteria and discussed them together with other features of globular clusters (IAU Symp. 67. 511). Frolov deduced a relation between the pulsating parameter $\Delta V/\Delta(B-V)$ and the metal abundances (A Cir USSR 941).

Bukhantsova and Nikolov investigated the group of galactic RR Lyrae variables with periods about $0\overset{d}{.}4$ (A Cir USSR 966). This group is well defined and might have special evolutionary aspects.

A system of criteria of two-dimensional quantitative spectral classification of RR Lyrae type variable stars were established by Fenina (Per Zv 20. 103). Using this method, absolute magnitudes of SW And, XZ Cyg and RZ Lyr could be determined (Fenina A Cir USSR 892).

Following a method advocated by van Hoof mean radii of some RR Lyrae variable stars were determined by Woolley and Davies (MN 179. 409). This method gives acceptable results and leads to mean absolute magnitude of $0\overset{m}{.}45$. The velocity ellipsoid of RR Lyrae stars has been computed and the best figures are roughly in the ratio $\bar{R}^2 : \bar{\Phi}^2 : \bar{\Theta}^2 = 3:2:1$ (Woolley MN 184. 311).

At the Sonneberg observatory extensive works were carried out on the variables discovered there. Gessner finished the processing of the 93 RR Lyrae variables discovered by Hoffmeister in the area η Arae (Veröff. Sternw. Sonneberg 8. 341). Meinunger derived the distances and the density distribution of RR Lyraes in the Sonneberg fields near the North Galactic pole (AN 298. 171).

During the past three years a great number of visual, photographic and photoelectric maxima have been collected for known and new RR Lyrae stars. These are mostly published in A. Cir. USSR, MVS, IB, Per Zv and Per Zv Sup. Accurate photoelectric observations indicate that HX Ara, originally classified as a W UMa star, is an RRc variable (Duerbeck, Walter IB 1224, AA 49. 471), while BC Dra, misclassified as δ Cep variable, is in reality an RRab star (Szabados, Stobie, Pickup IB 1197). Photoelectric observations and radial velocity measurements were published by Andrews (MN 172. 271) on the RRc variable HD 268892.

The study of period changes of RR Lyrae variables is still very popular. Szeidl gave a review of the observed phenomena of period changes (IAU Symp. 67. 545). He showed that both abrupt and smooth changes in period might occur. The dependence of the period changes on the evolutionary track through the instability strip has still found no explanation.

O-Cs were given for UY Cyg, SU Dra, CG Peg and AN Ser by Batyrev (Per Zv Sup. 2. 207), for TV Boo, TW Boo and RR Leo by Demjanovskij (Per Zv Sup. 2. 301), for AV Peg by Grigorev (Per Zv Sup. 2. 291) and for XX And, RS Boo, UY Boo, RZ Cep, RV CrB, XZ Dra, RW Dra, SV Eri, RR Gem, SZ Hya, Y LMi, RV UMa and SX UMa by Firmanyuk (IB 1152, A Cir USSR 924). He noticed that some of the O-Cs show possible periodicity in the period variations with a cycle length of 15000-40000 days. Oláh and Szeidl commenced an investigation of the period changes of field RR Lyrae stars. Their results on AT And, SU Dra, RR Leo, TT Lyn and AR Per have been published (Budapest Mitt. 71).

The Large and Small Magellanic Cluds have been searched for variable stars by Graham (PASP 89. 425 and 87. 461). Variables have been also found by Kinman, Sryker, Hesser (PASP 88. 393) in and around NGC 1841 and the extent of the Magellanic Clouds. A search for rapid variables and RR Lyrae stars in the LMC on ESO Schmidt plates is carried out by Geyer (IAU Symp. 67. 557).

Zinn and Searle investigated two RR Lyrae stars in the Draco system (ApJ 209. 734). The mean effective temperatures and the luminosities have been used to find their masses.

The sparsely populated globular cluster in Reticulum found by Sérsic has been investigated to V = 19.5 mag by Demers and Kunkel (ApJ 208. 932). Periods are given for 19 of 22 RR Lyrae variables found.

The investigation of multiple periodic RR Lyrae stars is still an outstanding problem.

Ultraviolet photometry of RR Lyrae was carried out from the Orbiting Astronomical Observatory-2. Hutchinson, Hill and Lillie (ApJ, 211. 207) used a hydrodinamical model to show the influence of atmospheric shock waves on the ultraviolet light curves. Their model indicates that visible light observations of the bump in the light curve before rising light and the stillstand in rising light are due to the energy deposition by two separate shock waves.

At the Konkoly Observatory, in Hungary RR Lyrae has been intensively observed and, its four-year cycle and the large phase shift in the 41-day period at the beginning of the new cycle was confirmed.

Szeidl (IAU Colloq. 29, Part I, 133) gave an observational review of RR Lyrae stars with Blazhko-effect. He discussed the problems of frequency and general characteristics of multiple periodic RR Lyrae stars. Kanyô (IAU Colloq. 29, Part II, 211) investigated the period-fluctuation of these stars and found out that the RR Lyrae stars of shorter secondary periods have more complex O-C structures than those of longer ones. Romanov (IAU Colloq. 29, Part II, 191) surveyed some characteristics of Blazhko-effect in RR Lyrae-type variables and discussed the hypothesis of the interference of radius oscillation first outlined by Tsessevich.

Fenina and Romanov's method of quantitative spectral classification of RR Lyrae-type variable stars (IAU Colloq. 29, Part II, 229) was applied to some variables with Blazhko-effect. The spectral pecularities of SW And was discussed (Fenina, Romanov, A Cir USSR 862) and a new period of $36\overset{d}{.}92$ of the Blazhko-effect in SW And was given (Lysova, Romanov and Firmanyuk, A Cir USSR 982). Romanov, Tsessevich, Shakun and Zgonyaiko (A Cir USSR 981) discovered some pecularities of light variation and anomalous Blazhko-effect in DR And. Recent observations (Lange and Firmanyuk, A Cir USSR 864) show the Blazhko-effect in ST Boo, too. A few papers (Fenina, A Cir USSR 862, Taylor, JAAVSO 4. 25, Blasberg, MVS 7. 31, Bezdenezhny Per Zv Sup 3. 125) have dealt with XZ Cyg. Photoelectric observations of this star were discussed by Kunchev (Per Zv 19. 477) and by Kunchev and Romanov (Per Zv 19. 471) Lebedev (Per Zv Sup 2. 313) observed XZ Dra photoelectrically but was unable to determine its secondary period. Romanov (Per Zv 20. 299) investigated the radial velocities in RZ Lyr and determined the radial velocity amplitudes at different phases of the secondary period. Romanov, Fenina and Klabukova (A Cir USSR 923) studied the variation of Δs and p indices of this star. AR Ser, which has anomalous Blazhko-effect, was thoroughly investigated at the Odessa Observatory (Firmanyuk, IB 1245; Tsessevich A Cir USSR 885, IAU Colloq. 29 Part I.3, Per Zv Sup. 3. 93). A detailed UBV photometry of the multiple periodic RR Lyrae star RV UMa was carried out by Kanyô (Budapest Mitt 69). Frolov called the attention to the interesting "ultraviolet" Blazhko-effect of X Ari (IB 1097). The multiple periodic RRc star TV Boo was observed by Firmanyuk (A Cir USSR 843).

Goranskij (Per Zv Sup. 2. 323) made an attempt at determining the secondary periods of RR Lyrae variables in globular clusters (No. 30 in M53 $P_s = 37\overset{d}{.}6$ and No. 2 in M5 $P_s = 132\overset{d}{.}4$).

AC And is unique in having three identifiable periods. Its fundamental and first and second overtone radial pulsation modes are all excited and nonlinearly coupled (Fitch and Szeidl, ApJ, 203. 616). This star provides a good opportunity for testing the pulsation theory and determining its mass (Cox, King and Hodson, IAU Colloq. 29 Part II. 233; BAAS 9. 360, ApJ 212. 451; Stellingwerf, IAU Colloq. 29. Part I. 153; Petersen, IAU Colloq. 29. Part I. 195, AA 65. 451). The U-B and B-V colours of this triple mode variable (Jakate ApJ 224. 603) are compatible with the idea that AC And is an RR Lyrae star.

AQ Leo has also complex light variation. The star is found to be an RR Lyrae variable with a $0\overset{d}{.}5497$ fundamental and $0\overset{d}{.}4101$ first overtone periods simultaneously excited (Wenzel, IAU Colloq. 29, Part II. 221; Jerzykiewicz and Wenzel AcA 27. 35).

9. VARIABLE STARS IN GALACTIC GLOBULAR CLUSTERS (H.B. Sawyer Hogg)

A. Introduction

Since the last report was prepared for the I.A.U. in 1976 more than a hundred papers have been published dealing with variables in more than two dozen galactic globular clusters. Obviously these can not all be discussed. The more significant results are summarized, with author's name and date. This will make easy access to the reference and summary in Astronomy and Astrophysics Abstracts. Also the writer is planning to publish the Fourth Catalogue of Variable Stars in Globular Clusters (David Dunlap Observatory Publications, vol. 3, No. 8) in time for the I.A.U. in 1979, with a full set of references.

A valuable and comprehensive review of the subject by L. Rosino appeared last summer in Vistas in Astronomy, volume 22, part 1, pp. 41-74 (1978). With 19 tables, 15 figures and more than a hundred references this paper brings us up to date with information available into 1977.

B. New Variables and New Periods (See also Binaries, X-ray clusters, Omega Centauri)

We draw attention especially to the following clusters, with considerable data or work: NGC 1261, 2419, 4833, 5986, IC 4499, NGC 6205, 6273, 6356, 6656.

NGC 288, Hollingsworth and Liller (1977) one new variable, RR Lyrae period; 362, Lloyd Evans (1977, 1978, letter) one new, P 130d; 1261, A. Wehlau, Flemming, Demers, C. Clement (1977) 6 new RR Lyrae stars, Wehlau and Demers, (1977) periods for 13 RR Lyrae and one 80 days; 2298, M. Liller (1976) one new variable, periods for 3 RR Lyrae; 2419 Pinto and Rosino (1976, 1977) elements and light curves for 32 RR Lyrae, one cepheid, one red SR; 3201, Lee (1977) 4 new variables; 4833, Demers and Wehlau (1978) 15 RR Lyrae periods, 9 of them new, 2 slow variables; 5272, Kholopov (1977 3 papers) 10 new variables, 2 new RR Lyrae periods, one changing period; 5286 Fourcade, Laborde, Puch, Colazo, Arias (1975) 6 new RR Lyrae periods, M. Liller and Lichten (1978) 2 new variables, 13 periods; 5466, Zinn and Dahm (1976) V 19, P $0\overset{d}{.}82$, anomalous cepheid; 5634 M. Liller and Sawyer Hogg (1976) periods of 5 RR Lyrae; IC 4499, Coutts Clement, Dickens and Epps Bingham (1979) periods and light curves for 52 RR_{ab} and 23 RR_c; 5897, Samus (1976) UV bright variable, Wehlau and Sawyer Hogg (unpublished) RR Lyrae periods; 5986, M. Liller and Lichten (1978) 7 new variables, 9 RR Lyrae periods; 6121, Lee (1977) 3 new RR Lyrae stars; 6205, Osborn and Fuenmayor (1977), Pike and Meston (1977) suspected red variables; 6235, M. Liller (1977) 3 new variables, periods for 3 RR Lyrae, 2 eclipsing; 6273, Samus (1976) one new variable, C. Clement and Sawyer Hogg (1977) periods for 5 Mira stars, one irr.; 6535, M. Liller and C. Clement (1977) one new variable, 2 RR Lyrae periods; 6638, Rutily and Terzan (1977) periods for 5 Mira stars and 15 RR Lyrae around cluster; 6656, A. Wehlau and Sawyer Hogg (1977) periods for 3 Mira cluster members and 6 probable field Mira's; 6752, Wesselink (1976) period for V, 1.378 days; 6838, Cohen (1977, letter) new red variable; 6934, A. Wehlau and Sawyer Hogg (unpublished) 10 new periods.

C. Period Changes

In NGC 5024 Goranskij (1976) has studied the period variability of 31 RR Lyraes and shows two types of variability, spasmodic or uneven variations and smooth, unperiodic variations, showing a dependence of period jumps on the period of the variable. In Messier 5 Wilkins (1976) has compared for 65 variables periods from many observers to fix their variations. In the same cluster C. Coutts and Sawyer Hogg (1977) show the 26 day period for the W Vir star V 84 has both increased and decreased markedly, but the similar period of V 42 has barely changed. In Messier 22 A. Wehlau and Sawyer Hogg (1978) have determined period changes for 14 RR Lyraes of which 8 are increasing and 5 decreasing, one with abrupt change. In Messier 15 Smith and Wesselink (1977) studied 12 RR Lyraes which, combined with earlier results show a significant excess of increasing over decreasing periods. Hoffleit (1976) confirms a relation found earlier between period length and abrupt change of period for Pop II Cepheids. Rastorgyev (1978) has compared the period changes of long period cepheids in 9 globular clusters to show that the cepheids with periods less than 10 days have more stable periods than those over 10 days. The time of life in the zone of instability is longer than theory predicts.

D. Red Variables

In addition to red variables mentioned in other sections we draw attention to the work by Eggen (1977) on the MARV and SARV stars in M4, M22, NGC 5897 which lie above the asymptotic giant branches and may be progenitors of halo population long period Cepheids. Mallia (1976) has barium characteristics for 2 red variables in M 22. White (1978) has made a determined hunt for variability of red giant and asymptotic branch stars. He concludes that photometric variability is a more generalized phenomenon throughout these branches than previously thought, with 7 such stars indicated in M 3, 9 in M 13 and 5 in M 15. In M 13 Osborne and Fuenmayor have studied the red variables, with results indicating all red giants redder than (B-V) 1.45 are variable.

E. Binaries

Very few binaries have been found within the area of globular clusters. Those considered for some years to be probably members have been gradually ruled out. Geyer (1965) showed V 65 in NGC 3201 a non-member. Geyer (1978) M. Liller and Lichten (1978) rule out, from velocity V 78 as a member of Omega Centauri. At the Globular Cluster Institute, Cambridge, England in 1978 M. Liller showed the velocity of V 3 in NGC 6838 indicates it is a non-member.

Meanwhile, because of the discovery of some globular clusters as X-ray sources, the hunt for binaries in them has been stepped up. Trimble (1976) conducted a hunt in the core of NGC 6809, but found an apparent absence of faint variable stars. In a further discussion of close binaries in globular clusters, Trimble (1977) following discussion of binaries by Hills (1975, 1976) states that while the number of binaries predicted does not conflict with available data, careful searches would provide a test of theories.

In Omega Centauri Niss, Jørgensen and Laustsen (1978) have conducted a systematic search for eclipsing binaries brighter than the turn-off point. They have found 22 variables (revised number, personal communication) one of which is almost certainly an eclipsing binary).

F. X-ray Clusters

In some clusters known variables lie within the error box of the X-ray source, but apparently no such variable has yet been positively identified as the source. In NGC 1851 Bolton and Mallia (1977) propose as the source a highly evolved star, not variable photometrically, but with radial velocity variation in a 15 day period. In this cluster A. Wehlau, M. Liller, Demers and C. Clement (1978) have periods for 19 RR Lyrae stars, 7 of them new variables. In NGC 6441 Hesser and Hartwick (1976) find that V 6, within the X-ray source error box, has an anomalous position in the C-M diagram. They also not two probable variables. In NGC 6624 M. Liller and W. Liller (1976) have one new variable and (1978) report that red V 3 is a probable member, but the one RR Lyrae star may be a field variable. In NGC 7078 Chu (1976, 1978) reports an ultrashort period variable of a new type, and Leroy (1976) reports that his work on the core has shown some new variables in the center.

G. Cluster Membership

The work on binaries in globular clusters emphasizes the importance of determining the probabilities of membership for variables in the cluster region. Proper motions, radial velocities, color magnitude diagrams in various wavelengths, and statistical probabilities are being used to decide on membership.

Cudworth, (1976) using proper motion in M 92 finds a 99 percent chance of membership for 5 RR Lyrae stars, with V 14 a field star. In M 13 (1978 letter) he has a 99 percent membership probability for 9 variables, with two more very probable. In NGC 6304 Hesser and Hartwick (1976) find that the RR Lyrae stars, even those closest to the center, may be field, but red variable V 15 may be a member.

In Omega Centauri Geyer proves non-membership for the very short period cepheid V 65.

In M 22 Lloyd Evans discusses the positions of 6 red variables on the giant branch.

In useful pieces of work R. Webbink (1978, private communication) has tabulated variables listed in the General Catalogue which are within the tidal radius of various clusters, and J. Nemec (1978 Master's Thesis, University of Victoria) has identified variables in the post horizontal branch stars of 19 globular clusters.

H. Omega Centauri, the most studied cluster

In addition to information in other sections of this report, we note that Freeman and Rodgers (1976) show the chemical inhomogeneity of 15 RR Lyrae stars. Sturch (1976) has photoelectric observations of three RR Lyrae stars confirming metal weakness. Glass and Feast (1977) with J, H, K photometry have classed the red variables as TiO absent or present. Butler, Dickens and Epps (1978) in a spectroscopic study have determined Fe/H values for nearly half the RR Lyrae stars in the cluster, showing a range in metallicity, and finding 2 new variables. The variation in metallicity is confirmed by Sturch (1978) with UBV photometry of 10 RR Lyrae stars. Fourcade, Laborde and Yurquina (1978) have a new long period variable.

I. Correlations

Auverge (1976) made a statistical study of RR Lyraes in 12 globular clusters with special reference to the period amplitude plane. Cacciari and Renzini (1976) present a graphical catalogue of RR Lyrae variables with numerous correlations. Samus (1976) discusses the relation between transition period and luminosity of RR Lyraes in globular clusters. Wilkens (1976) has derived the diameters of 66 globular clusters from their variables and Innanen and O'Brien (1977) studied the ellipticities of RR Lyrae systems in these clusters. Pike (1976) suggests NGC 6366 provides a chance to determine the absolute magnitude of a metal rich RR Lyrae. Gingold (1976) discusses the evolutionary status of Pop II cepheids with theoretical diagram of period distribution. In M 13 the mass of V 1 has been calculated by Osborn and Rosenzweig (1976), and by Pike and Meston (1977) in their improved colour photometry of the cepheids and RR Lyrae stars. Eggen (1977) in a classification of intrinsic variables from UBV and RI observations has derived parameters for long period cepheids in four globular clusters. In M 4 Smith and Butler (1978) show no star-to-star range in metallicity for the RR Lyrae stars, in sharp contrast to those in Omega Centauri.

10. RED VARIABLES (M.W. Feast)

The study of red variables is being pursued actively at the present time. This is very largely because new techniques (e.g. microwave work, infrared studies and diameter determinations) have revealed new phenomena and offer the possibility of deepening very considerably our knowledge of these stars. During the past three year period the accumulation of basic data has continued (e.g. UBVRI of Medium Amplitude Red Variables, Eggen, Ap.J. Supp. 34, 233, Ap. J. 213, 767, UBV of Miras, Celis A.A. Suppl. 29, 15, A.A. 63, 53. Discovery of Mira Variables etc. Rosino and Guzzi, A.A. Suppl. 31, 313, Friedemann et al. A.N. 298, 327, JHKL (Infrared) Photometry of Mira and other Variables Feast et al. (SAAO). Much remains to be done in this field if the statistical properties of the variables are to be properly studied and if the variables are to be used for studies in galactic structure (and also in the Magellanic Clouds). Consideration needs to be given to the completeness of discovery in various regions. In planning further photometric work considerable care will be needed in choosing the most effective wavebands (including band widths) at which to work.

Speckle interferometry opens up the exciting prospect of detailed studies of the diameters of Mira Variables and their variation with phase and wavelength. Variations of diameter by a factor of two are found for wavelengths inside or outside strong TiO bands (Labeyrie et al. Ap. J. 218, L75, 1977). Evidently not only the pulsation but much of the detailed structure of the atmosphere will eventually be studied by this method. The extent of circumstellar shells round Miras is indicated by the discovery of an optical emission region several arc seconds in size round R Leo (Lambert and van der Bout, Ap. J. 221, 854). Another powerful new tool, heterodyne interferometry, has been used to study the spatial structure of circumstellar shells of red variables at 11 microns (Sutton et al. Ap. J. 217, L97, 1977). In a different approach Barnes and Evans (M.N. 174, 489, 1976) have derived a relation between the V-R colour and surface brightness and applied it to the case of Mira Variable pulsations. Low resolution scans of Mira Variables in the infrared (Strecker et al. A.J. 83, 26, Puetter et al. P.A.S.P. 89, 320) also allow colour temperatures and diameters to be deduced. In addition they will be useful in interpreting broad band infrared colours of Miras.

Improved absolute magnitudes for Mira Variables remain an outstanding need. Work on statistical parallax etc. of Miras for infrared absolute magnitudes is in progress at SAAO. Eggen (P.A.S.P. 88, 715, 1976) has derived space motions for some Miras. However his absolute magnitude calibration is not in agreement with that of others. Aslan (Obs. 96, 149, 1974) has derived absolute magnitudes for SR Variables in groups whilst Upgren (A.J. 80, 828, 1974) has removed a number of discrepant parallaxes. Absolute Magnitudes for Miras have been predicted semitheoretically by Cahn and Wyatt (Ap. J. 221, 163) and it will be important to compare these predictions in more detail with observations.

Polarization studies hold out very considerable promise in understanding Mira Variables, especially when the results are combined with other work. It is known that Mira Variables with high polarization also have infrared excesses indicating circumstellar shells. However studies of time variability etc. suggest that the polarization occurs much closer to the star than the infrared excess. The increase in polarization in at least some Miras at phase $\sim$ 0.8 when emission lines first appear suggests the formation of particles in a shock wave in the atmosphere (c.f. Coyne and Magalhaes A.J. 82, 908, Shawl A.J. 80, 602, Kruszewski and Coyne A.J. 81, 641, Materne A.A. 47, 53, 1976). Shawl has discussed the polarization of Mira Ceti B which may indicate that a significant amount of the polarization of Mira A and B comes from a (common) circumstellar envelope. Mira B will certainly interact physically with such an envelope. Further light on this problem could well come from polarization work on symbiotic stars. The red star in these binary systems is either an M giant or a Mira (Feast et al. M.N. 179, 499). Extensive polarization work on these systems especially those with Mira components would be valuable for the general Mira problem. The OH and H_2O microwave emission from Mira variables may be anomalous in binary systems (c.f. Lepine et al. A.A. 56, 219) indicating a shell affected by the companion. In this connection it is worth noting that R Car which has anomolously weak H_2O (Lepine et al.) is a member of a binary system (Obs. 90, 25). Narrow band polarization work on Miras has recently produced some very important results (Landstreet and Angel Ap. J. 211, 825, McLean and Clarke M.N. 179, 293, McLean M.N. 1978, McLean and Coyne Ap.J. 1978). These observations show changes in percentage polarization and/or position angle across strong spectral features. This indicates that at least some of the polarization arises in the stellar atmosphere. However, the results obtained so far are quite complex, e.g. a decrease in polarization across a strong TiO absorption band in Mira Ceti but increased polarization across some ZrO bands in R And (with, in one band, an increased polarization at one time and a decrease at another). McLean and Coyne find a very high polarization (6.8%) of Hβ emission in Mira Ceti just before maximum. This could either be intrinsic (due for instance to emission phenomenon in a shock front) or produced within the stellar atmosphere (due to the large depth at which Hβ is emitted at that phase).

Some progress has been made with spectroscopic analysis of red variables (e.g. the C^{12}/C^{13} ratio in the Se variable χ Cyg (Hinkle et al. Ap. J. 210, 684) and in the SC star UY Cen (Catchpole, Leige 1978). Further work will undoubtedly be done on element and isotope abundances in red variables and should lead to a clearer understanding of evolutionary processes and mixing mechanism in these stars. C, N, O abundances relevant to Miras are discussed by Eggleton (Q.J. 17, 448) and Lequeux (A.A. 43, 71). An important spectroscopic result is the extreme phase dependency of SiO (4μ) absorption. This could indicate a chromospheric contribution to the infrared continuum and will obviously be studies further in more detail. The proper understanding of the role of TiO opacity remains of great importance. The problem has been studied theoretically by Krupp et al. (Ap. J. 219, 963). Maehara (P.A.S. Jap. 28, 135) has attributed line weakening in Miras to TiO opacity. The discovery that many red variables (especially Mira stars) are 18 cm OH emitted has given a very large impetus to the study of these stars. There is a need to know more about how OH emission is related to other properties of the variable and recent surveys of red variables for OH have been made by Kolena and Pataki (A.J. 82, 150) and Fix and Weisberg (Ap.J. 220, 836). Many OH sources can be recognised as probably very long period Mira stars from their 18 cm structure. These OH/IR sources are of great importance for galactic structure problems.

Amongst recent 18 cm surveys have been, Bowers (A.A. Suppl. 31, 127, A.A. 64, 307, 1978), Johanson et al. A.A. Suppl. 28, 199, A.A. 54, 323, Baud (Thesis, Leiden 1978). Some of the problems of these stars were reviewed at the 1977 Bamberg Variable Star Colloquium. Most OH/IR sources appear to be relatively young objects concentrated to the galactic plane. However, the nature of the OH sources near the galatic centre remains rather enigmatic. Work on optical Miras in the galactic centre is important for comparison with these OH sources cf Lloyd Evans M.N. 174, 169, Glass (infrared), Feast (radial velocities) in progress. Since the OH/IR sources are expected to be cool and/or heavily reddened objects they are best searched for optically in the infrared. In order that such search can be carried out successfully it is necessary to have good OH positions (e.g. Evans et al. Ap.J. 206, 440, 1976, Winnberg et al. A.A. 38, 145). So far there has been some success with finding IR candidates (e.g. Schultz et al. A.A. 50, 171 and 52, 475, Evans and Beckwith Ap.J. 217, 729, Glass M.N. 182, 93). However it is necessary to monitor these IR sources to establish them as (Mira) variables and to determine their periods. Schultz et al. (A.A. 63, L5) find that the 18 cm emission in some OH/IR sources has a phase lag in intensity between the two components of the emission line. Since the double emission components are usually interpreted as coming from the front and back of an OH shell they are able to calculate a shell diameter ($\sim 10^{16}$ cm). Clearly an extensive investigation of this idea would be very valuable. Simultaneous infrared light curves would also be useful to determine the way the OH intensities are related to the energy output of the star. A programme along these lines has been arranged between the Leiden Observatory and SAAO Cape. The integrated OH flux probably has little or no lag relative to the infrared flux (Fillit et al. A.A. 58, 281). In understanding Mira Variables it would be very useful to have some optical parameter which distinguished OH from non--OH Miras. Bowers and Kerr (A.A. 39, 473, A.A. 57, 115) find that the visual rise time is such a parameter at least statistically. This seems a valuable clue to a relation between the stellar pulsation and the OH intensity.

A detailed study of the structure of OH emitting red variables has been made by Reid et al. (Ap. J. 196, L35, 214, 60). Quite complex structures are evident. Both they and Fix (Ap.J. 223, 225) show that the OH maser lines are saturated. Reid et al. discuss a detailed model for OH masers (see also Elitzur A.A. 62, 305, Burdjuzha and Ruzmaikina A.A. 40, 233 Kwok J.R.A.S. Cand. 70, 49 Deguchi P.A.S. Jap. 29, 669, 1977. Olnon. Thesis Leiden 1977).

The discovery of other molecules in the microwave region (i.e. H_2O and SiO) together with OH and infrared observation allows us to make progress in studies of the detailed structure of the circumstellar shells of Miras. A good deal of work has gone into the discovery of microwave emitting Mira variables (e.g. H_2O sources, Lepine et al. A.A. 48, 269 Dickinson, Ap.J. Suppl. 30, 259, SiO sources A.S. 65, L7, Dickinson et al. A.J. 83, 36, Spenser A.J. 82, 706. and CO sources (Zuckerman et al. Ap. J. 211, 297, Lo and Bechis, Ap. J. 218, L27, Lambert and van der Bout, Ap. J. 227, 854). In the case of H_2O there is a double emission similar to that seen in OH but with smaller separation (Dickinson and Kleinmann, Ap. J. 214, L135). Blair and Dickinson (Ap. J. 215, 552) find that H_2O sources are much less frequent amongst S than M type Miras, due probably to the higher C/O ratio in the S stars. SiO microwave luminosity of Miras appear to be correlated to the period (Cahn Ap. J. 212, L135) though there is still some doubt about this (Dickinson et al. A. J. 83, 36) which will only be resolved by further observations.

Studies of the time variations of the microwave components will no doubt be carried out. The results presently available are rather tentative e.g. for Sio, probable variation of some Miras related to the light variation (Spenser et al. Ap. J. 199, L111) but in other Mira variation in a time scale of a few days (Balister et al. M.N. 180, 415). In some Miras non-maser SiO emission has been detected and its excitation has been discussed in detail by Morris and Alcock (Ap. J. 218, 687). A problem with the interpretation of H_2O microwave emission has arisen through the work of Wallerstein et al. (P.A.S.P. 89, 391). Working at 9400 A they fail to detect H_2O absorption in several Miras and deduce a much lower H_2O abundance than is required by current theories of H_2O masers.

It has been established by Reid et al. (Ap. J. 207, 784, 209, 595, 220, L113), that the true stellar velocities of Miras is mid-way between the two OH velocity components. Reid's work shows that this can be reconciled with optical results. He finds the interesting result that the "K" term derived using optical emission lines and a solution for galactic motion of Miras is different for those stars with measured absorption line velocities and those without. The most straightforward explanation of this is that it is related to the uncertainty in the phase at which optical observations were made, these being particularly uncertain for the fainter variables. The poor ephemerides coupled with the fact that emission lines are more difficult to detect before maximum than after could bias the optical sample to a mean phase somewhat after maximum. Since the emission velocity tends to decrease after maximum light, this would lead to the observed ffect. A more thorough investigation of this point might be useful. Whilst these results now considerably simplify the problems of model making so far as the shell is concerned they leave a number of problems still to be investigated, particularly; how are the optical absorption and emission velocities related to the OH velocities? Now that the stellar velocity can be defined from microwave observations the detailed study of optical velocities (both emission and absorption and in all spectral regions) becomes much more worthwhile since their interpretation should be considerably eased. The problem will require very extensive observations, as for istance the work of Wallerstein (Ap. J. Suppl. 29, 375) indicates.

11. THEORY OF STELLAR PULSATIONS (A.N. Cox)

This review deals with ten topics grouped in the two main headings: the interpretation of specific observations and theory with a general application. The purely observational aspects of red variables, Cepheids, RR Lyrae variables, β CMa variables, δ Ceti variables and dwarf Cepheids, and variable white dwarfs (ZZ Scuti variables), all of which pulsate, will be found elsewhere in this commission report. Not all references are given because this review is not a complete study of all pulsation theory but only of selected topics. Even for these selected topics the references are not necessarily complete.

We first consider the pulsation of our sun. The observations of Hill, Stebbins, and Brown seem to indicate that there are small amplitude variations in the position of the edge of the solar disc with periods of from 6 minutes to over an hour. The best reference to these data and a discussion of their significance is Brown, Stebbins, and Hill (Proceedings of the Solar and Stellar Pulsation Conference, Los Alamos LA-6544-C and 1978 Ap. J. 223, 324). These variations of only a few milliseconds of arc, a few kilometers in the solar diameter, have led to great arguments over whether they are real. This is especially so because different techniques of measuring these oscillations give very different results.

Severny, Kotov, and Tsap (1976, Nature 259, 87) and Kotov, Severny, and Tsap (1978, MNRAS 183, 61) at the Crimea Astrophysical Observatory have discussed an observed period between 120 and 180 minutes bydetecting Doppler shifts of the solar FeI line at 5123.7 A at the solar center relative to the outer rim. Their earlier periodic fluctuations in the radial velocity of the sun at 160 minutes were confirmed by Brookes, Issak, and van der Raay (1976, Nature 259, 92) who found two more periods of 40 and 56 minutes. However, Fossat and Ricort (1975 Ast. Ap. 43, 243), Grec and Fossat (1977, Ast. Ap. 55, 411), and Fossat, Harvey, Hausman and Slaughter (1977, Ast. Ap. 59, 279) could not confirm any of these periods with a sensitivity of better than one meter per second using similar radial velocity measurements of the sodium D line. They attribute the periods greater than 10 and less than 90 minutes to terrestrial atmospheric effects, and the maximum solar radial velocity is less than 1 meter/s, much smaller than the implied Hill value of about 40 m/s. Livingston, Milkey, and Slaughter (1977, Ap. J. 211, 281) find no fluctuations for periods longer than 5 minutes by observing the CI line depth which should be a temperature and luminosity indicator. Worden and Simon (1976, Ap. J. 210, L163) suggest that the 160 minute oscillations are due to the rotation of large scale solar velocity cells (supergranulation) ghrough the field of view of the detectors. Gough, Pringle, and Spiegel (1976 Nature 264, 424) suggest supergranule seiching may produce this 160 minute period. These sources of error have been discussed by Kotov, Severny, and Tsap and found not applicable.

The five minute oscillations of Leighton and Noyes are a completely different matter. They have been measured and interpreted extensively. The recent results of Deubner (1976, Proc. IAU Colloq. 36) and Rhodes, Ulrich, and Simon (1977, Ap. J. 218, 901) have been discussed in detail by Ulrich and Thodes (1977, Ap. J. 218, 521) who have been able to determine reasonably well the structure of the solar convection zone and how it oscillates in nonradial modes. The rotation of the outer solar layers as it decreases inward at first and then increases at depth is now being mapped using these short period oscillations.

Hill, Rosenwald, and Caudell (1978, Ap. J. 225, 304) have argued that the longer period very low amplitude oscillations are not so readily visible in measurements of the Doppler shifts of spectral lines or of intensity changes in the continuum and spectral lines. This conclusion which is based on linear theory solutions for eigenfunctions in the very uncertain outer layers of the sun, could also explain the lack of temperature fluctuations reported by Hill, Livingston, and Caudell (1977, Ap. J. (Letters) 214, L137) for the 5 minute oscillations and the lack of photospheric luminosity variations observed by Mussman and Nye (1977, Ap. J. (Letters) L95) and of Ca II K line variations reported by Beckers and Ayres (1977, Ap.J. (Letters) 217, L69).

The current situation regarding the longer period solar oscillations is confused. Three series of new limb observations obtained separately by Hill, Brown, and Stebbins seem to confirm the longer period oscillations (certainly one at 21 minutes) whereas disc observations do not. Dittmer (1978, Ap. J. 224, 265) sees no periodicity between 7 and 70 minutes in the Fe I 5123.7 line radial velocity. Detailed predictions of a large number of solar radial and nonradial mode periods are available (Christensen-Dalsgaard, and Gough (1976, Nature 259, 89), Iben and Mahaffey (Ap. J. 1976 (Letters) 209, L39), Goldrich and Keeley (1977, Ap. J. 211, 934), (1977, Ap. J. 212, 243), Rouse (1977, Ast. and Ap. 55, 477), but reconciliation among the various observations and between theory needs much further discussion.

The most understood pulsating star class, the Cepheids, has had a problem of uncertain masses. Evolution theory has consistently predicted that Cepheids are 3.5 - 18 $M_\odot$ stars core helium burning on blue loops in the H-R diagram. Recent tracks are given by Carson and Stothers (1976, Ap. J. 204, 461) and Lamb, Iben, and Howard (1977, Ap. J. 207, 209). Any mass loss more than 10 percent supresses the blue loops which have been discussed by Lauterborn, Refsdahl, and Roth and recently by Durand, Eoll, and Schlessinger (1976, M.N.R.A.S. 194, 671) and by Schlessinger (1977, Ap. J. 212, 507). Sreenivasan and Wilson (1978, Ast. Sp. Sci. 53, 193) have verifid the mass loss suppression at 15 $M_\odot$. Thus Cepheids should have masses (M_{ev}) given by the evolutionary theory mass-luminosity relation with no mass loss (Becker, Iben, and Tuggle (1977, Ap. J. 218, 633)). In extreme cases Buchler (1978, Ap. J. 220, 629) finds that pulsation might affect the evolution tracks, however. For 16 Cepheids with known distances, the known luminosity can give a mass which ranges from 4.8 to 12.4 $M_\odot$. These same Cepheids also have measured unreddened colors which can be converted to surface effective temperatures to evaluate radii. A period and radius used in the Eddington period mean density relation gives a pulsation mass which has been anomalously low until recently. Now new larger distances (Hanson, 1977 IAU Symposium 80) to the clusters with Cepheids and better unreddened colors give larger luminosities, radii, and masses, eliminating this pulsation mass anomaly.

There are, however, other mass problems with those Cepheids with bumps in their light and velocity curves and those which simultaneously display more than one pulsation mode, the double-mode or beat Cepheids (Cox and Cox, 1976 IAU Colloquium 29). For most of the Cepheids in these two classes. the luminosity is not known and, therefore, no evolution theory mass is available. However, use of the evolution theory mass-luminosity relation, the period-mean density relation, the linear theory pulsation constants (Q) for the proper pulsation mode, and the relation between luminosity, radius and T_e, one can solve for M, R, L, Q using only an observed period and an approximate T_e. The mass so obtained, termed a theoretical mass (M_T), is never more than 20 percent different from the evolution mass for those known ones. This theoretical mass used as a stand-in for the evolution mass allows quantitative comparisons of the ratio of the bump or beat masses with M_{ev}.

Bump masses are those derived from nonlinear hydrodynamic integrations which require anomalously low mass to predict the proper observed Hertzsprung relation between bump phase and period. The beat masses come from linear theory predictions that the ratio of the two periods is mass dependent. While the bump masses are low by a factor of almost two, the beat masses are as low in some cases as a factor of four.

Simon and Schmidt (1976, Ap. J. 205, 162) have indicated a method of getting bump Cepheid masses using only linear theory. The ratio Π_2/Π_0 of the second overtone to fundamental periods is correlated with nonlinear theory bump phase, being 0.53 for shorter period Cepheids with the bump far down on decreasing light and 0.46 when the bump is very early on rising light. At a given period the period ratio Π_2/Π_0 is obtained only for low masses of homogeneous models. Petersen (1978, Astr. and Astrophy. 62, 205) gets very low beat Cepheid masses by using the values of Π_1/Π_0 and log Π_0 on a theoretical plot of these two linear theory quantities with stellar mass as a parameter.

Cox, Michaud, and Hodson (1978, Ap. J. 222, 621) have shown that if a helium deficient Cepheid wind can enrich the helium content in the outer convection zones containing only 10^{-4} of the mass, the Cepheid bump phases, predicted from linear theory, and the beat Cepheid period ratios again from linear theory, can be made consistent with masses that are like the theoretical masses. Since the periods change only 5-10, percent the pulsation mass, and a mass obtained for some Cepheids from Wesselink radii (Balona, 1977, MNRAS 178, 231 and Evans, 1977, Ap. J. 209, 135), are only slightly changed. These inhomogeneous models seem to fit the available data, though their spectra might show some slight effects of the large helium (Y = 0.75) content. The principal problem with the enhanced helium is its instability against downward mixing. A little downward flow, however, has been used by Cox. King, and Hodson (1978, Ap. J. 224, 607) to explain the three simultaneous periods seen in AC And.

Other ideas about the beat Cepheid mass anomalies have been published by Cogan and Faulkner. Cogan (1977, Ap. J. 211, 890) proposes very extensive convection zones which can change the structure in the outer layers changing the Π_1/Π_0 period ratio. Deupree (1977, Ap. J. 215, 232) has found such large convection is unlikely. Faulkner (1977, Ap. J. 216, 49) proposes that the coupling between the fundamental and first overtone modes changes their ratio, but Cox, Hodson, and King (1978, Ap. J. 220, 996) found that the linear theory periods are only slightly changed in nonlinear calculations and presumably in real Cepheids.

Faulkner (1977, Ap. J. 218, 209) in a study of the periods of the double-mode Cepheid TU Cas. announced the occurrence of a third period, the second overtone. This has not been confirmed by Hodson, Stellingwerf, and Cox (1979, Ap. J. 229, 000) but the amplitude of the first overtone was found to be decaying from 0.4 to 0.25 mag. Faulkner has now found that another double-mode Cepheid, U Tr A seems to have a growing first overtone amplitude. These amplitude change rates are about as predicted by Stellingwerf in 1975, but the large number of double-mode Cepheids has led Stobie (1977, MNRAS 189, 631) to suggest that their double-mode life is as large as one third of their life as Cepheids.

Dwarf Cepheids show double-mode behavior also. Simon (1979, Astr. Astroph., in press) has discussed the observational data of AI Vel and has proposed, as others before, that its mass is like 0.4 $M_\odot$. Stellingwerf (1978, Ap. J., in press) was able to get the observed period ratios using modern opacities whereas previous studies go the ratio too low for masses 1 - 2.5 $M_\odot$ expected for normal evolution tracks. Cox, King, and Hodson (1979, Ap. J. 229, 000) then derived theoretical masses for 9 of the dwarf Cepheids. SX Phe has a mass of 1.1 $M_\odot$ with Z = 0.001, as observed, and a decreased Y = 0.1 in the surface layers due to downward settling and some remixing discussed by Vauclair (1976, Astr. Astroph. 50, 435 and 1977, Astr. Ap. 55, 147). The idea that SX Phe is very old near the main sequence after a mass loss as a red giant has been pursued by Saio and Takeuti (1978, PASJ, in press).

The question then of whether the stars which show metallicism can pulsate was addressed by Valtier, Baglin, and Auvergne (1979, Astr. Ap., in press) and by Cox, King, and Hodson (1979, Ap. J., in press). The two papers disagree on how the pulsation can occur, the first paper suggesting that hydrogen alone causes the driving and the second alleging that there is enough residual helium after the lower convection zone, and its severe mixing effects, disappears to cause the pulsation in at least the red third of the δ Scuti instability strip.

Turning now to general theory we consider first the problem of calculating full amplitude pulsations. Since the last IAU General Assembly it appears that only three nonlinear calculations have been published apart from the 2D convection studies by Deupree. The bump Cepheid problem for a 10 day homogeneous composition Cepheid has been calculated by a dynamically zoned mesh program and published by Davis and Davison (1977, Ap. J. 221, 929). Vemury and Stothers (1978, Ap. J. 225, 939) considered the pulsations of homogeneous composition Cepheid models using the unpublished opacities of Carson. Bumps at the correct phases were found with the expected evolution theory masses at amplitudes just a bit larger than for actually observed Cepheids. Opacity variations (mostly due to helium) between 25,000 and 100,000 K which are different from those in the Los Alamos opacities (Cox and Tabor 1976, Ap. J. Suppl. 31, 271), seem to cause the bumps. The accuracy of these calculations depends on the accuracy of the Carson opacities which are still being questioned.

Nonlinear calculations of Cox, Hodson, and King (1978, Ap. J. 220, 996) address the even more difficult question of modal selection. Use of the Baker-von Sengbusch-Stellingwerf relaxation method to get strictly periodic full amplitude solutions has resulted in a situation where no double-mode behavior can now be predicted except for mode switching at the transition line between the fundamental and first overtone modes. In principle, if one could find two full amplitude periodic solutions which, by linear analysis, tend to decay to each other, then double-mode behavior should occur. Searches for the physics of this behavior are continuing.

One hopeful idea about what causes bumps and double-mode pulsations has been discussed by Simon (1977, Ap. J. 217, 160; 1978, Astr. Ap., in press). The Cepheid light curve bumps discussed earlier in this review could be a resonance between the fundamental mode and the second overtone, possibly even when the second overtone is not unstable in the linear theory. Double-mode Cepheids are predicted to occur when the interaction frequency which is the sum of the fundamental and first overtone freouencies is resonant with the third overtone. Double-mode dwarf Cepheids are predicted to occur when this sum frequency is resonant with the fourth overtone. Some application of these ideas is being published by Simon and Cox, but there has been some problem of finding the resonance for nonlinear periods which differ a percent or so from the linear theory ones.

Another idea about the double-mode Cepheids has been suggested by J.P. Cox in his review of Cepheids at the Goddard Conference on stellar pulsation instabilities. Perhaps there is a small admixture of nonradial modes in these variables to alter the period ratios just enought to imply anomalously small masses. Osaki (1977, PASJ 29, 235) and Dziembowski (1977, Acta Astronomica 27, 95) have shown that nonradial p mode pulsations with $\ell \geqslant 4$ can exist in Cepheid and RR Lyrae variables.

Extensive convection calculations for RR Lyrae variables have been made by Deupree (1977, Ap. J. 211, 509; 1977, Ap. J. 214, 502; 1977, Ap. J. 215, 232; 1977, Ap. J. 215, 620) to investigate time dependent convection in two space dimensions. The main result is that the cause of the red edge is a conversion of the hydrogen driving zone to a damping zone when the full amplitude case is considered. Normally, the γ and κ effects operate on the radiative luminosity to phase its flow in order to excite the growth of pulsations. At maximum compression, for example, the luminosity is absorbed locally in the hydrogen and helium ionization zones by a low γ (ionization occurring) and a high κ (reducing energy flow). Then when expansion occurs the energy is released to drive the pulsation. However, the convection in the hydrogen ionization layers is so strong that it actually produced more luminosity at maximum compression and causes pulsation damping.

Gough (1977, Ap. J. 214, 196) discusses the linear theory pulsation with convection.

With red edge data for both the fundamental and first overtone modes Deupree has considered the width of the RR Lyrae strip as a function of helium content. Even though the blue edges, as mapped by Iben and Tuggle in 1972 and Cox, King, and Tabor in 1973, indicate, a helium mass fraction Y of between 0.20 and 0.25, the total width, which is independent of reddening corrections, indicates Y between 0.25 and 0.30. Evolution theory in comparing the relative life times in the red giant and horizontal branches to the relative numbers observed in these H-R diagram regions suggest a Y of 0.25 or less, according to 1972 results of Demarque, Mengel, and Sweigart. This dilemma of why the red edge is predicted a bit too hot, making the strip too narrow for Y = 0.25 in Deupree's convection calculations, is being investigated.

Another convection result, not directly related to pulsation, is a recipe for the ratio of mixing length to pressure scale height given by Deupree (1977, Ap. J. 215, 620) for use in 1D models of pulsating stars.

A general theoretical result has been given by Renzini and Swigart (in press) to explain the positive and negative period changes for RR Lyrae variables. The semiconvection that seems to occur in the nuclear burning core of these horizontal branch population II stars might adjust itself in small jumps. These changes in the central structure cause a change in the radius of the model which then changes the period. The average rate of change of the period is given by the rate of evolution along the H-R diagram track, but both positive and negative radius and period changes of the sizes observed can occur as the semiconvection adjusts itself.

Mira variable pulsations have been studied in the presence of mass loss. The earlier calculations of Wood (1975) showed that Miras were pulsating in the first overtone mode; the fundamental mode is dynamically unstable. Hill and Wilson (1979, Ap. J., in press), however, think that observations indicate fundamental modes. At lower masses, however, Keeley earlier did get fundamental modes. Wilson (1976, Ap. J. 205, 172) and Slutz (1976, Ap. J. 210, 750) have discussed shock waves which probably produce a puff of mass loss at each pulsation cycle. More details are given by Willson and Hill (1979, Ap. J., in press). The fate of the red giants as they lose mass has been discussed by Wood and Cahn (1977, Ap. J. 211, 499). They predict that stars with masses up to 4.5 $M_\odot$ become Mira variables or planetary nebulae and avoid being supernovae. Further work has been reported by Cahn and Wood (1978, Ap. J. 221, 163). Wood has reviewed this work at the Goddard stellar pulsation instability conference, and he has shown that the multiple shocks differ from the single shock of Slutz and cannot cause appreciable mass loss.

12. FLARE STARS (P.F. Chugainov)

Important progress has been made in investigations of individual flare stars and those which are contained in stellar aggregates as well as so called BY Draconis stars (that is, variables showing flare activity and slow quasi-periodic light variations). Current problems have been discussed at the Symposium held in 1976 in Bjurakan Observatory (see "Flare Stars", Armenian Academy of Sciences, Erevan, 1977). Review papers have been published by Mullan (Irish AJ, 12, 162; 12, 278), Gurzadyan (ASS. 48, 313) and Mirzoyan (IAU Coll. 42, 106).

New flare stars of UV Cet-type have been discovered by Bond (IB, 1160), Lovas (IB, 1345), Veeder and Hansen (IB, 1266), Byrne (IB, 1407). Photoelectric observations of known stars of this type are reported by Jarrett and Gibson (IB, 1105, 1112), Sanval (IB, 1180, 1210), Contadakis et al. (IB. 1181), Slovak (IB, 1271), Hansen and Veeder (IB, 1292), Kareklidis et al. (IB, 1354, 1355, 1356), Ichimuza and Schimizu (Tokyo Astr. Bull. No. 255). Some conclusions from the study of UV Cet, YZ CMi and AD Leo at the Uttar Pradesh Observatory in 1971-75 are made by Sinvhal and Sanval (IB, 1263). Photoelectric observations of the double flare star EQ Peg with the area scanner are presented by Rodono (AA, 66, 175). The relation between ultraviolet, red and infrared amplitudes of flares of EV Lac and UV Cet has been investigated by Kilyachkov and Shevchenko (Letters to AZ, 4, 224). 32 flares of EV Lac and 89 flares of UV Cet were observed in the U, 7100 and 8050 A filters. It is found that the flares in the red and infrared are preceded by decrease of the star light in 1/3 of cases.

Results of coordinated X-ray, optical and radio observations of YZ CMi have been discussed by Karpen et al. (ApJ, 216, 479, see also Ap. J. 225, L35). The upper limit of the ratio of soft X-ray to B-band luminosity during the flares is obtained and it is shown that the flare stars contribution to the diffuse X-ray background is negligible. Simultaneous radio and optical observations of UV Cet-type stars are reported by Spangler and Moffett (ApJ, 203, 497). The relationship is studied between the radio emission at two frequences (196 and 318 MHz) and 62 optical flares most of which were weak. In the previous studies this relationship was investigated mostly for large optical flares. Davis et al. (Nature, 273, 644) have worked out the new method of measurement of the radio emission (408 MHz) from flare stars based on use of the interferometer. In December 1977 two radio events on YZ CMi were recorded by them. Moffett at al. (PASP, 90, 93) have made a search for centimetric wavelength emission from flare stars. The upper limit of 1420 MHz emission during a typical flare is found.

In the paper of Mirzoyan at al. (Af, 13, 205) are presented photographic observations carried out mainly in 1973-74 and the discussion of all relevant data on flare stars in the Pleides. The paper is the fifth of this series. In previous papers the distributions of flare stars according to the number of observed flares were represented by the sum of two Poisson distributions. Nevertheless the authors conclude that observations accumulated to the present time can be better represented by the Poisson distribution with a continuous function of flare frequences of stars. This means that large differences are now revealed in activity of the Pleiades flare stars. In the other work Mirzoyan and Ohanian (Af, 13, 561) have shown that in the period 1957-1975 only one half of the low luminosity probable members of the Pleiades possessed flare activity. This conclusion contradict the previous conclusion made by Mirzoyan at al. that all or almost all members of Pleiades with V $\geqslant$ 13.3 are flare stars. However Mirzoyan and Ohanian have presented some evidences that this discrepancy can be explained on the assumption of cyclic recurrence of the flare activity in the low luminosity Pleiades stars. Results of the study of flare stars in several stellar aggregates (including Pleiades, Orion, Cygnus and other) have been surveyed by Haro (Bol. Inst. Tonantzintla, 2, 3) and by Ambartsumian and Mirzoyan ("Flares Stars", p. 63, Erevan) from the point of view of an observational approach to stellar evolution.

Observations of BY Draconis stars provide much information about the physical properties of active regions on stellar surfaces and the nature of flare activity as well. Now in common use is the idea that quasi-periodic light variations reflect the rotation of these stars and are caused by spots. Bopp and Fekel (AJ, 82, 490) have shown that relatively rapid rotation ($V_{rot} \geqslant 5$ km/sec) is the necessary condition for the ocurrence of BY Dra phenomenon. Direct evidences have been presented by Bopp and also by Chugainov (ICAO, 55, 85; 56, 24), Anderson et al. (ApJ, 216, 42), Walker (PASP, 89, 874) and Rucinski (PASP, 89, 280) that rotational velocities of these stars are in agreement with their photometric periods and at least some representatives of this group are young contractive stars.

Survey of the BY Draconis phenomenon has been undertaken by Bopp and Espenak (AJ, 82, 916). 22 dwarf K-M stars are studied and light variations on a time scale of several days are found in 9 of them (Gl. 15B, 176, 182, 277A, 494, 867A and BD+48°1958). It is concluded that variations of two kinds exist, periodic and unperiodic as well. The later kind of variation was observed in two late-type Me-stars (Gl. 15B and 277A). Probably unperiodic variations are due to the growth and decay of active regions near the rotational pole. Quasi-periodic light variations are found by Chugainov (ICAO, 54, 89; 61, in press) in dG-dK stars ξ Boo, BD+30°448, BD+27°4642 and HD 1835 and by Busko et al. (IB, 1275) in dM3e star CoD-38°11343. BD+30°448 and BD+27°4642 showed also some evidence of flare activity. Photoelectric observations of known BY Dra stars have been obtained by Davidson and Neff (ApJ, 214, 140), Bopp et al. (IB, 1443), Chugainov (ICAO, 54, 85). Oskanyan et. al. (ApJ, 214, 430) confirmed the existence of variations of the photometric period of BY Dra and conclude that at the present time light variations of this star are caused by bright spots. Contrary, Chugainov (ICAO, 54, 85; 54, 89) from observation of colour variations concludes that only dark spots present on BY Dra and on other three stars of this type. Different amplitudes of colour are found by Chugainov from one star to another corresponding to the temperature differences between photospheres and spots from 400 k to 1400 K. A similar conclusion is made by Bopp and Espenak (AJ, 82, 919) from their observations.

The flare activity of BY Dra variables AU Mic, HDE 319139, CC Eri, TW PsA and AT Mic together with their quiscent variability has been studied by Busko and Torres (AA, 64, 153). Considerable change in flare activity of AU Mic between 1970 and 1974 is found. The known relation is confirmed that larger flare durations correspond to higher star luminosities. The mean level of flare activity of the stars studied, with the exception of TW PsA, corresponds to about 1% of the bolometric star luminosity which is a typical value for usual flare stars. Further continuous observations are desirable for TW PsA, a dK5e-star with low flare activity, as well as for other dwarf G-K stars showing quasi-periodic light variations but little known as flare stars.

Much efforts are devoted in order to find out in BY Dra stars the existence of variations in emission lines, particularly those ones which presumably are caused by the rotational modulation. No variations of Ca II K line intensity in BY Dra during the period of rotation has been revealed by Bopp and Ferland (PASP, 89, 69) from precise photoelectric observations; a similar result was also obtained by Fix and Spangler (ApJ, 205, L63). The lack of variation indicate that Ca II emission is uniformly distributed on the surface of BY Dra. However, changes of emission Balmer lines have been observed in BY Dra stars by Ferland and Bopp (PASP, 88, 451), Bopp and Schmitz (in press), Busko et al. (AA, 60, L27). These changes only partly can be attributed to the rotational modulation. Sudden variations of H intensity and (or) profile also exist due to the flares and radial gas movements in envelopes (see Chugainov, ICAO, 57, 31; Hartmann and Anderson, ApJ, 213, L67).

Attempts have been made to detect the magnetic field in BY Dra by means Zeeman observation of the Hα emission line (Anderson, Hartmann and Bopp, ApJ, 204, L51) or polarimetric observations (Koch and Pfeifer, ApJ, 204, L47). Koch and Pfeifer have found variable linear polarization in BY Dra rising toward shorter wavelengths. It is interesting also to note that ξ Boo A in which Merchant Boesgaard et al. (PASP, 87, 353) have observed the variable magnetic field is found to belong to the BY Dra-type (ICAO, 54, 89).

IB = IBVS, Budapest. Letters to AZ = Letters to Astron. Zh.
Af = Astrofizika. ICAO = Izvestia of the Crimean Ap. Obs.

13. NOVAE AT OUTBURSTS (E.R. Mustel and L.I. Antipova)

A large number of papers have been devoted to the V1500 Cyg - the brightest nova during the last 33 years. The amplitude of its outbursts was larger than 19 mag. The character of light variations and the spectral evolution correspond to a fast nova; the MacLaughlin parameter $t_3 \sim 4$ days. The absolute visual magnitude at maximum was

M_o = -10 mag. The expansion velocity of the main envelope was about 2000 km/s. It was noticed (IAU Circ. No. 2839, Chin. Astron., I, 194, 1977) that the brightness of V1500 Cyg increased gradually prior to its outburst. At maximum. a strong uv-excess was observed in the spectrum (IAU Circ. No. 2827). The ultraviolet radiation decreased with time (Jenkins et al., Ap.J., 212, 198) and due to that the bolometric luminosity decreased, as opposed to the case of FX Ser which remained constant during the first 100 days (Jenkins et al., Ap.J., 212, 198, Wu, "Novae and related stars", p. 184). There was no infrared radiation which would indicate the formation of dust around nova (Kawara et al., Pub. Astr. Soc. Japan, 28, 163, Kiselev and Narizhnaya, Astr. Circ., No. 893, 5). The observed polarization is of interstellar origin (Kemp and Rudy, Ap.J., 203, 131, Shenavrin et al., Astron. Zh., 54, 629). The absence of variable polarization also indicates that, unlike the case of other novae, the was no formation of a dust envelope. Within 10 days after maximum the short period light variations were discovered with P $\sim$ 0.137^d and amplitude A $\sim$ 0.15 mag. (Tempesti, IBVS No. 1052). Further observations showed slight variations of the amplitude and the period (0.141 - 0.137^d) Tempesti, IBVS No. 1098, Semeniuk and Kruszewski, IBVS No. 1157). Investigations showed that the alternating maxima are of different form so that the period is more likely 0.282^d (Abramenko and Prokofeva, Astron. Zh., 54, 510, Young et al., Pub. Astr. Soc. Pacific, 89, 34, Kemp et al., IAU Circ. No. 2981). Photometric and spectroscopic observations demonstrated that the light variations are connected with the continuum and not with the emission lines (Young et al., Pub. Astr. Soc. Pacific, 89, 34, Prokofeva and Abramenko, IBVS No. 1219, Ferland et al., Nature, 264, 627). It was found that the profile of the Hβ emission line shows intensity variations (P = 0.141^d) with a phase delay depending on the wavelength within the line (Hutchings and McCall, Ap.J., 217, 775). The origin of these light variations is not fully understood. The following hypotheses have been advanced: (a) V1500 Cyg is a close binary with the orbital period of $\sim$ 6 hours (Abramenko and Prokofeva, Astron. Zh., 54, 510, Fabian and Pringle, M.N.R.A.S., 180, 749, Marcocci et al., Astron. Astrophys., 55, 171). In this case the analysis gives the mass of the red component $M_2 \sim$ 0.5 M$_\odot$, the mass of the white dwarf $M_1 \sim$ 0.8-1.4M$_\odot$, and the orbital radius (8.7-9.8)x10^{10} cm. (b) V1500 Cyg is a single white dwarf with an expanding envelope and with the light variations being due to either rotation or pulsation (Young et al., Pub. Astr. Soc. Pacific, 89, 34, Starrfield et al., Ap.J., 208, 23). (c) V1500 Cyg is an ellipsoidal variable (Kemp et al., Ap.J., 211, L71. It was also discovered that the intensity of the peaks of the emission lines is also variable on a time-scale of the order of minutes (Kemp et al., Ap.J., 211, L71, Ruchi and Thompson, Ap.J., 211, 184). Observations at radio wavelengths indicated that there was no strong emission from relativistic electrons during the outburst (Pynzar et al., Astr. Circ., No. 893). From an upper limit to the γ -radiation it was concluded that novae do not contribute to the galactic cosmic rays (Stepanian et al., Astron. Zh., 54, 515). The curve of growth analysis resulted in an overabundance of CNO and a normal content of Si (Boyarchuk et al., Astron. Zh., 54, 458).

The transient X-ray source A-0620-00 was identified with an eruptive star V616 Mon. From a survey of old spectrograms it was found that the star erupted also in 1917 and is probably a recurrent nova (Eachus et al., Ap.J., 203, L17). The spectrum of V616 Mon is that of a hot source with weak emission lines (IAU Circ. No. 2840).

The UBV colors suggest the presence of a red star (MO) and a hot (~ 20,000K) gas (Lutyi, Astron. Zh. Letters, 2, 402). The X-ray data give period of 7.8 d (IAU Circ. No. 2949), while the spectrophotometric data indicate P = 3.92d (Duerbeck and Walter, Astron. Astrophys., 48, 141). This period is interpreted as the orbital period. No rapid light variations have been detected (IAU Circ. No. 2854).

The curve of growth analysis of HR Del resulted in an excess of CNO both before maximum (Raikova, Astron. Zh., 54, 55) and during maximum (Antipova, Astron. Zh., 54, 68); prior to maximum this CNO excess increased with time. The spectral analysis of HR Del led to a conclusion about continuous outflow of material during the long, pre-maximum phase (Antipova, Astron. Zh., 54, 68). In addition there were ejections of clouds producing short-lived absorption components (Antipova, Astron. Zh., 55, 531). It is estimated within this model that the total amount of mass ejected was unusually large: ~ $0.01 M_\odot$ (Antipova, Astron. Zh., 54, 68). On high dispersion spectrograms of HR Del taken during the nebular stage it was found that the emission lines consist of many components which illustrates the complexity of the outburst (Gallagher and Anderson, Ap.J., 203, 625). The four strongest components are, as usually, connected with the polar condensations and polar belt. The physical conditions in these regions are different. On the early stages of the outburst the spectrum of HR Del contained also narrow emission components (of half-widths ~ 30 km/s). They were observed also after maximum, their half-widths increasing with time to about 140 km/s (Raikova, Astron. Zh., 55, 540). It was found (Antipova, Astron. Zh. Letters, 4, 177) that the expansion velocity of the main envelope decreased with time. These facts can be explained by the presence - prior to the outburst - of an extensive circumstellar envelope. The mass of the main envelope obtained from the emission lines in the nebular stage was about $10^{-4} M_\odot$ (Anderson and Gallagher, Pub. Astr. Soc. Pacific, 89, 264).

Nova Vul 1976 (NQ Vul) was subject of extensive spectroscopic studies (Cordoni et al., Astron. Astrophys., 55, 307, Fahrenbach and Adrillat, C.r.Acad. Sci., 284, B149, Karetnikov and Medvedev, Astr. Circ., No. 935, Rafanelli and Vittone, IAU Coll. No. 42, Volf, ibid.). Short period light variations have also been noted (Cordoni et al. op. cit.). Polarization is of interstellar origin (Martin et al., Nature, 265, 314). The infrared data showed during two months after outburst the dust was formed in the envelope (IAU Circ. No. 3000).

The increasing infrared luminosity, implying the dust formation in the envelope, was observed in two fast novae: Nova Aql 1975 (Vrba et al., Ap.J. 211, 480) and Nova Sgr 1977 = HS Sgr (IAU Circ. No. 3082). From the analysis of the emission lines in the nebular stage of Nova Cep 1971 it was found that He, O, and N were overabundant (de Freitas, M.N.R.A.S., 181, 421).

An analysis of the observational data pertaining to the nova outburst (Friedjung, "Novae and related stars", 61) seems to imply that the most plausible model is that which consists of a major ejection of material near the maximum, followed by a continuous outflow. A model with a continuous outflow (thick stellar wind) due to radiation pressure was also considered (Bath et al., M.N.R.A.S., 175, 305, Bath, M.N.R.A.S., 182, 35).

The recurrent novae RS Oph, T Pyx, and V1017 Sgr were investigated to confirm the correlation between the amplitude and interval between outbursts found earlier by Kukarkin and Parenago. It turns out that this relation involves the time interval between the present and the next outburst, so that the latter can be predicted from the amplitude of the present one (Steiner, Astron. Astrophys., 62, 273).

From a statistical material including 37 novae a relation was found between the amplitude of the outburst and the absolute magnitude at minimum (Steiner, "Novae and related stars", 93). Sharov (Var. Stars, 1978) investigated the relation between the luminosity at maximum and the rate of decline and found that either the dispersion in this relation is higher than estimated earlier, or there are distant novae in galaxies located in the halo or even in the extragalactic space. Kholopov and Efremov (Var. Stars, 20, 277) found that there is no physical identity between the cycle-amplitude relations for the recurrent novae and U Gem stars.

Ford (Ap.J., 219, 595) considered the problem of the number of outbursts and concluded that all novae should be of the recurrent type with the minimum number of outbursts between 160 and 660.

14. MAGNETIC VARIABLES AND RELATED OBJECTS (K. Stepien)

The report given here does not cover the results presented at the IAU Colloquium No. 32 held in Vienna in September 1975, and published in 1976.

In the last three years the theoretical investigations concentrated mainly on the origin of stellar magnetic fields, models of rotating stars with magnetic fields, model atmospheres and origin of chemical peculiarities. Although the fossil theory has still more observational support the dynamo theory has recently gained a more firm basis by demonstrating the ability both to produce magnetic fields of the observed strength and to diffuse them to the surface of a star in a time short enough. On the other hand the models of rotating stars with perpendicular magnetic and rotation axes (Moss, M.N.R.A.S. 181, 747) showed that the anti-correlation between rotational velocity and surface magnetic field, considered as an important argument against the dynamo theory, disappears in perpendicular rotators. The most serious difficulty for the dynamo theories remains in explaining the existence of an oblique decentered dipole which best fits to observations. The attempt by Oetken (Astr. Nachr. 298, 197) to fit the magnetic field symmetric in respect to equator (being the mixture of a dipole and a quadrupole) was shaked by the observations of HD 215441 obtained by Borra and Landstreet (Ap.J. 222, 226) Also Deridder, van Rensbergen and Hensberge (preprint) showed that the Oetken´s model does not give a satisfactory agreement with observations.

Thanks to a series of papers on rotating models of magnetic stars by Mestel and Moss (the references can be found in Moss op.cit.) we understand now much better the interaction of the magnetic field with rotation for different magnetic fields and finite conductivity.

A careful analysis of several magnetic stars carried out by Hensberge, van Rensbergen, Goossens and Deridder (preprint) showed that the observations do not support conclusively any hypothesis about a preferred inclination of the magnetic axis to the rotation axis.

Some authors looked for evolutionary effects in Ap stars. Earlier suggestion by Wolff (Ap.J. 202, 121) that one can observe effects of megnetic breaking operating when a star is on the main sequence has been criticized by Hartoog (Ap.J. 212, 723) who analyzed magnetic stars in four open clusters and could find no evidences for magnetic breaking. An important question in this respect is the determination of radii of peculiar stars. The radii and effective temperatures for individual Ap stars were determined by Schallis (preprint) on the base of IR photometry and model atmospheres. The results indicate that the radii of these stars are about two times larger than on the zero-age main sequence. These results are in contradiction with the results by Weiss (private communication) who, using essentially the same method, obtained radii similar to normal A type stars. Similar result was obtained by Babu and Rautela (Astr. and Sp. Sci., in print) on the basis of spectrophotometry of forty Ap, Am and A stars.

Early UV observations of Ap stars showed that they are flux-deficient in ultraviolet due to numerous and strong metal lines in this region. Line-blanketed model atmospheres are necessary to compare observations with theory and to derive meaningful physical parameters. The situation is, however, much more complicated than in case of normal stars. The latter have a uniform chemical composition and a series of standard models can directly be compared with individual stars. In case of Ap and Am stars no such standard chemical composition exists. Instead, each star has the chemical composition of its own and should be modelled individually. Using an average, most representative composition one can model only some general differences between these and normal stars. Such a series of models with one "peculiar" chemical composition was obtained by Muthsam (Astr. and Ap., in print). The results give a good agreement with observations producing the UV deficit and many other features characteristic of Ap stars. In case of the best observed magnetic star α^2 CVn Muthsam and Stepien (preprint) obtained models which show that the star has to change the effective temperature and the visible disk in course of rotation if the observed light variations are to be explained.

Model atmospheres including the magnetic forces into the equation of hydrostatic equilibrium showed (Stepien, Astr. and Ap. 70, 509) that the vertical structure of an atmosphere is very little changed in the presence of the magnetic field. The star can, however, be slightly distorted and have variations of effective temperature over its surface, contributing to the observed light variations.

Of two most widely known theories of the origin of chemical peculiarities the diffusion theory receives presently more observational support. Following Michaud et al. (Ap. J. 210, 447) who demonstrated that most of the overabundant elements in Ap stars will be pushed upwards in a relatively short time scale Alecian (Astr. and Ap. 60, 153) was able to reproduce satisfactorily the chemical composition of χ Cnc using the diffusion theory. Vauclair, Vauclair and Michaud (Ap. J. 223, 920) refined this theory by including turbulence while Vauclair, Hardorp and Peterson (preprint) were able to demonstrate that the magnetic field can play an important role in forming silicon spots on the surface of a magnetic star.

It is now generally accepted that helium-poor and helium-rich hot stars form an extension of magnetic stars into the hot domain. These stars have chemical peculiarities (even if restricted only to helium), they rotate slower than normal stars, and they have apparently a spotty structure as phtometric and spectroscopic observations show (Pedersen and Thomsen, Astr. and Ap. Suppl. 30 No. 2). Some of them have strong magnetic fields. They probably are losing matter which can be trapped in the magnetic field above the atmosphere (Landstreet and Borra, Ap.J. Letters 224, L5). But do classical magnetic stars lose matter? Some authors claim that they observe stellar winds in peculiar stars, e.g. Rakosch, Jenkner and Wood (preprint). It has also been reported in the literature that Vega - a sharp-lined normal star haing the spectral type right in the middle of the interval covered by Ap stars (and believed to be very "quiet" in respect to loss of matter) - may also have a stellar wind. If Ap stars do lose matter it would seriously affect the mechanism producing chemical peculiarities.

The diffusion theory seems to have more problems with explaining Am phenomenon. There, due to the presence of two convective zones, constraints put on the diffusion theory are more severe. Hence the suggestion by Vauclair (Astr. and Ap. 55, 147) that Am phenomenon is transient in the life of a main-sequence star.

A large amount of photometric. spectrophotometric and spectroscopic data have became available in the last three years. Here we report only on photometry and spectrophotometry. Jamar, Macau-Hercot and Praderie (Astr. and Ap. 63, 155) analyzed 77 Ap stars observed with TD-1 satellite. Their conclusions are essentially similar to those earlier obtained by Leckrone. HgMn stars are least flux deficient in UV while other types of Ap stars have strong absorption features masking the true continuum. Most of these features could be identified with lines of iron peak and lighter elements. Autoionization lines may contribute in some cases. It seems that the role of rare earth elements in forming the UV pseudocontinuum of Ap stars has been overestimated in the past.

Van Dijk et al. (Astr. and Ap. 66, 187) discussed UV phtometry of 79 Ap and 26 Am stars obtained with the ANS satellite. It is demonstrated that the UV flux deficiency decreases with decreasing temperature and is absent for Am stars. A new flux of data is expected to come soon from the IUE satellite.

Many new ground-based photometric observations have became available. The complete list of these is given in Astronomy and Astrophysics Abstracts andwe mention here only the most extensive or representative studies: observations obtained by the Potsdam group, published in Astr. Nachr., vol. 297, Rakosch and Fiedler (Astr. and Ap. Suppl. 31 No. 1), Renson and Manfroid (Astr. and Ap. Suppl. in print), Hensberge et al. (Astr. and Ap. 48, 383; 54, 443), Hensberge et al. (Astr. and Ap. Suppl. 34, 67), Blanco et al. (Astr. and Ap. Suppl. 31 No. 2). They supply us with periods and amplitudes of light variations. Several new peculiarity indices have also been proposed, like flux deficiency at 1400 A (Jamar et al., op.cit.) or those based on a particular photometric systems, e.g. vilnius system (Straizys and Zitkevicius, Astr. Zhu., 54, 987) and Geneva system (Hauck, Astr. and Ap., 69, 285). It is interesting to note in this place that Preston and Wolff (Pub. A.S.P., 90, 406) proposed to use variations of Ca II lines when looking for periods of cool Ap stars.

Although it is almost generally accepted that HgMn and Am stars do not possess strong magnetic fields and hence are not variable, their variation is reported from time to time in the literature e.g. Rakosch and Kamperman (Astr. and Ap. 55, 53) in case of α And, or Boehm-Vitense (Ap. J. 225, 514) who suggests that some Am stars may be variable on a long time scale. The problem of variability of these stars requires more attention in the future.

15. VARIABLE STAR SURVEY WORK (W. Wenzel)

A. Introduction

In the author's opinion the meaning of the word "Survey" has changed appreciably during the last decades, and to-day it seems difficult always to find the correct definition in a Report like the present. We remember the great genuine surveys in test fields of the previous decades, for instance the Harvard plan of the "Milton Fields" and the "Plan of Sonneberg Fields for the Statistics of Variable Stars of the Northern Milky way". By these two enterprises, which were expanded several times, more than 10 000 new variable stars of all types were detected in each. Tremendous stellar-statistical and astrophysical findings resulted from those work. Unfortunately, processing of the data of all the variables discovered at that time, i.e. determining of type, period, brightness etc., is not yet completed at present, because other "more modern" matters of research are taken up. This tendency is also connected with the fact that the present "surveys" are often devoted to the search for special classes of variable stars only. The Palomar-Groningen Durchmusterung for instance was mainly initiated to find a representative sample of RR Lyrae stars of the galactic halo. A third kind of photographic survey consists in searching on routine plates, for example of the sky patrol, for new variable stars, with or without giving preference to special galactic regions, or in searching on plates which originally were taken for other purposes. Finally there exists a fourth possibility, namely of performing a photoelectric survey by looking for special types of variables in certain groups of stars, for instance for β CMa variables in a sample of B stars.

B. General Surveys

As already mentioned, general surveys at present are of only small scope each, but are carried on systematically:

At the Maria Mitchell Observatory Miss Hoffleit and her collaborators regularly investigate variable stars of the Harvard lists for improvement of previous results, for finding period variations and for discovering cases of variability which are of special astrophysical importance. Results are published for instance in IBVS 958; 1207; 1349; 1431, mostly concerning stars of the Sagittarius cloud.

G. Romano at Reviso and his co-workers continued several surveys on Asiago plates. This work dealt with the fields around β Per (IBVS 999; Mem. Soc. Astr. It. 47, p. 229), M 33 (where a new Hubble-Sandage variable was discovered; IBVS 1421 = 1433; Astron. Astrophys. 67, p. 291) and M 31 (Astron. J. 82, p. 319). The number of recorded GR stars is now 292. This work will be continued in several fields in Perseus, Lyra, in high galactic latitudes, and together with L. Rosino, at M 33.

Further Asiago material including a lot of infrared plates (Kodak IN hypersensitized + RG 5) was used by L. Rosino, Padova, to search for new variables in the area around the peculiar nebula NGC 7635 Cas (Astron. Astroph. Supp. 24, p. 1) and by P. Maffei, Catania, of the field γ Cyg (IBVS 1302). Out of 62 new variables found near γ Cyg, 48, are to be classified as Miras or SR types; a previous search on blue plates of that region by G. Romano had yielded 21 such objects only, but obviously the increase of the number of variables by using infrared techniques is not quite as large as in the M 16-17 field (see Report 1973).

At Sonneberg Observatory the troublesome processing of the 488 variables discovered there by C. Hoffmeister in the field of η Ara was finished by H. Gessner (Sonneberg Veröff. 8, No. 7). The main contents of the field consists of 121 eclipsing stars, 149 RR Lyrae variables, 119 slowly variing semiregular and irregular stars and 57 Miras. The work on the fields of 20 Vir (E. Splittgerber) and α Per (H. Busch and associates) is completed, too (forthcoming no. of Sonneberg Veröff.). 5 new variables were discovered by H. Gessner in the course of a systematic search on Sonneberg Sky Patrol plates covering parts of the constellations Coma, Bootes, Corona Borealis and Hercules (Mitt. Veränd. Sterne 8, p. 65). Good progress is achieved by L. Meinunger in surveying the surroundings of M 92 on Tautenburg 52 inch Schmidt plates and on Sonneberg 40 cm astrograph plates; by this program the work described already in Reports 1973 and 1976 is continued.

By E.H. Geyer and F. Gieseking at the Observatory Hoher List of the University of Bonn, surveys in the regions of some stellar associations yielded a number of new variable stars of different kinds in the fields of μ Cep, o Per and IC 1396 (IBVS 967; 1008; 1145). During that work the old experience has been confirmed that in the course of a search amining at the finding of a special kind of variables the latter often are in the minority among the detected objects, if not very specific methods are used.

At the Bamberg Remeis Observatory the discovery, patrol and processing of variable stars of the southern sky on plates of patrol type were continued by W. Strohmeier, R. Knigge and collaborators (Bamberg Veröff. 10 to 12; IBVS 994); the number of variables discovered for finally confirmed at Bamberg (BV) has reached 1736.

C. Surveys on Special Types

Survey work on special types of variable stars cannot be quoted completely at this passage of the Report. We will rather give here an enumeration of the variety of methods and illustrate them by some characteristic examples, this brevity being necessary also to avoid overlaps.

Some of those investigations were already touched upon in the previous chapter because of their extensive importance. The work of P. Maffei and L. Rosino et al. mentioned above led for instance to a certain specialisation with respect to the discovery of Mira variables by using photographic infrared techniques. The paper in Astron. Supp. 31, p. 313 = Asiago Contr. 377 is especially important in this connection: In a Milky Way field in Sagitta surrounding the globular cluster Palomar 10, by blinking infrared plates of the 67 cm Schmidt telescope of Asiago, 123 new long-period variables have been found, the number of such stars previously known for that region having been 2!

In section B also the Bonn survey work on stellar associations was mentioned. In a special study on NGC 7000 F. Gieseking and J.D. Schumann (Astron. Astroph. Supp. 26, p. 367) continued to check the Hα emission stars announced by several previous authors. As already stated by R. Hudec et al. (for instance Mitt. Veränd. Sterne 6, p. 171) in a high percentage of those stars the presence of the H emission couldnot be confirmed. Theremaining objects, if not distant Be stars, are supposed to be T Tauri variables and related stars. The search for further emission line stars in and near NGC 7000 with the aim of objective prism plates taken with the Byurakan 40 inch Schmidt telescope has been continued by M.K. Tsvetkov and co-workers in Sofia (Astrofizika 11, p. 579; IBVS 1447). 77 new Hα stars were found by this work; of course they should be checked individually for their real membership in the T Tauri or flare type class.

At Sonneberg the novae of the halo of M 31 (see for example Report 1976) were statistically investigated by W. Wenzel and I. Meinunger (Astron. Nachr. 299, p. 239). The comparison of the number of novae in the halo of M 31 (corrected for the probability of discovery) with the density gradient of the "normal" halo stars of our Galaxy points to the conclusion that a high percentage of novae - if not all of them - belong to population II. Special surveys of M 31 and its surroundings, as initiated for instance by A.S. Sharov et al. (Astron. Tsirk. 869), and in fields of the galactic halo by means of wide-angle exposures, are urgently recommended. A test program of the latter kind was the survey of H. Gessner in Bootes and adjacent constellations mentioned above (section B); this Durchmusterung indeed yielded, among other variables, one probable nova Bootis 1962 with a maximum brightness of $10^{m}.5$ (IBVS 1428).

Occasionally the general surveys have the disadvantage of a rather coarse determination of the magnitudes of the variables found. So for example I. Meinunger by using Tautenburg 52 inch Schmidt plates had to determine a new the B magnitudes of 136 RR Lyrae variables discovered previously on Sonneberg plates in order to render possible a statistical investigation of those stars for the completion of existing similar surveys (Astron. Nachr. 298, p. 171).

A search for long period Cepheids in the direction of the Norma spiral arm (discovered radio-astronomically) was undertaken by A.L. Cabrera et al. by means of San Juan and Cerro Tololo plates (IBVS 1299). 8 new variables have been found; the type of variation however must still be investigated.

Surveys for finding new faint flare stars in star clusters and associations (Plejades, Hyades, Praesepe a.o.) are going on in a number of observatories by means of the method of multi-exposures, mostly in the U region. Without claiming to be complete we quote the observatories of Tonanzintla (G. Haro et al., Tonanzintla Bol. 2, p. 95), Byurakan, Budapest (Jankovics, Bamberg Veröff. 121, p. 120) and Heidelberg (C. Paulakos, Acta Astron. 27, p. 87).

Finally we make reference to the surveys which are performed photoelectrically in order to look for small amplitude variables of population II by R.S. McMillan et al. (Astron. Soc. Pac. Publ. 88, p. 495) at McDonald Observatory or in order to search for new β Canis Majoris stars by Baloma (South African Astr. Obs.) (RAS Mem. 84, p. 101). By the former work was intented to find short-period variables among evolved stars of low mass in the lower RR Lyrae instability strip; the result was negative. The latter work resulted in recognizing 8 new probable or certain β CMa variables out of a sample of 31 candidates.

J. SMAK
President of the Commission

29. STELLAR SPECTRA (SPECTRES STELLAIRES)

PRESIDENT: M. Hack
VICE-PRESIDENT: W. K. Bonsack
ORGANIZING COMMITTEE: Y. Andrillat, A. A. Boyarchuk, C. O. R. Jaschek, J. Jugaku, R. J. Kovachev, D. C. Morton, J. P. Swings, K. O. Wright.

This report has been written with the cooperation of the vice-president, the members of the O. C. and of several other members.

Commission 29 has co-sponsored IAU Coll. 35 on Compilation, Critical Evaluation and Distribution of Stellar Data (Strasbourg, August 1976), IAU Symp. 80 on The HR Diagram (Washington, November 1977), and IAU Symp. 83 on Mass-loss and Evolution of O-type stars (Vancouver, June 1978). Other meetings of interest for the Commission: Conference on Novae and Related Stars (Paris, September 1976) 21st Liège Symp. on Laboratory and Astrophysical Spectroscopy of Small Molecular Species (Liège, June 1977); IAU Coll. 42 on The Interaction of Variable Stars with Their Environment (Bamberg, September 1977); 1st Latin-American Regional Meeting (Santiago, January 1978); 22nd Liège Symp. on Elements and their Isotopes in the Universe (Liège, June 1978); 4 th Trieste Coll. on High Resolution Spectrometry (Trieste, July 1978); IAU Coll. 47 on Spectral Classification of the Future (Vatican City, July 1978); 4 th European Regional Meeting (Uppsala, August 1978).

Commission 29 has supported the publication of the Atlas of Be stars by Drs. A. M. Hubert-Delplace and H. Hubert as a project of the Commission.

Because of the very limited space available for this report, it is impossible to include all the papers published since the last IAU meeting. Two research areas have received special attention:1) the physics of outer stellar layers:UV observations have revealed the existence of hot plasma and stellar winds in all early-type supergiants, in a few dwarfs and in many close binary envelopes, and have permitted detections of chromospheres and coronas of late-type stars. 2) the chemical composition gradients and the chemical evolution in our galaxy, and in nearly outer galaxies.

WR, O, Of, B, Be, early-type supergiants (written with the collaboration of Y. Andrillat, A. B. Underhill, H. G. Groth and T. P. Snow).

WR, O, Of Stars. It is not clear if all WR stars are binary systems; however, the majority of them are binaries. Computations of evolutionary sequences taking into account mass loss by stellar winds suggest that WN non-binary stars can be the result of the evolution of O-type stars of mass 80-120 M$_\odot$ (Loore et al. AA 67, 373); WC would be the result of the evolution of stars of 40-60 M$_\odot$. Relations between WR and Of are discussed by Conti (20th Liège Symp. p. 193). 13 WR stars have been studied individually: the binaries γ^2 Vel (a cloud of CS gas with radius 10^{15} cm, much larger than that of the binary orbit, has been revealed by radio and IR observations; mass loss of $3.9\ 10^{-5}$ M$_\odot$/y has been estimated by Morton, MN 182, 47P), θ Mus, HD 193793, V 444 Cyg (IR observations of this WN5 star show the presence of stellar wind escaping at constant velocity, Hartmann Ap J 223, 193), CV Ser, HD 90657, and the non-ascertained binaries HD 50896, HD 192163, HD 156385 (characterized by a very strong extinction at λ 2200, in spite of the fact that no nebulosity has been found around it, Willis et al. AA 59, 133), HD 93162 (Moffat, AA 68, 41), HD 151932 (Seggewiss, IAU Coll. 42, 633), HD 191765, HD 192103. Velocity gradients in the atmospheric regions where the absorption lines as well as where the emission lines originate were found by Niemela (IAU Symp. 83) in two Of stars and in 4 WN stars. The gradient appears to be spectral-type dependent. An atlas of WR stars for the region 3500-6700 A has been produced by Bappu and Scaria (Kodaikanal Obs. Bull. Ser. A, Vol. 2 No 2), who studied the violet absorption edges of CIV lines at 5807 A. Spectrophotometry of WR stars brighter than 8.5 mag is underway in the range 3400-11000 A.

The extended series of studies of O-type stars and O-type binaries by Conti and collaborators is continuing (Ap J 212, 728; 213, 438; 214, 759; 215, 561; 218, 431; 224, 558). Cassinelli, Olson and Stalio (Ap J 220, 573) studied the structure of the winds and coronae of O stars applying the Sobolev method to the Hα profile of ζ Ori, O9.7 Ib.

Bohannan and Garmany (Ap J 223, 908) have made a search for binaries and stellar winds among the O stars, in order to verify if the high mass-loss rate indicated by UV observations is found only in binary systems or is a general property of O stars. A velocity gradient in the Balmer lines is observed in most O supergiants and Of stars, and is certainly present in 3 of 6 late O-type giants and main-sequence stars. This gradient and the presence of photospheric mass motions complicates the problem of searching for spectroscopic binary systems. Walborn (Ap J 205, 419) has found that all of the O9.5-B0.7 supergiants in the Orion belt and in NGC 6231 are systematically nitrogen-deficient with comparison to the majority of the supergiants of the same spectral type. Bolton and Lars Rogers (Ap J 222, 234) have measured RV of OBN and OBC stars in order to determine binary frequency. At least 50% and possibly 100% of the OBN stars are members of short period binaries; none of the OBC is member of short period binaries. Masses and luminosities are normal for both groups; there is no evidence that the two groups are kinematically distinct. Walborn (Ap J 215, 53) has studied the brightest O, WR and B supergiants in the Magellanic Clouds. Further evidence for large metal deficiency in the SMC was found. Moreover the supergiant nature of Of stars has been confirmed in both Clouds.

The spectrum of the high galactic latitude O star HD 93521 has been studied (Hack and Yilmaz, ASS 48, 483) and several possibilities examined for explaining its distance from the galactic plane (z > 750 pc) and its slight He excess.

B stars. Several studies of the UV spectra of B stars have been published: three extensive lists giving identifications of the lines visible on Copernicus tracings of the UV spectra of B stars (Underhill and Adelman, Ap J Sup. 34, 409, Hill and Adelman, Ap J Sup. 37 No 2, Selvelli, Stalio and Crivellari AA Sup. 27,1) an Atlas of τ Sco (Rogerson and Upson, Ap J Sup. 35, 37) at Copernicus resolution 0.05 A and representation of the UV spectra of 17 O and 41 B stars (Snow and Jenkins Ap J Sup. 33, 269) at Copernicus resolution 0.2 A. Lamers and Snow (Ap J 219, 504) have interpreted the shapes and strengths of the resonance lines of Si IV, N V, and O VI in O and B stars and they suggest that the electron temperature in the coronal layer ranges from $7 \pm 3 \cdot 10^4$ (presence of Si IV) to $4 \pm 2 \cdot 10^5$ (presence of N V and O VI). Cassinelli and Olson (Wisconsin preprint No 71) have proposed that these ions originate from Auger ionization due to X-rays coming from a thin coronal zone at the base of a cool stellar wind. It is clear that absorption lines seen in the spectra of main sequence O and B stars as well as O and B supergiants indicate the presence of a plasma having a degree of ionization greater than can be maintained by the radiation field of the star. No ultimate source of the extra energy needed in O and B stars has yet been established.

Although asymmetric resonance lines of some high ions are seen in the UV spectra of a few main sequence stars (Snow and Morton, Ap J Sup. 32, 429; Lamers and Snow, Ap J 219, 504), rarely is material seen moving with a velocity great enough to permit it to escape unless the material is proved to be at more than 4 stellar radii. The same is true for Be stars. The available analyses suggest that the higly ionized, outward moving layers observed for some B and Be stars lie inside the cool shell. Thus these layers will not escape unless the material is given additional acceleration when it is in a state of ionization not observed. The broad emission wings of Hα in shell stars may be revealing this material. In normal B stars the outer cool shell is not seen; it is visible only when it is enhanced during the Be shell episodes.

Abundance analyses, using standard techniques, of γ Peg, B2 IV, and τ Her, B5 IV, by Peters (Ap J Sup. 30, 551) and Adelman (MN 181, 667) respectively, have given further confirmation that normal B stars have essentially solar composition. Line blocking factors in the UV spectra of 35 B6-A0 stars have been measured by Underhill and van der Hucht (AA 54, 393) using data from S 59 on TD 1. Rajamohan (Pramana 7, 160) has derived rotational velocities and H and He line intensities for 112 members of the Scorpio-Centaurus Association. For stars fainter than M_V = 0.0 the distribution of rotational velocity closely resembles that of field stars. Stars brighter than M_V = 0.0 which are found in the dense upper Scorpio region rotate faster than the field stars and the members of the Pleiades and α Persei clusters. Analyses of high resolution spectra of 5 members of the association give normal helium abundance.

Pulsating B stars. Line-profile observations at high resolution using an electronic detector have shown that several sharp-lined B stars have line-profile asymmetries which vary in a matter of hours and which are reminiscent of the line doubling observed in β Cephei stars (Smith, Ap J 215, 574; Smith and Karp, Ap J 219, 522; Smith and McNall, Ap J 223, 221). These variations can be interpreted in terms of travelling waves associated with nonradial pulsations. Smith and McNall (Ap J 221, 861) interpret the sharp components of γ Peg for 4567 Si III in terms of shell gas in a radial pulsating atmosphere. Line-profile variations in time scale of hours (Si IV, Fe III near λ 1120 A) and in time scale of days (C IV 1550 A) have been observed in β Cep (Hutchings and Hill Ap J 213, 111).

He-variable, He-strong stars. It is not clear if they are stars which have lost much of their hydrogen rich matter or if they are young stars. Spectral changes can be explained by the oblique rotator model (Bond and Levato, PASP 88, 905; Wolf and Wolf Ap J 203, 171; Pedersen AA 49, 217; Hardorp and Megessier AA 61, 411), but this does not explain particular cases like a Cen or σ Ori E. Lester (Ap J 210, 153) has made a differential abundance analysis of the He-rich star HR 3089 compared to the normal star λ Sco.

A spectroscopic search for B-stars with enhanced He^3 abundances has been made by Hartoog and A. P. Cowley (Lick Bull. No 814). 8 new stars of this class were found. They have helium abundances 5 to 20 times lower than normal B stars and He^3/He^4 ranging between 2.7 and 0.47. They occupy a narrow strip in the log g-log T_e plane between the He-strong and a group of He-weak stars, which present no He^3 enhancement. These results indicate a link between He-w and He-s stars and the possibility that the diffusion theory can explain both.

Be stars. The spectra of peculiar Be stars with IR excess have been studied by Allen and Swings (AA 47, 293) and three mechanisms of mass-loss which may explain the formation of the shells are discussed. UV observations (950-1450 A) of 6 Be stars with a range of v sin i obtained with Copernicus show radial velocity components up to a few hundred Km/s for stars with large v sin i. Stars with low v sin i do not show a similar behavior (Marlborough Ap J 216, 446). The region of Mg II lines of some Be and shell stars has been observed by Morgan et al. (Ap J 216, 457); evidence for expanding layers has been found.

The spectral variations of the shell star 48 Lib from 1965 to 1977 have been studied by Aydin and Faraggiana (AA Sup. 34, 51). The spectrum variations of Pleione from 1938 to 1975 have been studied by Gulliver (Ap J Sup. 35, 441). The same star has been studied by Hirata and Kogure (PASJ 28, 509; 29, 477 and 30 in press) and by Higurashi and Hirata (PASJ 30, in press). The mass of the shell is of the order of 10^{-10} $M_\odot$ and the mass loss 5 10^{11} $M_\odot$ /y. An atlas of high dispersion intensity tracings of Pleione is in press (Kodaikanal Obs. Bull. Series A, Vol. 2, No 3). Hirata and Asada (PASJ 28, 713 examined 15 suspected Be stars and Kogure, Hirata and Asada (PASJ 30, 385) studied the formation of the shell absorption lines in the Balmer series. Variations of γ Cas from 1965 to 1970 have been studied by Ivanova (Soobschenia Bjurakan Ap Obs 50). Significant variations of the equivalent widths of Hα with time scale of several minutes have been found for the Be stars φ Per, γ Cas, 28 Tau (Mamatkazina, Trudi Ap Inst Alma Ata, 31, 73); variations of Hα and Hβ during a period of 10 minutes have been found by Luud (AZ Lett, in press) in the spectra of γ Cas, γ Dra, 11 Cam and π Aqr.

Early-type supergiants. Considerable UV data have been accumulated during the last three years, providing information on stellar parameters such as flux distributions temperatures, line identification, and especially on the winds from early supergiants. For the O4 If star ζ Pup a complete Copernicus high resolution spectrum has been published (Morton, Underhill Ap J Sup. 33, 83), the flux distribution analyzed by Holm and Cassinelli (OAO-2 data, Ap J 211, 432), the mass-loss rate studied by Morton (Ap J 203, 386) and later by Lamers and Morton (Ap J Sup. 32, 715). Another well-studied supergiant is β Ori, B8 Ia; an extensive line list was compiled by Selvelli et al. (AA Sup. 27, 1) and line-blocking factors were measured and new determination of the temperature made by Stalio et al. (AA 60, 109). For the A2 Ia supergiant α Cyg line identification based on Copernicus data was compiled by Underhill (Ap J 217, 488) and by Barbier et al. (AA Sup. 32, 69) and an investigation on mass-loss from α Cyg, β Ori and η Leo (AO Ib) was made by Lamers et al. (Ap J 223, 207) using data from BUSS VII and VIII. The two Ia supergiants present shifted components of the resonance lines of singly ionized metals while the Ib supergiant does not. Other data on mass-loss in early-type supergiants have been presented by Chentsov (Izv Spets Ap Obs 8, 128) using Orion 2 data; by Kondo et al. (Ap J 209, 489) using BUSS data, by Morton (IAU Symp. 83) using high resolution Copernicus scans of the O VI doublet at 1032, 1037 A. The physical conditions in stellar winds have been discussed by several authors in view of their importance as clues to the wind origins. The conclusion is that the high ionization observed in stellar winds precludes any possibility that pure radiative equilibrium exists and the nature of the non-radiative processes has been widely debated (Lamers and Morton, Ap J Sup. 32, 715; Lamers and Snow, Ap J 219, 504, Underhill, IUE data, private communication; Lamers and Cassinelli, PASP in press; Hearn, Lamers, Castor and Cassinelli, and Thomas in IAU Symp. 83). The mass-losing prototype P Cygni has been studied by Ambartsumian et al. (Ap J in press) using Copernicus data and found not to show evidence for non-radiative heating. Probably dynamic instabilities cause the outflow and it cannot be considered a prototype of mass-losing supergiants. UV indications of time-variability in stellar winds were found by: York et al. (Ap J Lett 213, L 61 for ζ Pup and δ Ori A in time scales of hours; Snow (Ap J 217, 760) for a number of stars over a three-year period; Lamers et al. (Ap J 223, 207) for β Ori and α Cyg. More detailed studies on time-scale variations in supergiant winds have been carried out by Snow and Hayes (Ap J in press) and by Wegner and Snow (Ap J Lett in press). New observations of α Cam, O9.5 Ia were carried out with IUE and ground based observations were obtained at the same time over a continuous period of 3 days.

Ground based observations of early-type supergiants also give important evidence of mass-loss rates. Hutchings (Ap J 203, 439) calculated mass-loss for 65 OB stars from RV measurements; Conti and Frost (Ap J 212, 728) calculated mass-loss from Hα for 20 O stars and compared measured intensities and profiles with those computed for non-LTE models. Groth (unpublished) derived a time-dependent Balmer progression including a reversal, and considerable variations of the Hα profile of HD 92207. Similar results were obtained by Wolf and Sterken (AA 53, 355) for HD 91619 and HD 96919. High resolution echellograms of α Cyg confirm the existence of asymmetric emission core in the K line that may be due to chromospheric emission or to scattering from a CS shell (McClintock and Henry, Ap J 218, 205).

An attempt to explain the variable line profiles of α Cyg during the period 1962-1974 is made by Inoue and Uesugi (PASJ 29, 149); a change of the microturbulence field cannot explain them, at least within the limits of plane parallel geometry and LTE assumptions. Luud et al. (A Z Lett in press) found that Fe II lines belonging to multiplet 42 in α Cyg spectrum sometimes have P Cygni profiles while at the same time Hα has normal pure absorption profile.

A-F-type stars, Ap and Am stars (written with the collaboration of C. R. Cowley and C. van't Veer Menneret).

Improved model atmospheres for early-type stars and availability of high resolution observations from the Lyman limit to the IR make it increasingly important to have accurate measurements of absolute energy distribution and attempts to detect chromospheric indicators. Energy distribution of A-type stars from 1100 to 8000 A was compared with thoretical predictions by Panek (Ap J 216, 747); Joshi and Rautela have determined the absolute energy distribution of 14 A-F stars lying in the δ Scuti region of the HR diagram (ASS 45, 199); absolute energy distribution of α Lyr and 109 Vir in the range 3295-9040 A is obtained by White and Lockwood (AA 61, 679) by comparison with black body sources. Line-blanketing was determined from high dispersion spectrograms for ten stars of type F5 by Climenhaga and Wright (PASP 88, 213). Line blanketing is considerably higher for the supergiants than for the dwarfs. A complete identification list of the UV spectrum of α Lyr (0.2 and 0.4 A resolution, Copernicus data) has been published by Faraggiana, Hack and Leckrone (Ap J Sup. 32, 501). Morton et al. (Ap J 212, 438) have presented UV spectra (2100-3200 A) of the normal stars α Aql, A7IV-V and α CMi, F5IV-V obtained with Copernicus and compared with the solar spectrum obtained with a rocket by Broadfoot.

High resolution studies of Sirius have been made by Millard et al. (AA 54, 869, v sin i measurements), Kurucz et al. (Ap J 217, 771; an overabundance of B_a by a factor of 50 was found). Freire et Al. (AA 61, 785) have measured high resolution profiles of the Ca II K line, with the spectrograph of the Meudon solar tower, for α CMa, A1 V, γ Gem, A0 IV and α Aql, A7 IV-V. A solar abundance for Ca has been found and no evidence for chromosphere. Asymmetries in the K line point to the existence of velocity fields in the photosphere. This result contrast with those of R.E.M. and R. F. Griffin (AA in press) who found it symmetrical and non variable. R.E.M. and R.F. Griffin are studying also α Lyr in the range 3100-6900. The emission lines observed by Johnson and Wisniewski (PASP 90, 139) were not confirmed by the Griffins' observations (PASP in press). Boyarchuk and Snow (Ap J 219, 515), using the Copernicus data, have found that α Lyr and α CMa have no excess of heavy elements such as Tc, Pt and Hg in comparison with the solar abundance. Castelli and Faraggiana using the Copernicus data and the Kurucz, Peytremann and Avrett models derived the atmospheric abundances for α Lyr. They found that if it is required that the abundances have the same value in the visual and UV range and that the ionization equilibria be respected, it is necessary to admit a depth-dependent microturbulence (4 th Trieste Coll. in press). The visual and photographic spectrum of ε Aur, F0 Ia was studied by Castelli (ASS 49, 179; AA 69, 23) who found spectrum variability in time-scale of 20 minutes, and depth-dependent microturbulence, increasing inward. IUE observations in the high resolution mode show the presence of chromospheric emissions in the two resonance lines of Mg II (Hack and Selvelli, Nature in press, AA in press).

Boesgaard (Ap J Dec. 15, 1978) has determined the abundance of boron from Copernicus scans of 1362.46 B II in 16 normal stars of spectral types A and B. A cosmic value $B/H \sim 2\ 10^{-10}$ is found, not dependent on the temperature. The cosmic values of Li, Be and B agree with the predictions of galactic-cosmic-ray spallation reactions for the origin of light elements.

Ap stars. Several abundance studies, especially of some particular atomic species, have been made, in order to check the diffusion theory. Of the possible suggestions for the origin of the Ap-abundance anomalies, the diffusion hypothesis is most compatible with established notions of stellar structure, stellar evolution, and nucleosynthesis (even if the stability required for the stellar atmosphere represents the greatest difficulty with the theory). The appeal of this theory lies here rather than in the results of detailed comparisons of theoretical predictions with observational results which are often of low accuracy. In its most recent formulation, diffusion theory provides the basis for an observational probe of numerous processes in stellar interiors from an examination of surface abundances. This positive aspect of the theory has yet to be exploited.

The proceedings of three conferences on Ap stars have appeared: the Shemaha conference (AZ 51, English abstracts), The IAU Coll. 32 and the Prague Conference (Publ Czech Ac Sci Astr Inst. No 54). Several papers by C. R. Cowley, M. S. Allen and coll. (Ap J 206, 196, Ap J 213, 121, Ap J Sup. 32, 631, PASP 89, 386,

PDAO 15, 37) are devoted to the systematic study of very heavy elements in the spectra of Ap stars (Y, Ba, Lanthanides, Actinides, U and Th) and of Cr, Mn and Fe. Aikman and Cowley have a continuing program to make element identification in Ap stars using 2.4 A/mm plates (22nd Liège Symp.). Identification lists and occasional abundance determinations have been prepared for: γ Equ (Adelman et al. Preprint), HD 25354 (Pyper, Ap J Sup. 31, 249), HD 215038 (Rice, PASP 89, 770), ε UMa (Mallama and Molnar, Ap J Sup. 33, 1), α^2 CVn (Leckrone, private comm.) HD 161733 (Levato and Malaroda, PASP 89, 84), HD 34452 (Pavlova et al. Sov Astr 21, 554), α And (Aydin, Hack AA Sup. 33, 445). The Hg-Mn stars were studied by Kodaira and Takada (Ann Tokio Astr Obs 2nd Ser 17, 79) who discuss abundance work on 12 stars. Heacox (preprint) made abundance determinations for 21 elements in 21 stars. Properties of Hg-Mn stars were discussed by Aikman (PDAO 14, 379) and by S. Wolf and Preston (Ap J Sup July 1978) who derived the distribution of rotational velocities for a sample of 256 late B-type stars, and found that 16% of the non-magnetic late B-type stars are Hg-Mn stars and that these objects are not distinguished from normal ones by their age, duplicity, rotation, or any combination of these characteristics.

A summary of the properties of the hotter Ap stars (Bp-He-w) is given by Vilhu et al. (Report No 2 of Ap Lab Univ Helsinski). Hartoog and A. P. Cowley (Ap J in press) have studied the He^3 phenomenon in B stars. A. M. Boesgaard (PASP 88, 353) treats Li, Be, B abundances in Ap and other stars. C. Cowley and Etzler (PASP 89, 524) find the presence of weak CN bands in the cooler Ap stars. Several individual Ap stars have been studied for abundances by a large number of authors. Three papers have appeared dedicated to the very peculiar rare-earth star HD 101065 (Przybylski, IAU Symp. 72, 135; MN 178, 71 and 735). High resolution (0.086 A) polarization observations inside spectral lines of the magnetic star β Cr B have been made by Borra and Vaughan (Ap J 216, 462). The observations can be approximately explained by a decentered dipole model. Variability of spectrum and magnetic field have been investigated for several magnetic stars at the Zentralinstitut für Astrophysik in Potsdam (53 Cam, HD 9996, 52 Her, ε UMa, 73 Dra, β CrB) and the distribution in spots of the elements is studied by several Soviet astronomers (e. g. Astrophysics 12, 631). Babu has derived effective temperatures, radii and bolometric magnitudes of several Ap and Am stars (ASS 43, 57; Babu and Rautela, ASS in press). Raiamohan and Babu (MN 182, 773) have studied the correlation between rotational velocities and colors of magnetic variables, both singles and members of spectroscopic binaries.

Am stars. The main trend is (a) to improve spectral resolution for a more efficient comparison of the spectroscopic results with the prediction of the diffusion theory; (b) to study the interaction between the metallicity phenomenon and duplicity, rotation, evolutionary stage and other physical parameters affecting the stellar atmosphere. Disagreements appear between high and low dispersion spectroscopic results, as well as photometric work; recent UV observations give results in contrast with the observations in the visible range.

C. van't Veer and collaborators are continuing detailed analyses of Am stars; M. Kondo and Kitamura and collaborators (Ann TAO 2nd Ser 16; TAB 2nd Ser No 246) have made extensive studies of the eclipsing binary Am stars and found that the two components of RR Lyn have a different compositions while WW Aur and δ Cap may show a non-uniform distribution of metallicism on the surface components.

Kurtz (Ap J Sup. 132, 651) found that of seven δ Del stars analyzed, five showed similar abundances to evolved Am stars. Fracassini and Pasinetti (AA 66, 37) studied three suspected δ Sct variables (σ And, θ Peg and 2 Lyn) which are progressively far from the hot limit of the instability strip of the δ Sct variables. 2 Lyn may be an Am star.

Late-type stars (written with the collaboration of G. Cayrel De Strobel).

The role of late-type stars in astrophysics is of paramount importance, because in these objects evolution has not depleted the initial stellar content of the Galaxy and the full span of ages is still present. Hence these stars are used to study the chemical evolution of stellar populations, the abundance gradients across the galactic disc, the constraints on nucleosynthesis imposed by the chemical composition of extremely metal-deficient objects, and the connection between chemical, kinematical and dynamical evolution of the Galaxy.

High dispersion analysis of G and K standard stars. A critical survey of the solar abundances has been made by Holweger (22nd Liège Symp). Martin is carrying out a detailed non-LTE analysis of the sun; Hardorp (AA 63, 383) is comparing the energy distribution of the sun with solar-type stars by photoelectric scanning. Hardorp, Cayrel de Strobel and Perrin (IAU Symp 80) analyzed some of the stars belonging to the solar-type "boxes" of Golay (AA 60, 181). The found that, once analyzed in detail, stars belonging to the same "box" do not have identical atmospheric parameters. R. F. Griffin has reported on high resolution spectrometry of solar-type stars (4th Trieste Coll). Two new analyses of the iron spectrum of ε Vir (Hearnshaw et al. MN 185, in press and PASP 89, 797) and Cayrel de Strobel (AA 54, 797) are in good agreement with the former ones. β Gem, KO III, is analyzed by Griffin (MN 175, 225) and by Holweger and Griffin on high resolution FTS spectra; β UMi, K4 III by Williams, α Tau, K5 III, by Tsuji (AA 62, 29) and by Foy and de Batz on FTS spectra. The problem of the gravity of Arcturus is debated: log g = 0.9 ± 0.35 (Mackle et al. AA 38, 239) or log g = 1.6 ± 0.20 (Ayres and Linsky, Ap J 200, 600; Ap J 201, 212; Blackwell and Willis MN 180, 169; Ayres and Johnson Ap J 214, 410; Martin, AA 31, 591). High resolution studies of the Ca II and Mg II lines in Arcturus have been made by Adams et al. (Ap J 211, 453) and by Kondo et al. (Ap J 207, 167) and van der Hucht et al. (AA in press). The region 2700-3400 of the Arcturus spectrum has been studied at 0.1 A resolution by Stencel and van der Hucht (Ap J Sup. Sept. 1978). CO bands in Arcturus give evidence of inhomogeneous chromosphere (Heasley et al. Ap J 219, 970).

G and K field stars. A series of studies on the abundances of Li, Be and B has been made by A. M. Boesgaard (Ap J 210, 466 and 475; PASP 88,353), by Luck (Ap J 218, 752) and by Kipper et al. (Publ. Tartu Obs. 44, 271). CNO abundances which are essential for understanding the different nucleosynthesis processes have been investigated by several authors: Bell and Branch (MN 175, 25), Clegg (MN 181, 7); Bell et al. (MN 181, 31),Lamber and Ries (Ap J 217, 508),Luck (Ap J 219, 148), Sneden and Peterson (Ap J 222, 585). The C^{12}/C^{13} ratio in 38 giants and supergiants has been determined by Tomkin et al. (Ap J 210, 694) and by Lambert and Sneden in the most metal-deficient star HD 122563. A solar C/Fe ratio has been found in 12 metal-poor dwarfs and subgiants (Peterson and Sneden, Ap J in press) and in the metal-rich K dwarf 14 Her (Peterson, Ap J 224, 595). On the other hand, a range of CNO/Fe abundance ratio in metal-poor field stars is suggested by Pilachowski (Ap J in press). Oxygen abundance has been studied by Eriksson and Toft (AA in press) and by Kjaergaard and Gustafsson (Stromgren Symp. p. 127). Kameswara Rao (MN 185, in press) has measured the strengths of the G band in 10 weak-G-band stars and found a correlation of the weakness of the G band to enhancement of NH-band strength. Kollatschny has analyzed the K5 V star ε Indi. Abundance anomalies very close to those of the Ba and CH stars have been found. This study confirms the trend of decreasing T_{min}/T_{eff} with decreasing T_{eff}. This value for ε Indi is 0.76. Hartmann and Anderson (Ap J 215, 188) have analyzed high resolution red spectra of three late-type dwarfs and two emission-line flare stars. The abundances are similar to the solar ones.

SMR stars. Contrasting results have been found. Peterson (Ap J Sup. 30, 61) has explained the SMR phenomenon as an effect of difference in surface temperature. Several analyses indicate that the SMR stars or some of them, have normal abundances (Kuroczkin and Wisniewski, Act Astr 27, 145; Hearnshaw AA 51, 85) while others found an overall metallic overabundances of 0.3 dex relative to the sun (Luck Ap J 212, 743). Weak metallic lines in the near IR are very sensitive to metallicity (Campbell and Garrison). Hearnshaw (AA 51, 71) in analyzing 8 high-velocity dwarfs found that the spectroscopic gravity is about a factor of 10 lower than the gravity derived from the trigonometric parallax. This effect was not confirmed by Cayrel de Strobel (AA 54, 797) in a new analysis of one of these stars, HD 103095.

Metal-poor stars. Foy (AA in press) shows that high-velocity stars are only moderately metal-deficient and that metal-deficiency and kinematic properties are weakly correlated. Wallerstein et al. (Ap J in press) have found a field red-horizontal branch star HD 175305. Spite and Spite (AA in press) have found two extreme metal-deficient objects (like the only one known up to now: HD 122563) with [Fe/H] = -2.6 : HD 84903 and HD 184711. From a sample of 11 very metal-deficient stars Spite and Spite (AA 67, 23) ascertain the following trends: i) no enhancement of the odd-even effect; ii) no overdeficiency of Eu (r-element); iii) when Fe/H $\lesssim$ -1.5, Ba (s-element) is overdeficient, and to a lesser degree also Y (s-element). Pagel (22nd Liège Symp.) has carried out a survey of abundances in unevolved cool stars. There seems to be good evidence for variations in N, O and Ba abundance and for zero or very small variations in C and members of the iron group. The state of evolution of dwarfs and subgiants is discussed by several authors (e.g. see Cayrel de Strobel, Stromgren Symp. p 205). Greenstein has observed G and K stars of large proper motion with the multichannel spectrometer on the 5-meter telescope. The existence of a genuinely sub-luminous sequence with appreciable internal scatter is established for the high velocity stars. Some of these objects have no lines visible except Balmer lines, G-band and H and K, indicating a metal-defect by a factor of 1000 with respect to the sun. Some objects present retrograde motion with respect to the galactic center, a about 500 km/s with respect to the sun. These must be among the oldest metal-poor objects known.

G and K stars in galactic and nearby extragalactic systems. Owing to progress in instrumentation interesting results have been obtained for individual stars belonging to stellar systems. Significant differences in chemical composition have been found between members in the Arcturus group (Cayrel de Strobel, Martin, Pasinetti, AA 61, 591). Their result could raise doubts about the physical reality of old moving groups. Bonsack and Pilachowski (IAU Symp. 80) are conducting a low dispersion survey of stars in the clump and nearby regions of the giant branch of open clusters of intermediate age, in order to identify stars of peculiar composition and look for abundance variations within clusters and between clusters. A sensitive method of deriving metallicities of young G stars has been developed by Barry et al. (Ap J 212, 462; Ap J 219, 942; Ap J 222, 1032, PASP in press); they show that the metallicity difference between coeval clusters Hyades and Coma is greater than between old (Sun and M 67) and very young (NGC 2264) objects. The Hyades metal enrichment of 0.3 dex appears to be excluded (Tomkin and Lambert, Ap J 223, 937). Abundance determinations in globular clusters are being carried out by Bell et al.(Stromgren Symp. p. 249); Ap J Sup., submitted); Wallerstein and Pilachowski 22nd Liège Symp.); Zinn (Ap J 218, 96), Mallia, (AA 60, 195). Pagel et al. are obtaining spectra at 5 and 10 A/mm of globular cluster giants. Cohen et al. (Ap J 223, 487) have analyzed giants in M3, M13, M92 and M 67, Griffin in M 13 (MN in press), Kraft in M 15, M 13 and M 3. Kraft and Carlson (IAU Symp. 80) have found large N/Fe values and wide variation (1.30 dex) in these values among stars on the subgiant branch in M 92, suggesting that the CNO-cycle processing is initiated much earlier than had been thought until now.

Vilhu, Tuominen and collaborators (Report No 3, of Ap Lab Univ Helsinski) have studied high resolution spectrograms of several K giants in Wolf 630 moving group. μ Her, γ^2 Del and η Eri form a homogeneous sample, more metal-rich than the standard ϵ Cyg. Two peculiar stars of the group, 56 Peg (a barium star) and HD 148897 (a marginal CH star) have been studied. The latter is probably not a true member of the group because it is slightly metal-poor.

Metal abundance in the LMC cluster NGC 2209 is in reasonable agreement with that found for the LMC IS-medium (Gustaffson et al. Ap J Lett 216, L7); Foy is studying the chemical composition gradient in the SMC and Coehen (Ap J in press) has determined gradients of line strengths within M 31, M 81, NGC 4472 and NGC 3115.

Norris and collaborators have performed spectroscopic studies in M 13, M 92, NGC 6397, M 15, M 22, M 55, ω Cen, 47 Tuc, and discussed the spectroscopic similarities of dwarf spheroidal galaxies with ω Cen.

M, C, S stars. Maehara and Yamashita (PASJ, 28, 135) observed the energy distribution of M stars and found that it is an indicator of the TiO band intensity.

Yamashita and Maehara (PASJ 29, 319) and Yamashita et al. (PASJ 30, 219) observed the companion of Mira Ceti near light minimum, and estimated the mass-loss rate from Mira (PASJ 30, 409) to be about 2 10 ^{7} M⊙/y. F. and M. Querci (AA, submitted) have observed λ 10830 He I in R CrB in a pre-maximum phase. It appears with a P Cygni profile; the violet-displaced absorption gives an RV of 240 km/s. This line confirms the presence of He in R CrB and states the physical conditions in high temperature and high density prevailing in the CS cloud. F. and M. Querci (21th Liège Symp.) discuss the information that molecules in C star spectra can give on nucleosynthesis processes undergone by carbon stars and on their hydrodynamics. Fujita (JRASC 71, 115) examined the red spectrum of nine carbon stars for the presence of HCN, CH_2 and C_2H_2. Fujita and Tsuji (PJA 52, 296; PASJ 29, 711) determined the isotopic ratio C^{12}/C^{13} for nine carbon stars and found that this ratio is abnormal only in peculiar carbon stars (see also PJA 54, 207). The broad absorption band at 3 μ of 22 carbon stars was studied by Noguchi et al. (PASJ 29, 511). They found that the band depth correlates with the abundance ratio CN/C_2. The continuum between 1 and 4 mm from the carbon-Mira-Ceti star V Cyg is interpreted by M. Querci et al. (AA submitted) on the basis of the near IR flux. Shock waves and mass-loss should support the assumption of a flare-like event.

Kipper et al. (Publ Tartu Obs 44, 271) have determined the abundance of lithium in the atmospheres of 14 K and M giants. Dragunova et al. (AZ 52, 6) have investigated the blanketing effect of M0-M5 stars. In order to clarify the problem of the origin of lithium in the galaxy, de la Reza and F. Querci (AA 67, 7) are studying the importance of non-LTE effects in the Li abundance determination in Super-Li-Rich Carbon stars. de la Reza and Torres are also investigating the possibility of enrichment of Li by the mass-loss of the numerous flare stars. For this the Li abundance in the M dwarf Gl 182 is being calculated.

Mould and Wyckoff (MN 182, 63) found that the predicted properties of FeH do match the observed characteristics of the unidentified molecular band near 9900 A discovered in spectra of late M dwarfs by Wing and Ford in 1969.

Wyckoff and Clegg (MN 184, 127) have prepared an atlas of pure S stars near minimum light and a list of the positions of more than 200 molecular band heads. Model atmosphere analyses indicate that pure S stars have $0.98 \lesssim C/O \lesssim 1.02$.

Spectroscopic evidence of motion or variability in late-type photospheres and chromospheres. The main result is the detection of chromospheres and coronas in the UV. Separation of macroturbolence from rotation is being made on several K giants (Smith Ap J in press, Gray Ap J in press). A radial-tangential velocity distribution for macroturbolence fits the observations better than does a gaussian (Gray Ap J 202, 148; 211, 198; Smith et al. Ap J 207, 308), but supergiants may be an exception (Luck, Ap J 218, 752). SMR K-giants appear to show a different turbulence compared to normal K giants (Gray and Martin in preparation). Correlations between microturbolence, macroturbolence, chemical composition, and age are sought for (Smith, Ap J 224, 584; Foy AA 67, 311).

Wilson (Ap J Dec. 1 1978) and Vaughan et al. (PASP 90, 267) study the flux of Ca II emissions and find evidence for cycles similar to the solar one. Absolute fluxes of K chromospheric emission on the HR diagram have been evaluated by Blanco et al. (AA 48, 19) and the intensity effect in Wilson-Bappu relation was investigated on 248 G, K and M stars by Gleboki and Stawikowski (AA 68, 69).

The UV spectra of late-type stars have been studied with Orion 2 (Epremyan, Byurakan Obs Bull), Copernicus, BUSS. Observationally, new spectra obtained with IUE (Stencel, private communication) demonstrate that there exist two types of mid-UV emission line spectra for G and K stars: the early G bright giants, all G and K0-K1 giants and seemingly all G and K dwarfs present high excitation lines, including C II to N V; whereas the late G bright giants and the mid- and late-type K giants have only low excitation lines (Si II). This dichotomy appears to parallel the locus of evidence for strong mass-loss described by Reimers (AA 57, 395) and the Ca II V/R asymmetry changeover described by Stencel (Ap J 223, L37). The successful flight of BUSS X in May 1978 recorded the mid- UV spectrum of the active chromosphere star β Dra (G2 Ib-II). Mg II profiles present strong emission with possible multiple CS features superimposed.

Bappu and Mekkaden have studied K-emission in Canopus and find fluctuations in intensity. Bappu, Raveendran and Mekkaden have in progress a survey for the detection of the He I 5876 line of chromospheric origin. The absolute energy distribution in ρ Cas, F8 Iap (Joshi and Rautela, MN 183, 55) indicates that it is appreciably modified by the CS shell surrounding the star.

Spectral evidence of stellar chromospheres in late-type stars is being investigated at the Observatorio Nacional, Brazil: for carbon stars by studying the line formation of emission lines, by de la Reza; for M giants by means of the Ca II absorption lines in the infrared, by Batalha.

Binaries - (Written with the collaboration of M. Plavec and K. O. Wright).

Semi-detached systems of the Algol type: lines of the secondary component of Algol were finally seen by Tomkin and Lambert (Ap J 222, 119). Popper and Plavec discovered that some apparently normal Algol primaries show symptoms of a shell (weakness of Mg II, Si II); the phenomenon is quite pronounced in TT Hya.

Ultraviolet excesses in U Cep, U Sge, SX Cas and several RS CVn stars were found by Rhombs and Fix (Ap J 209, 821; 212, 446; 216, 503). These are probably due to free-free and bound-free emission associated with the hot components in Algols, but with the cool components in RS CVn stars. York, Flannery and Bahcall (Ap J 210, 143) found an H II region around V Pup. The RS CVn binaries became objects of great interest when it was discovered that they tend to be radio (and even X-ray) sources. Weiler (MN 182, 77) from a study of H, K, Hα emission suggest an exstensive chromospheric activity on the later-type component.

Perhaps significant for our understanding of He-rich stars is the suggestion that Sigma Ori E might be an interacting close binary (Hesser et al. Nature 262, 116; Groote and Hunger AA 56, 129).

Intensive observations of the far ultraviolet spectrum of β Lyr continued (Hack et al. Ap J Sup. 34, 565; Kondo et al. Ap J 208, 468) with both Copernicus and IUE. The peculiar emission spectrum is interpreted as arising in a supercorona; matter associated with the companion is detected through the absorption components of 1175 C III and a mass ratio $m_2/m_1 \simeq 3$ is derived. The He I line at 10830 in the spectrum of β Lyr was observed by Girnjak et al. (Iz Cr Ap Obs 58, 75) and shown to be formed in the gas stream and envelope. Leushun et al. (Iz Sp Ap Obs 9, 3) have estimated the He/H ratio in β Lyr to be equal to 1.55. The spectrum of β Lyr was believed to be unique until Plavec using IUE discovered a similar spectrum in W Ser, W Cru, and V 367 Cyg, and Koch in RX Cas and SX Cas (IBVS). υ Sgr was studied by Duvignau, Friedjung and Hack using Copernicus data (AA in press) and again by Hack with IUE. Evidence of an expanding hot plasma is found. N V, Si IV, C IV resonance lines present terminal velocity of about -700km/s. The photographic spectrum of V 367 Cyg was studied by Aydin et al. (ASS 53, 345) and a peculiar behavior of the lines at some phases detected. In the supergiant binary μ Sgr Plavec and Polidan found a hotter component (B1?) than the observed B8 Ia star, using Copernicus and IUE. Hack and Selvelli observed Epsilon Aur with IUE (IBVS 1439, Nature in press) and discovered a faint blue companion, V$\sim$10, Sp. type $\sim$ B8). R. F. Griffin and Radfoot (Obs 97, 169) and Griffin (MN in press) have shown that the G5Iab-Ib star HD 187299 has a B companion. Among the ζ Aur type stars, VV Cep (M2Ia+OB) was observed before and during atmospheric eclipse by Wright (IAU Circ 2811, JRASC 77, 152) and Faraggiana (AA 46, 317). IUE spectra in the region 1150-3200 A were obtained by Selvelli and are now being analyzed by Faraggiana. Saito and Kawabata (ASS 45, 63) observed the Ca I line at 6572 in spectra of ζ Aur and 31 Cyg; it is probably due to CS gas expanding from the systems.

T Tauri stars. (Written with the collaboration of W. K. Bonsack).

One of the main problems is the location of T Tau stars in the HR diagram: they probably represent the earliest evolutionary phase for which stellar spectra are observable. Many problems are posed by their envelope and photospheric spectra which present very complex and variable line profiles.

A review of the relationship between the T Tauri stars and star formation is given by S. E. Strom in Star Formation (T. de Jong and A. Maeder eds., P. 179, Reidel Publ.). Rydgren, Strom and Strom (Ap J Sup. 30, 307) obtained spectra and photometry of 49 T Tau stars. The veiling superimposed on the stellar spectrum has the characteristics of H free-bound emission. The veiling, emission lines and IR excess can be accounted for by a CS envelope with T near 20000 K; is some cases a dust envelope must be added to account for silicate emission at 10 and 20 μ. Kuhi (Protostars and Planets conference, Tucson 1978) emphasizes the breadth of the absorption lines, implying v sin i up to 100 km/s. Kuhi noted that pre-main-sequence stars without H emission do not have unusually broad lines, suggesting the need for a rotational braking mechanism to act between T Tau phase and subsequent pre-main-sequence evolutionary phases. True P Cyg profiles are rare, and inverse P-Cygni profiles more so. Herbig (Ap J 214, 747) has made a radial velocity survey of T Tau stars and finds that the stars are essentially at rest with respect to the clouds with which they are associated. Frequency distribution of photospheres is peaked near MO. Schneeberger et al. (AA 62, 447) discuss the intensity ratio of He triplet lines to He singlet lines and point out the usefulness of this datum in discussing physical conditions in the CS cloud. Kuhi (Tucson conf.) pointed out that rapid variability of the emission lines (time scales of a few hours for the Na I lines, 24 hours for Fe II, weeks or months for Balmer lines and blue continuum, while forbidden lines are essentially constant) is common to many T Tau stars.

The subgroup YY Ori —type is characterized by the presence of inverse P Cyg profiles. S CrA, a bright member of this group, has been observed by Rydgren (PASP 89, 557 ; Appenzeller and Wolf AA 54, 713 and by Wolf et al. AA 58, 163).

FU Ori objects are characterized by an increase in brightness of four to six magnitudes in a time on the order of a year. The post-maximum spectrum is similar to an A or F supergiant. Three objects of this class are known: FU Ori, V 1057 Cyg and V 1515 Cyg. It has been suggested that the FU Ori phenomenon represents the clearing of the dust envelope of a protostar. A critical review by Herbig (Ap J 217, 693) points out that it is more likey that the FU Ori is an event which occurs several times during the lifetime of most T Tau stars.

An observational study of the Ae-Be Herbig stars, which are considered to be pre-main-sequence objects, has been made by Garrison and Anderson (Ap J 218, 438). The infall rotation model proposed by Ulrich (Ap J 210, 377) for T Tau stars, may be applicable to these variables.

Flare stars. (Written with the collaboration of R. E. Gershberg).

For understanding the mechanism of stellar flares it is very important to have simultaneous spectroscopic and photometric observations, like those performed by Moffet and Bopp (Ap J Sup. 31, 61). These data show that continuum emission dominates at the spike phase, while line emissions dominate during the phases of slow rise before maximum and of slow decay after maximum. Helium emissions are seen only symultaneously with strong flare continuum. High dispersion spectra of two faint flares of AD Leo have been studied by Schneeberger et al. (in press). Spectra of flare stars in quiet state have been studied by several authors. Bopp and Fekel (A J 82, 490) have analyzed the spectra of 13 BY -Dra -stars and found that 8 of them are SB2. one is SB1 and 4 are single stars. R. F. Griffin did not find close companions for 3 other flare stars (PASP 89, 65). It seems that the presence of a close companion is not necessary for spottedness to appear, and that the ultimate cause is relatively rapid rotation. Bopp and Shmitz (in press) have discussed the Hα variability in 15 dMe stars, and Worden and Peterson (Ap J 206, L145) have studied the profile of the Hα emission in 8 dMe stars and found that the central reversal is a common feature of Hα profiles of dMe stars, but that 5876 He I does not show such characteristics. Hartmann and Anderson (Ap J 213, L67) have found a short-lived absorption feature near Hα in the spectrum of EQ Vir which may be interpreted as large scale infalling gas with a velocity of 70 km/s. Zeeman, observations of 15 dM, dMe and BY Dra stars (Vogt BAAS 8, 535; 9, 593) show no evidence of magnetic fields.

A positive correlation between the variation of the emission of $H\alpha$ and the star luminosity has been found for the star CC Eri (Busko, Quast and Torres, AA 60, L27). These authors are continuing the same kind of research for other variable stars of the type BY Dra.

Late stages of stellar evolution. (Written with the collaboration of Y.Andrillat and J. P. Swings).

Hot evolved stars. Five eruptive blue variables at a distance from the plane of 0.5 to 5 kpc have been observed by Greenstein et al. (PASP 89, 741) including the X-ray variable AM Her. High velocity stream motions in the lines are reported from high resolution spectrograms of AM Her. Several of the "ultraviolet objects" found with S2/68 on TD-1 are studied by groups in Liège and Strasbourg with an objective prism (Barber et al. Ap J Lett 66, L9). One halo star, F 86, which in the visible is very similar to the young Bp star 3 Cen A (Baschek and A. I. Sargent AA 53, 47) has been observed by Hack with IUE. The UV spectrum presents an energy distribution typical of an early B - type star. The same kind of peculiarity observed in the visual seems to be present also in the UV: strong P, Si and metallic ions, weak C and He. Two high galactic latitude stars, HZ 43 and F 24 have been observed with the Apollo Soyuz experiment in the far UV (100-900 A) (Margon et al. Ap J 209, 525; Liebert et al. Ap J 216, 18) and coude spectrograms of F 24 were studied by Thorstensen et al. Ap J 223, 260).

A survey of suspected white dwarfs has been made by Wickramasinghe, Bessel and coll. A magnetic field of 3.6 10^7 gauss was found for the DA star BPM 25114 (MN 183, 533). LOS 785 A has an $H\alpha$ absorption line, the first example in a DB, with the result that H/He = 10^4 to 10^5 by number (MN 178, 11P). LDS 687 B is an intermediate DA with no evidence of hydrogen absorption, so that H/He $< 10^{-4}$ by number (MN 182, 53 P).

Symbiotic stars and peculiar emission line stars.An interesting paper on the nature of symbiotic stars is that by Feast et al. (MN 179, 499). To define the nature of the objects-BQ] , symbiotics, proto-planetary nebula, planetary nebulae-is often quite difficult as Allen and Swings (AA 47, 293) remark in their survey of 65 peculiar emission-line stars with IR excess. Some of the results on peculiar emission line objects are to be found in the following publications:J. P. Swings (NASA TM-X73, 190), Y. Andrillat and J. P. Swings (Ap J 204, L123), J. P. Swings, Y. Andrillat (AA submitted), Johansson (MN 178, 17P), Thackeray (Mem RAS 83, 1). Future work would be desirable in the following direction: simultaneous spectroscopy and photometry of the variable objects (see e. g. Klutz et al. IAU Coll. 42, Bamberg, 126); extension of the observations towards the UV with the help of IUE (Hack and Selvelli have observed the symbiotic M6 III star CH Cyg in April and July 1978, during a period of activity and found evidence of a flat continuum in the range 1150-1900 and a strong 1302 O I emission indicative of an extended envelope of H II); use of the FTS in the near IR, e. g. to try to detect absorption bands and thus to confirm the presence of a late-type companion, which is just suspected in several cases, or to study Zeeman patterns or the velocity structure of lines formed in different layers of the atmosphere.

Nuclei of planetary nebulae. The study of the spectra of the nuclei of planetary nebulae is fundamental for understanding the evolution of the nebulae. Observations are difficult because stars are faint and nebular lines interfere with the stellar lines.However, the new technical developments have produced a quantity of new data. Several reviews on these nuclei have been given at the IAU Symp. 76 on Planetary nebulae. The main problems are: 1) to find the cause of the different spectral types of nuclei; 2) to discover what is their evolution; to look for possible correlations between the characteristics of nuclei and those of the nebulae. Ten planetary nebulae have a binary star as nucleus. This is a very important problem correlated to the evolution of nebulae, and it is strange that no systematic search for binariety among the nuclei of PN has been made.

Observations in the UV of 39 PN show that only 5 of them present variations on the order of 15% at maximum in the UV (Gilra et al. AA 63, 297; IAU Symp. 76 p. 210). Fast variations of low amplitude have been observed in several central stars of PN.

The spectrum of the unique object FG Sge has been observed by Chalonge et al. (Astroph 13, 437; Smolinski et al. PASP 88, 67; Kipper T. A. Sov Astr Lett 4, 280). The results of the observations of several central stars of PN (spectral type, population-membership, binariety, peculiarity) have been reviewed by Lutz (IAU Symp. 76, 185) and by Aller (20th Liège Symp. p. 271).

UV observations have permitted the derivation of the energy distribution and of the temperature of the central stars of PN and the comparison of them with those derived by the Zanstra method (Pottasch et al. AA 62, 95). The disagreement is strong in some cases; and the reasons for it are discussed by Heap (Ap J 215, 864).

Abbreviations.

AA = Astronomy and Astrophysics
AA Sup = Astronomy and Astrophysics Supplement
Act Astr = Acta Astronomica
AJ = Astronomical Journal
Ann TAO = Annals of the Tokyo Astronomical Observatory
ApJ = Astrophysical Journal
ApJ Sup = Astrophysical Journal Supplement
ASS = Astrophysics and Space Science
Astroph = Astrophysics
AZ = Astronomicheskij Zhurnal
IBVS = Information Bulletin Variable Stars
IRASC = Journal Royal Astronomical Society of Canada
MN = Montly Notices of the Royal Astronomical Society
Mem RAS = Memoirs of the Royal Astronomical Society
Obs = Observatory
PASJ = Publications of the Astronomical Society Japan
PASP = Publications of the Astronomical Society Pacific
PDAO = Publications of Dominion Astrophysical Observatory
PJA = Proceedings Japan Academy
Sov Astr = Soviet Astronomy
TAB = Tokyo Astronomical Bulletin

M. HACK
President of the Commission

30. RADIAL VELOCITIES (VITESSES RADIALES)

PRESIDENT: A.H. Batten
VICE-PRESIDENT: M. Duflot
ORGANIZING COMMITTEE: R.F. Griffin, A.G.D. Philip.

General

The Commission suffered severe losses by death during the last three years. In 1976 October, a few weeks after he had been re-elected to the Organizing Committee, J.F. Heard died as a result of a heart attack. He had already served several terms on our Organizing Committee and was President of the Sub-Commission on Radial-Velocity Standards from 1955 until 1961, when all sub-commissions were eliminated in a reorganization of commission structures. In 1978, another distinguished member of our Commission, A.D. Thackeray, was killed in an accident caused by a freak whirlwind. He had been Commission President from 1952-58, and was actively corresponding with me about our affairs shortly before his death. We also note with regret the death of R.K. Young in 1977 December. He was a member of the Commission from 1935 to 1952.

Two Working Groups were set up by the Commission during the Grenoble Assembly. The first, under the chairmanship of M. Barbier-Brossat, dealt with the problem of defining a mean radial velocity from heterogeneous sources of material. All members of the Commission have had a chance to contribute to the deliberations of this Group, and its report and recommendations will be a major item of discussion at Montreal. The Organizing Committee has formed the other Working Group which has been concerned with various problems of standard-velocity stars. No firm recommendations have yet been agreed upon, except that standards should be observed more thoroughly. Many observers, however, clearly feel that present lists of standards are inadequate. Hube emphasizes that the velocities of some stars adopted as standards may be variable or different from the accepted value. Andersen remarks that there are not enough standards in the southern sky, and he points out that if early-type standards are to be adopted they should include stars of widely differing rotational velocities. Beardsley has drawn attention to the possible existence of rapid velocity-variations in B-type stars that should not be overlooked if standards of that spectral type are chosen. Some of these problems have been discussed in more detail in a recent article (Vistas Astron. 22, p. 265.) and further discussion of them will be an important part of our meetings in Montreal.

The remainder of this report consists of edited versions of the various reports sent to me by observatories active in the field of radial velocities. I am grateful to Ch. Fehrenbach, A. Baranne, M. Duflot, M. Imbert, E. Maurice, M. Mayor, J. Andersen, G.A. Harding, I.D. Kharachentsev, and L. Prévot, who have each provided sections or sub-sections of the report, as well as to other members of the Commission whose short reports I have summarized in the concluding section. Much of the work done by members of this Commission has applications to other topics and interested readers will sometimes find more detailed discussion of radial-velocity work in the reports of other commissions, especially 26, 27, 28, 29, 33, 34, 37 and 42.

Observatoire de Haute Provence

L'activité de l'Observatoire de Haute Provence pour la mesure des Vitesses

Radiales a continué dans trois domaines:

1. Mesure de V.R. avec le spectrographe à fente.
a) d'étoiles standard IAU - résultats communiqués avant publication à Batten (publication en cours).
b) d'étoiles de tous types pour des magnitudes 9-10 pour étalonnage des champs de P.O. (publication en préparation).
2. Le groupe de Genève-Marseille a installé au télescope de 1 m helvétique de l'Observatoire de Haute Provence un spectrographe a corrélation. Ce groupe fait son propre rapport.
3. Les prismes objectifs de 40 cm de diamètre de l'Observatoire de Haute Provence et de l'E.S.O. continuent à fonctionner normalement. Les résultats sur le Grand Nuage de Magellan sont publiés dans le rapport de l'Observatoire de Marseille.

Mme G. Amieux et R. Burnage ont étudié l'amas galactique NGC 3114. Sur une centaine d'étoiles, 80 étoiles font partie de cet amas.

G.P.O. DE 62 CM DE DIAMETRE

Un prisme objectif de 62 cm de diamètre, monté devant le télescope de Schmidt Liège-CNRS à l'Observatoire de Haute Provence est utilisé à mi-temps pour la mesure des Vitesses Radiales.

Cet instrument donne une dispersion de 200 A mm^{-1} vers 4220 A, le domaine spectral va de 3900-5900 A, les spectres sont d'excellente qualité. Avec les plaques IIaJ hypersensibilisées, la magnitude limite est de 12, avec une hauteur de spectre de 250 μ en 1 heure de pose (Fehrenbach-Burnage-Compt. Rendus Ac. Sci. Paris, B 281 p. 481).

Un effort a été fait pour la mesure des clichés par corrélation. Cette méthode est absolument impersonnelle; le pointage, l'enregistrement, la mesure et le calcul nécessitent 2 minutes.

Nous avons comparé nos résultats à ceux obtenus par le spectrographe à fente classique ou à corrélation. Pour tous les types spectraux aucune différence systématique n'apparaît dans cette comparaison.
(Compte Rendu Acad. Sci. Paris B 286 p. 289).

L'erreur standard sur l'ensemble des mesures de plusieurs clichés est de ± 8,5 km s^{-1}.

PROGRAMMES ACTUELLEMENT EN COURS OU ARCHIVES (P.O. 62)

Etude systématique de champs situés à + 30° et -30° de latitude galactique pour l'étude des répartitions des V.R. et recherches d'étoiles du halo, analogues à celles détectées dans la direction du Grand Nuage de Magellan.

champ de 2 x 4 degrés	Centre 1950	Etat
champ Ia	$\alpha = 1^h00^m04^{sec}.3$	$\delta = 31°32'11".0$
II	$\alpha = 2^h09^m27^{sec}.9$	$\delta = 30°04'13"$
IIa	$\alpha = 3^h17^m25^{sec}.1$	$\delta = 23°30'37"$
III	$\alpha + 4^h11^m52^{sec}.1$	$\delta = 9°53'12"$

Chaque champ comprend environ 250 étoiles de magnitude 9 à 12.

Observatoire de Marseille

PRISME-OBJECTIF

L'équipe du prisme-objectif a terminé la mesure des clichés pris en direction du Grand Nuage de Magellan (GNM). Complétant le catalogue publié en 1974, un deuxiéme catalogue sera prêt à la publication dans le premier semestre de 1979. Il comprendra environ 1500 nouveaux résultats de vitesses radiales répartis approximativement

en 300 vitesses radiales d'étoiles du GNM
et 1200 vitesses radiales d'étoiles de la Galaxie

Une étude de la qualité de ces vitesses radiales, très variable avec l'étoile mesurée, est préparée par Ch. Fehrenbach et M. Duflot.

CORAVEL

Le projet CORAVEL a été mené à terme par le Laboratoire d'optique de l'Observatoire de Marseille (équipe Baranne et al.). Il s'agit d'un spectromètre photoélectrique destiné à la mesure des vitesses radiales. Le spectre stellaire obtenu (appareil dispersif: réseau échelle + prisme à champ normal) est déplaceé devant la grille correspondante par l'oscillation d'une lame à faces parallèles.

Les parties électronique et informatique ont été réalisées à l'Observatoire de Genève, les parties optique et mécanique à Marseille.

Un premier appareillage CORAVEL fonctionne depuis un an sur un télescope genevois de 1 m à l'Observatoire de Haute Provence. Un deuxième appareillage sera installé sur le télescope danois de 1,5 m à l'ESO-Chili.

Les résultats obtenus depuis un an à l'OHP montrent que depuis que le "CORAVEL" est en fonction, les conditions et les résultats d'observation sont complétement changés en ce qui concerne les vitesses radiales d'étoiles froides (F5 → M).

L'ordre de grandeur des deux paramètres, quantité et qualité, est différent:

La 11e magnitude s'atteint en 3 minutes avec souvent moins de 0,5 km s^{-1} d'erreur. (A. Baranne).

Détermination d'éléments orbitaux d'étoiles doubles spectroscopiques et à éclipses dans les hémisphères nord et sud.

Les vitesses radiales sont obtenues à partir de spectres à 20 A mm^{-1} à l'Observatoire de Haute Provence et de l'ESO et récemment avec le spectromètre à corrélation CORAVEL.

Les premiers résultats obtenus avec CORAVEL montrent une amélioration considérable de la rapidité d'accès à la vitesse radiale et de la précision sur les éléments orbitaux. Pour quelques étoiles observées avec CORAVEL et dont la couverture de l'orbite est presque complète on peut déterminer la vitesse radiale du centre de gravité à 0,1 km s^{-1} et les masses ou fonctions de masses avec une erreur relative de 0,02, l'écart type des résidus étant inférieur à 0,5 km/s et ceci quelle que soit la magnitude de l'étoile.

Actuellement plusieurs dizaines d'étoiles doubles spectroscopiques et à éclipses sont à notre programme d'observation au CORAVEL. Plusieurs d'entre elles ont été découvertes avec un spectre double. (M. Imbert).

Les vitesses radiales d'une dizaine de supergéantes M de la région de h et

χ Persée ont été mesurées avec le spectromètre photoélectrique CORAVEL.

La comparaison de nos mesures avec celles obtenues par G.F. Gahm et J.G. Arkling (Astron. Astrophys. 10, p. 414) avec une dispersion de 16 A mm^{-1} montre une différence systématique de 0,1 km s^{-1}. La grande précision des mesures obtenues avec le spectromètre CORAVEL nous a incités à entreprendre une étude systématique et détaillée de toutes les supergéantes rouges de la région de h et χ Persée. (L. Prévot et E. Rebeirot).

SLIT-SPECTROGRAPH VELOCITIES

Maurice in collaboration with Ardeberg (Lund Observatory, Sweden) has published (Ardeberg and Maurice, Astron. Astrophys. 54, p. 223, Astron. Astrophys. Supp. 28, p. 153 and 30, p. 261) stellar and interstellar radial velocities in the Small Magellanic Cloud and in the galactic cluster IC 2944 in the direction of the inner side of the Carina Spiral feature. All these velocities are measured on intermediate dispersion spectrograms at 74 Amm^{-1}. In the SMC, radial velocities are given for 51 supergiants; the Cloud and Galactic interstellar Ca II K line velocity is given for 6 and 3 stars respectively. Velocities of forbidden lines of ionized oxygen have been measured in the spectra of 3 stars. In IC 2944, 58 stellar velocities are given together with 28 velocities for interstellar calcium and 17 forbidden lines of ionized oxygen. In these papers measurements of radial-velocity standard and reference stars are given and discussed; the accuracy and zero-point of radial velocities obtained with Cassegrain spectrographs are discussed. (E. Maurice).

MINOR PLANETS

A program of spectrographic observation of minor planets as radial-velocity standards has been undertaken by M. Duflot and E. Maurice. Ceres and Vesta were observed, at intermediate dispersion, during five nights from La Silla (June 1978) and the plates are being measured. M. Imbert has measured radial velocities of Ceres, Eunomia and Vesta with the (French-Swiss) correlation spectrometer "CORAVEL" installed in Haute Provence (OHP) with an accuracy (standard error) of 0.5 km s^{-1}; these test observations were made respectively in August 1977, January 1978 (1 measurement for each) and June 1978 (2 measurements).

AUTRES OBSERVATIONS

D'autre part l'Observatoire de Marseille est engagé dans l'étude de nombreuses galaxies extérieures et régions H II galactiques. Ces études s'appuient sur des mesures de vitesse radiale, obtenues par des techniques très différentes (spectrographe nébulaire, interférometrie sur le gaz ionisé).

Observatoire de Genève

Depuis le printemps 1977 un spectrophotomètre CORAVEL pour la détermination des vitesses radiales par corrélation est en activité sur le télescope de 1 m de l'Observatoire de Haute-Provence. Le principe optique de ce spectrophotomètre a réseau échelle a été brièvement décrit (A. Baranne, M. Mayor, J.L. Poncet, 1977). Une description de l'ensemble de l'appareil va être publiée ainsi que l'analyse des mesures de divers échantillons de comparaison, en particulier les étoiles standard IAU (A. Baranne, M. Mayor, J.L. Poncet, 1978). Durant la première année de mesure, environ 6000 mesures ont été effectuées.

Un programme important de mesures d'étoiles d'amas ouverts a été commencé. (Voir le rapport de la Commission 37.)

Des programmes d'observations ont été entrepris dans le champ des étoiles binaires visuelles, des étoiles (m_B < 13) des zones de McCormick et de la micro-variabilité des supergéantes froides.

Dominion Astrophysical Observatory

Selection and testing of suitable wavelengths for radial-velocity measurements at various dispersions has continued. Two sets of wavelengths for spectral classes A0-F2 and B0-B9 at dispersion 15 A mm^{-1} have been published (Batten Pub. Dom. Astrophys. Obs. 14, p. 367), together with radial velocities for 12FK4 stars of these spectral types. The IAU standard-velocity stars are being observed with dispersions of 6.5, 10, and 15 A mm^{-1} in order to check their constancy and their suitability as standards. As a further check, both on the standards and the adopted wavelength system, a programme of observations of the brighter minor planets has been begun. Preliminary results are encouraging, although there is some evidence that small systematic differences between our spectrographs can develop if care is not taken to keep everything in correct adjustment.

Crampton has continued his studies of the radial velocities of O-type and B-type stars in H II regions. He and Georgelin have compared stellar velocities with those of the gas in the corresponding regions (derived from Hα Fabry-Perot measures) they have shown that velocities of early-type stars are being obtained on nearly the same system as those of other objects. They found $V_{gas} - V_{star}$ to be -1.9 ± 1.5 km s^{-1} for the B-type stars and -0.2 ± 2.1 km s^{-1} for the O-type stars (see Astron. Astrophys. 40, p. 317).

Preliminary measures by Crampton and Morrison of spectrograms at 30 A mm^{-1} obtained with an ITT 4089 image tube indicate that there is no significant systematic difference between these results and those obtained from ordinary spectrograms made with the same spectrograph, but that random errors with the image tube are about twice as large as for the photographic spectra.

Hill, in collaboration with Aikman, Barnes (Kitt Peak) and Hilditch (St. Andrews) is continuing his studies of the dynamics of A-type and F-type stars in the galactic pole regions in order to determine the local mass density. So far at least three spectrograms (80 A mm^{-1}) each of some 550 stars brighter than 10^m within 15° of the NGP and some 350 stars brighter than 9^m within 15° of the SGP have been obtained. Photometric results (ubvy, β) are available for all of the stars.

Morbey has developed a method of digitally averaging the line profiles in a stellar spectrogram and in the accompanying comparison spectrum simultaneously. Averaging the results for a selected list of wavelengths gives a measure of the radial velocity of the star. The method is analogous to that employed directly at the telescope in a radial-velocity scanner. The technique is particularly useful in sorting out multiple spectra where low contrast or close blending of different sets of lines makes it difficult to measure the components of individual lines. It was developed for the triple system HD 165590 (Morbey et al. Pub. Astr. Soc. Pacific 89, p. 851), and has since been successfully applied to the detection of marginally visible secondary spectra. Comparisons of the technique with more conventional methods are being made, and the results indicate that at a dispersion of 6.5 A mm^{-1} the plate-to-plate scatter is less than 0.5 km s^{-1}.

Copenhagen University Observatory

The observations for the programme by J. Andersen, B. Nordström, and P. Grosbol, comprising some 740 bright southern B - F4 stars, have been completed. About 2800 plates (ESO 1.5 m coudé, 20 A mm^{-1}) have been taken, including some 350 plates of standard stars, and the measurements have been started. The programme will complete the radial velocities for the entire FK4 catalogue and the Bright Star Catalogue to spectral type F4, and is designed to enable a full discussion of available uvby β photometry and proper motions of these stars.

A programme by E.H. Olsen and B. Strömgren to define the group of intermediate Population II F stars on purely photometric grounds has resulted in a candidate list of some 400 stars. J. Andersen and B. Nordström, in collaboration with B. Mayor, Geneva, have started an observing programme to determine radial velocities of these stars, combining one coudé plate and several photoelectric measures in order to check for spectral peculiarities at the same time. Observations are starting October 1978.

Royal Greenwich Observatory, Herstmonceux

1. Pike, Stickland and Lloyd have investigated techniques for obtaining radial velocities using the PDS microdensitometer. Two methods have been evolved:
   a) scanning individual lines and fitting profiles to determine the centroids.
   b) cross-correlating segments of spectrum to find velocity changes.

The results for α Equ and HR 6684 are contained in Mon. Not. Roy. Astron. Soc. 184, p. 265 and Acta Astronomica (in press).

2. Stickland has used the Isaac Newton telescope to obtain about 40 spectra of Hg-Mn stars at 8.5 A mm^{-1}. The programme is continuing and being extended to the Southern Hemisphere.
3. Walker has observed the Ap Si star HR 3413 photometrically and for radial velocity in order to monitor variations. Radial velocity changes appear to be small except for the λ 3954, 4200 and 4261 lines of silicon which show a total amplitude of $\sim$ 10 km s^{-1}.

Observatories in the U.S.S.R.

Radial velocities have been measured for 340 Markarian and Arakelian galaxies. The observations were made in the Crimea (125 cm aperture, 220 A mm^{-1}) and at Alma-Ata (70 cm aperture, 90 A mm^{-1}). Results have been published in several papers in volumes 11 and 12 of Astrofizika by Arakelian, Denisyuk, Doroshenko, Kopylov and others, and also by Denisjuk et al. (Pysma v Astr. Zh. 3 p. 7) and Arakelian et al. (Astr. Circ. No. 914, p. 7).

Radial velocities for 190 galaxies in groups and interacting systems have been obtained at the Crimean Astrophysical Observatory (260 cm aperture, 360 A mm^{-1} and as given above), Alma-Ata (as above), Tautenberg (200 cm, 180 A mm^{-1}) and ESO (100 cm, 280 A mm^{-1}). The results have been published by several authors in a variety of journals (Astronomocheskii Zhurnal, Astronomy and Astrophysics, Astronomical Circular, Astronomische Nachrichten, and Astrofizika) during the past three years.

At Zelenchuk (Special Astrophysical Observatory of the U.S.S.R. Academy of Sciences), Karachentsev has measured 500 radial velocities of double galaxies in his Catalogue of Isolated Pairs. The work was done with 6m telescope at a dispersion of 90 A mm^{-1} and the results are in press. A Catalogue of Galaxy Red-shifts, containing over 5,000 entries, has been compiled on magnetic tape.

Miscellaneous Short Reports

Sahade reports from the Institute de Astronomia y Fisica del Espacio (Buenos Aires) that V.S. Niemela is continuing her studies of Of stars and Wolf-Rayet stars. She finds velocity gradients both in the regions of the atmospheres where the absorption lines are formed and in those where the emission lines are formed.

She has studied three objects of these classes in the Carina association HD 93129A, HD 93131, and HD 93162) and can find no periodicity in their velocity variations. R.H. Mendez has studied the central star of VV 1-7 and found no clear evidence of radial-velocity variations, although his value for the mean velocity from 12 plates disagrees with a single earlier observation.

W. Buscombe reports from the Lindheimer Astronomical Research Center (Northwestern University) on a study of high-velocity stars in the southern hemisphere by H.J. Augensen. Careful comparison of measurements on direct photographic and image-tube spectrograms shows the latter to give reliable results consistent with those obtained from the former. Some stars have been found to possess interesting kinematical properties. E.M. Hendry, also at Lindheimer, continues her radial-velocity work on spectroscopic binaries.

Griffin reports that his programmes discussed in earlier Commission reports are continuing, with observations being made both from Cambridge and Palomar.

Hube (Calgary) continues work on standard-velocity stars, Ap stars, and binaries, mainly at the Dominion Astrophysical Observatory in Victoria. Gieseking (Bonn) has developed a new method of determining radial velocities with the objective prism and is applying it to spectroscopic binaries. Oetken reports only limited activity in the field of radial velocities at Potsdam. The chief results are new orbital elements for some spectroscopic binaries containing magnetic stars.

A. H. BATTEN
President of the Commission

35. STELLAR CONSTITUTION (CONSTITUTION DES ÉTOILES)

PRESIDENT: B. Paczyński
VICE-PRESIDENT: R. J. Tayler
ORGANISING COMMITTEE: D. J. Faulkner, P. Giannone, C. Hayashi, I. Iben, R. Kippenhahn, A. G. Massevitch, G. Ruben, J-P. Zahn.

1. INTRODUCTION

Following the example set by the previous President of the Commission (L. Mestel), we have arranged for this report to consist of detailed reports on a limited number of topics of current interest. The subjects discussed are: (1) Developments in the Theory of Evolution of Single Stars (Dr. Ergma); (2) Close Binary Stars (Drs. Bath and Pringle); (3) Effects of Opacity and Composition on Stellar Evolution (Dr. Cox); (4) Effects of Mass Loss on Stellar Evolution (Drs. Dearborn and Schramm). Because these reports have been produced on a tight schedule by authors from different countries, there is some unavoidable overlap in their contents. We are grateful to the authors for providing the reports.

Study of the evolution of stars continues to be central to astrophysics although research in galaxies and cosmology may have appeared a more glamorous subject in the recent past. To some extent this is because the subject of stellar evolution has come of age and most of the easy and straightforward problems have been solved. There is now reduced emphasis on the study of single spherical stars in quasi-static stages of evolution. Instead there is much work on the more difficult problems of non-spherical stars, either single or interacting, and of mass loss from stars, either steady or explosive. Many of these problems are too complicated to be solved exactly with the present generation of computers even if the physical processes involved are fully understood. One particular problem in which uncertainties in the laws of physics, departures from spherical symmetry and extremely short timescales are probably all involved is the explosion of a supernova and it is perhaps not surprising there is not agreement on what causes a supernova explosion. In saying this it must not be forgotten that there are still uncertainties in the structure of single spherical stars. The solar neutrino discrepancy is still not resolved, there are disagreements about the form of stellar opacity, the maximum mass of a neutron star is not settled and further resonances may yet be found in nuclear reactions which are important in nucleosynthesis.

Since the 1978 Grenoble General Assembly and the symposia and colloquia associated with that General Assembly, meetings of interest to members of Commission No. 35 have included IAU Symposium 76 "Planetary Nebulae", Ithaca, U.S.A.,June 1977, IAU Symposium 80 "The HR Diagram. The 100th Anniversary of Henry Norris Russell", Washington, D.C.,U.S.A., November 1977, IAU Symposium 83 "Mass Loss and Evolution of O-type Stars", Vancouver Island, Canada, June 1978, IAU Colloquium 42 "The Interaction of Variable Stars with Their Environment", Bamberg, G.F.R., September 1977 and IAU Colloquium 52 "Protostars and Planets", Tucson, U.S.A., January 1978. Meetings planned in conjunction withthe General Assembly are IAU Symposium 85, "Star Clusters", Victoria, B.C.,Canada, August 1979, IAU Symposium 88, "Close Binary Stars", Toronto, Canada, August 1979 and IAU Colloquium 51 "Convection and Turbulence in Stellar Atmospheres", London, Ontario, Canada, August 1979.

2. DEVELOPMENTS IN THE THEORY OF EVOLUTION OF SINGLE STARS (E. V. Ergma)

1. General Problems

Evolution of massive stars with mass loss have recently been considered in several countries. [C de Loore et al. (17), Chiosi et al. (2), Massevitch et al. (3), Sreenivasan and Wilson (4)]. It was shown that the effect of mass loss causes a broadening of the hydrogen core burning phase in the H-R diagram. It was pointed out by Chiosi et al. (2) that the core He burning phase is rather sensitive not only to mass loss in the acoustic-flux regime (in the red giant stage) but also to mass loss occurring in the core H-burning phase. To explain an evolutionary scenario from O-B stars to W-R stars a very large mass loss is required ($\dot{M}>10^{-4} M_\odot\, yr^{-1}$). A mass loss rate $\dot{M} \geq 4 \times 10^{-6} M_\odot\, yr^{-1}$ may lead to a decrease of the luminosity and to the production of an OBN star near the main sequence (5).

Besides the mass loss mechanism the observed widening of the upper part of the M-S may be explained by two other mechanisms: Firstly there can be an increase of the convective core by convective overshooting. Such an increase may lead to a widening of the M-S band so that not only blue and yellow giants but also more massive red giants remain in the phase of hydrogen core burning. The time of H-burning in the core also increases (3). The second mechanism of the widening of the upper M-S band is connected with the new Carson opacities. Stothers and Chin (6) have found that a pronounced difference exists between evolutionary tracks calculated with Cox-Stewart's and with Carson's opacities. It is to be regretted that Carson's opacity tables are still not available to the astronomical community.

The physical picture in Cas A has been analyzed by S. Lamb (7) with the result that the progenitor of Cas A could be a massive star that has lost considerable part of its envelope during the early supergiant phase.

The special evolutionary code proposed by B. Paczyński has been developed by Rozyczka (8,9) by a replacement of the Lagrangian coordinates by Eulerian coordinates which significantly reduces the computing time when investigating the helium and carbon flashes or accretion processes on to white dwarfs. With the help of this program the position of the helium flash as a function of rotation,chemical composition and screening factor has been investigated.

Calculations for 46 stars have been performed from the subgiant branch to the onset of helium burning ($0.10 \leq Y \leq 0.4$, $0.7 \leq M/M_\odot \leq 2.20$, $10^{-5} \leq Z \leq 0.04$). The tracks are particularly sensitive to changes in the heavy-element abundances (10).

The effect of heavy-element abundance on stellar evolution has been investigated by Alcock and Paczyński (11) with the result that in a 7 $M_\odot$ star carbon is consumed non-violently if $Z = 4.10^{-4}$ but is ignited in a highly degenerate core for $Z = 0.03$. This indicates a possible different behaviour for massive Population 2 and Population 1 Stars.

Evolutionary calculations for small super-metal-rich stellar masses ($0.6 \leq M/M_\odot \leq 1.5$, $Z = 0.10$) have shown that Z values as large as 0.10 could account for gaps in evolutionary tracks near the main sequence as large as $\Delta M_b \sim 0^m5$ independent of the He content (12). Evolution of extremely metal-poor stars remains sensitive to initial heavy element abundance and evolution is significantly faster than that of pop I stars. Core helium ignition takes place prior to reaching the red giant branch (13). Evolution of massive helium stars with the new opacities given by Carson has been followed by Stothers and Chin (14). If neutrino losses are neglected, the phase of carbon burning in the core occurs in the red-supergiant region but with neutrino losses carbon burns near to the He M-S.

Several authors (15, 16) have studied advanced evolution of massive stars. According to Arnett (15), helium stars more massive than 12 $M_\odot$ avoid the silicon flash. For M_α about 6-8 $M_\odot$ there is a mild silicon flash. For less massive helium stars silicon starts burning in highly degenerate conditions.

Ergma and Vilhu (16) have found that for 9$M_\odot$ stars carbon starts burning at some distance from the centre, so that the flash moves inwards removing finally the degeneracy. No mixing has been found.

The problem of occurrence of loops (one or two) in the H-R diagram is still being discussed by several authors. The factors that affect the length of the first loop are as follows: the mass fraction in the helium core, the opacity in the radiative envelope, the location of the hydrogen-burning shell at core helium ignition, the location of the outer boundary of the zero-age M-S convective core, the depth of convective mixing at the peak of the giant branch (if deep mixing occurs), the duration of core helium burning, the fraction of the energy produced by the hydrogen burning shell and the extent of intermediate convective mixing (17). The behavior of the second loop in the H-R diagram has been studied by Höppner et al. (18).

Evolution of rotating stars with $7M_\odot$ and $10M_\odot$ has been followed by Endal and Sofia (19). They found that evolved stellar cores develop nonaxysymmetric instabilities corresponding to the bifurcation of the Maclaurin sequence before they reach critical (Keplerian) rotational velocities. For $7M_\odot$ the instability point is reached prior to carbon ignition, indicating that catastrophic carbon detonation may be avoided by rotating stars.

2. Chemical evolution

A most interesting problem is the variation of surface chemical abundances caused by the evolution of stars. In this field most promising investigations are connected with s-process elements in the helium burning shells and consequent mixing.

Paczyński (20) has studied helium shell flashes in models of Pop I stars with $M_{CO}=0.8\ M_\odot$ for $M_{tot} = 8\ M_\odot$ and $3M_\odot$. If no over-shooting occurs at the base of the convective envelope, the features of the helium shell flashes do not depend on the mass of the envelope. If convective overshooting is taken into account, then following the peak of the helium flash the massive envelope penetrates the core and mixing occurs. Several papers by Iben and Truran (21, 22) and Iben (23) are devoted to the s-process nucleo-synthesis in thermally pulsating stars, based on Iben's mechanism for the production of neutrons $^{14}N(\alpha,\gamma)^{18}F(,\beta^+\nu)^{18}O(\alpha,\gamma)^{22}Ne(\alpha,n)^{25}Mg$. The main results of these studies are: 1) significant enhancement of s-process elements does not occur for luminosities less than M_B ~-6. 2). For stars of a given initial composition the surface abundance of ^{12}C exceeds that of ^{16}O at roughly the same surface temperature independent of mass. 3) $M<1.5M_\odot$ stars do not as a consequence of thermal pulses produce $^{12}C/^{16}O>1$. 4)$M<2.5\ M_\odot$ stars do not achieve convective shell temperatures sufficiently high for the production of s-process elements. 5) For $M>3M_\odot$ the surface enhancement of s-process elements ~2-20, maximum $M{\sim}5M_\odot$. 6)stars with $3\lesssim M/M_\odot\lesssim 8$ are the major sources of s-process elements in the galaxy.

The distinguishing feature of the ^{22}Ne neutron source studied in (21) is that, under a wide variety of conditions in the convective helium shells, s-process elements ($70\leqslant A\leqslant 204$) are invariably produced in solar system proportions. This neutron source fails to produce in solar proportions elements in the range $60\leqslant A\leqslant 70$ and the heavy lead isotopes.

Neutron capture nucleosynthesis in the helium-burning cores of massive stars studied by Lamb et al. (24) showed that: 1) a significant amount of elements ^{25}Mg, ^{26}Mg, ^{36}S, ^{37}Cl, ^{40}Ar, ^{40}K, ^{58}Fe are produced in stars with $M\gtrsim 15M_\odot$. 2) Nuclei in the mass range $60\leqslant A\leqslant 70$ are built in stars more massive than $15M_\odot$. 3) The helium cores of massive stars cannot be the major source for formation of the heavier ($A>90$) s-process nuclei. For the explanation of the origin of CH and carbon stars, the core helium flash (off-center) has been proposed (25). It was shown that the amount of carbon produced during the flash increases with the mass fraction of the off-center core helium flash and at least some R-type carbon stars may be explained by this effect.

Fujimoto (26, 27) studied the possibility of hydrogen mixing during helium flicker and concluded that the helium convective zone grows extensive enough to trigger the Schwarzschild mechanism $^{12}C(p,\gamma)^{13}N(,\beta^{+}+\nu)^{13}C(\alpha,n)^{16}O$ for two cases: 1) when the mass fraction of the hydrogen-rich envelope is very small; 2) when the helium burning luminosity L_{He} grows larger than $10^{8}L_{\odot}$.

Hot-bottom convective neutron and s-process element production has been studied by Despain (28).

References

1. De Loore, C., De Greve, J. P., and Lamers, H. J. G. L. M.: 1977, Astron. Astrophys. 61, p. 251.
2. Chiosi, C., Nasi, E., and Sreenivasan, S. R.: 1978, Astron. Astrophys. 63, p.103.
3. Massevitch, A. G., Popova, E. I., Tutukov, A. V., and Jungel'son, L. R.: 1978, to be submitted to Astrophys. Space Sci.
4. Sreenivasan, S. R., and Wilson, W. J. E. 1978, Astrophys. Space Sci. 53, p.193.
5. Dearborn, D.S. P., and Eggleton, P. P.: 1977, Astrophys. J. 213, p.448.
6. Stothers, R., and Chin, C-W.: 1977, Astrophys. J. 211, p.189.
7. Lamb, S. A.: 1978, Astrophys. J. 220, p.186.
8. Rocyczka, M., 1977, Acta Astron. 27, p.415.
9. Rocyczka, M., 1978, Acta Astron. 28, p.19.
10. Sweigart, A. V., and Gross, P.: 1978, Astrophys. J. Suppl. 36, p.405.
11. Alcock, C., and Paczyński, B.; 1978, Astrophys. J. 223, p.244.
12. Castellani, V., and Tornambe, A.: 1977, Astrophys. Space Sci. 46, p.195.
13. Wagner, R. L.: 1978, Astron, Astrophys. 62, p.9.
14. Stothers, R., and Chin, C-W.: 1977, Astrophys. J. 216, p.61.
15. Arnett, D.W.: 1977, Astrophys. J. Suppl. 35, p.145.
16. Ergma, E., and Vilhu, O.: 1978, Astron. Astrophys. 69, p.143.
17. Schlesinger, B. M.: 1977, Astrophys. J. 212, p.507.
18. Höppner, W., Kähler, H., Roth, M. L., and Weigert, A.: 1978, Astron. Astrophys. 63, p.391.
19. Endal, A. S., and Sofia, S.: 1978, Astrophys. J. 220, p.279.
20. Paczyński, B.: 1977, Astrophys. J., 214, p.812.
21. Truran, J. W., and Iben, I.: 1977, Astrophys. J. 216, p.797.
22. Iben, I., and Truran, J. W.: 1978, Astrophys. J. 220, p.980.
23. Iben, I.: 1978, Astrophys. J. 217, p.788.
24. Lamb, S. A., Howard, W. M., Truran, J. W., and Iben, I.: 1977, Astrophys. J. 217, p.213.
25. Paczyński, B., and Tremaine, S.: 1977, Astrophys. J. 216, p.57.
26. Fujimoto, M. Y.: 1977, Publ. Astron. Soc. Japan 29, p.331.
27. Fujimoto, M. Y.: 1977, Publ. Astron. Soc. Japan 29, p.537.
28. Despain, K. H.: 1978, Astrophys. J. 212, p.744.

3. THEORETICAL PROBLEMS IN CATACLYSMIC BINARIES AND RELATED OBJECTS (G. T. Bath and J. E. Pringle)

The cataclysmic variables (dwarf novae, nova-like variables, and AM Her-type objects) provide a testing ground for theoretical studies of mass transfer and accretion processes in close binaries. Recognition of this in the period prior to this report (IAU Symposium 73, Eggleton et al. 1976) has led to many theoretical developments in the subject.

The presence of a disc around the white dwarf component is well established. The optical luminosity of a steady-state accretion disc depends primarily on the mass transfer rate through it, and the temperature distribution over the surface. The temperature distribution may be estimated assuming black body emission, since for any viscosity likely to be generated by subsonic turbulence the disc is optically thick, at least in the brightest inner region. Tylenda (1977, Acta Astron. 27, p.235) compares the theoretical spectrum of a steady state disc from the soft x-ray to the infra-red region with the observed spectrum of SS Cygni at outburst. The fit is good for an accretion rate 6 x 10^{17} g s^{-1}. Agreement with the UV spectral distribution of two old novae, V603 Aql and RR Pic, is satisfactory for accretion rates 8 x 10^{17} g s^{-1} and 4 x 10^{18} g s^{-1} respectively, confirming the widespread suspicion that old novae are similar to the outburst state of dwarf novae, but 'stuck' permanently at a higher accretion rate.

A more accurate flux distribution is needed to calculate colours. Schwarzenberg-Czerny and Rozyczka (1977, Acta Astron. 27, p.429) fit steady-state discs with a grid of normal stellar atmospheres but ignore the important contributions of the 'hot- spot' and emission lines from optically thin regions. They find the observational agreement in the two colour diagram is poor. Not surprisingly, only the brightest systems with highest accretion rates and disc temperatures show reasonable agreement. Rather than interpolating over a grid of published stellar atmospheres, as in the previous study, Mayo (Ph.D. Thesis, Cambridge, 1978) has constructed models covering the correct range of g and T_e, including absorption line formation, and finds good agreement with the outburst spectrum of SS Cygni, but the quiescent state is less well fitted.

Two sources of soft x-ray emission in dwarf novae have been proposed - non-thermal radiation from the shocked 'hot-spot' region and thermal radiation from the boundary layer region of the disc/white dwarf interface, where about half the accretion energy is liberated where the disc grazes the stellar surface. Pringle (1976, Monthly Notices Roy. Astron. Soc. 178, p.195) concludes that the thickness of the shocked region in the 'hot-spot' is very much less than the radius of the incoming stream, and the spot cannot be optically thin to soft x-ray photons; whereas much of the boundary layer energy can be emitted as soft x-rays (0.15>KeV) if the boundary layer is optically thick. These conclusions receive support from the recent observation of a 9 sec periodicity at outburst in SS Cyg at both soft x-ray (Cordova et al., 1978, IAU Circ. 3235) and optical wavelengths (Patterson et al.Astrophys.J. Letters in press) clearly produced close to the boundary layer region. A further, and more detailed, study by Pringle and Savonije (Monthly Notices Roy. Astron. Soc., in press) has shown that for low accretion rates ($\lesssim 10^{16}$ g s^{-1}) the boundary layer can be optically thin and emit x-rays with energies up to about 20 KeV. The recent discovery of hard x-ray emission in SS Cyg in the inter-outburst state (Ricketts, King and Raine, Monthly Notices Roy. Astron. Soc., in press) indicates strong non-thermal radiation can be important in the quiescent state.

Of direct relevance to the discovery of x-ray emission from cataclysmic variables have been a number of studies of the x-ray emission processes associated with accreting white dwarfs. Katz (1977, Astrophys. J. 215, p.265) has calculated the emitted spectrum of radially infalling material under the assumption that the outward flux does not interact with the accreting gas. Fabian, Pringle and Rees (1976, Monthly Notices Roy. Astron. Soc. 175, p.43) estimated the importance of absorption in the accreting gas and argued that this leads to an upper limit to the luminosity of about 10^{36}erg s^{-1}. This is confirmed by Kylafis (Ph.D. thesis, Illinois, 1978). When radial infall takes place along magnetic field lines, cyclotron cooling can be important. The effects of this have been estimated by Masters et al. (1977, Monthly Notices Roy. Astron. Soc. 178, p.501). Because of the importance of this process in AM Her variables more detailed calculations of this process have been undertaken (see Tuohy et al.:1978, Astrophys. J. Lett. 226, p.L17 and references therein).

The cause of bursting flows through the disc in dwarf novae during eruptions remains the most speculative and uncertain area in the subject. Madej and Paczyński (IAU Colloquium 42, p.313, Kippenhahn et al.) following the suggestion of Osaki, argue that the optical brightness of the 'hot-spot' in U Gem favours a disc storage mechanism, and construct a model with the incoming stream forming a torus with constant angular momentum per unit mass and vanishingly small viscosity. They suggest eruptions are due to an ad hoc increase in viscosity leading to sudden accretion. The cause of the sudden viscosity increase and the factors which determine a repetition timescale of ~30 days remain unknown. Sunyaev and Shakura (1977, Soviet Astron. Letters 3, p.138) discuss the formation of disc reservoirs around a magnetised central star. If the magnetosphere exceeds the corotation radius, matter storage may occur in circumstances which would otherwise lead to steady accretion.

The alternative outburst scheme for dwarf novae associates enhanced accretion with bursting mass transfer by the companion. Webbink (1977, Astrophys. J. 211, p.486; 1977, Astrophys. J. 211, p.881) finds that main sequence stars going through case A evolution can experience 'dynamical' phases of transfer, a result confirmed by other workers. Dynamical instabilities can only be properly studied with a full hydrodynamic code. A non-linear hydrodynamic code, which includes time-dependent convection using a mixing length formulation, has been developed by Wood (1977, Astrophys. J. 217, p.530). He confirms the existence of dynamical instability in some models, and shows, as in earlier work, that this can produce bursting mass transfer. Mass transfer rates and repetition times fit those required in dwarf novae, but Wood confirms earlier results that the existence of the instability is sensitive to the treatment of convection. Given the complexity of the geometry near the Lagrangian point, the lack of a realistic theory of convection, and uncertainties about the correct value of the mixing length to scale height ratio in the theory which is available, the existence of instabilities remains unclear. Detection of hard (>2 KeV) x-ray emission from SS Cyg throughout the quiescent phase (Ricketts, King and Raine, Monthly Notices Roy. Astron. Soc., in press) argues in favour of continuing accretion and against a simple disc storage mechanism.

Further work has been done on investigating the conditions needed for an equilibrium (time-independent) accretion disc to be stable. Shakura and Sunyaev (1976, Monthly Notices Roy. Astron. Soc., 175, p.613) have made a detailed study of the inter-relationship of secular and thermal instabilities, showing that in general both are present (if not all) and that the fastest growing mode is the thermal one. Pringle (1976, Monthly Notices Roy. Astron. Soc. 177, p.65) demonstrated how a simplified instability criterion could be obtained by considering just the thermal mode. Piran (1978, Astrophys. J. 221, p.652) has generalized previous work to consider arbitrary viscosity laws and cooling mechanisms.

An understanding has been reached of what happens to the angular momentum of matter that is transferred in a binary system in which an accretion disc is formed. Numerical experiments had earlier showed that almost all the transferred material is accreted and the disc more or less extends out to fill the Roche lobe of the accreting star in an equilibrium situation. Investigations by Paczyński (1977, Astrophys. J. 216, p.822), Papaloizou and Pringle (1977, Monthly Notices Roy.Astron. Soc. 181, p441) and Weber (Ph. D. Thesis, Berkeley, 1978) have shown that the outer edge of the disc is curtailed by tidal processes which remove angular momentum from the transferred material and return it directly to orbital angular momentum. In this way the excess angular momentum of the transferred material is removed, allowing essentially all the transferred matter to be accreted.

Both Ritter (1976, Monthly Notices Roy. Astron. Soc. 175, p.278) and Robinson (1976, Astrophys. J. 203, p.485) have determined component masses for cataclysmic binaries assuming the secondary obeys a main-sequence mass-radius relation and fills its Roche lobe. A method based on particle trajectories and the position of the 'hot-spot' has been used by Smak (1977, Acta Astron. 26, p.271). The difficulty with the last method lies in determining the size of the disc, whilst the first method relies on the assumption of a normal main-sequence structure for the companion.Despite these difficulties, the methods are sufficiently accurate to show that early suggestions that the white dwarf mass is close to 1.2 $M_{\odot}$ in almost all systems are incorrect.

Further observations have been made of coherent short period oscillations seen in cataclysmic variables during outburst. These oscillations are highly variable and lie in the range of 8-50 seconds. The more recent observationsl results can be found in papers by Patterson, Robinson, and Nather (1977, Astrophys. J. 214, p.144) Warner and Brickhill (1978, Monthly Notices Roy. Astron. Soc. 182, p.777) and Nevo and Sadeh (1978, Monthly Notices Roy. Astron. Soc. 182, p.595) and references therein. Robinson and Nather (Astrophys. J., in press) have recognised a new phenomenon of quasi-periodic oscillations which can be seen in the light curve but which have sufficient phase-jitter that they do not show up well in power spectrum analyses.

Theoretical work by Papaloizou and Pringle (1978, Monthly Notices Roy. Astron. Soc. 182, p.423) suggests that much of what is observed can be interpreted as non-radial oscillations of a rapidly rotating white dwarf, whose rotation period is comparable to the period of the observed oscillations. They argue that the amount of mass involved in the oscillation is exceedingly small (around 10^{-8} to $10^{-10}M_{\odot}$).

One of the most startling recent discoveries has been the AM Her-variables (at present AM Her, AN UMa, and VV Pup). These stars are characterized by displaying strong circular and linear polarization which varies in a regular manner with periods of 2-3 hours. The best studied of these, AM Her itself, has been known for decades as a semi-regular variable. It is now known to vary with a 3.09 hr period, both photometrically, in all regions of the spectrum up to 60 KeV, and spectroscopically. The nature and relative phasing of these variations are highly unusual. A large body of observational data has now been published, most of which can be traced through papers by Priedhorsky and Krzeminski (1978, Astrophys. J. 219, p.597), Chanmugam and Wagner (1978, Astrophys. J. 222, p.641) and Liebert et al.(1978, Astrophys. J. 225, p.201). From these papers emerges a plausible theoretical model in terms of a close binary system consisting of a lobe-filling red dwarf and a strongly magnetic (10^8 gauss), accreting white dwarf. The strong field ensures that the white dwarf is phase locked to the red companion, and so rotates with the binary period. It also ensures that accretion takes place only on to the magnetic pole(s) of the white dwarf, giving rise to the x-ray emission as well as the strong optical polarization. The relative positions of the polar axis and of the gas stream of the accreting matter can be adjusted to obtain reasonable agreement with observations.

The origin of cataclysmic variables remains problematic. The presence of a white dwarf in a system with an unevolved companion and a separation of a few solar radii indicates extensive angular momentum loss. In particular the specific angular momentum of cataclysmic binaries is so low that, during an earlier, more widely separated phase, an efficient mechanism must have existed for transferring angular momentum from the stellar core(s) of the stars to outer envelope material which is finally lost. Ritter (1976, Monthly Notices Roy. Astron. Soc. 175, p.278) discusses anevolutionary scheme involving case B or C mass transfer combined with extensive mass and angular momentum loss, but does not consider the problem of angular momentum exchange between the stellar core and the envelope. Lin (1977, Monthly Notices Roy. Astron. Soc. 179, p.265), Flannery and Ulrich (1977, Astrophys. J. 212, p.533) and Nariai and Sugimoto (1976, Publ. Astron. Soc. Japan 28, p.593) have studied various aspects of binary stability and evolution with mass and angular momentum loss, but the use of the outer, L_2, Lagrangian point to define a critical surface, analogous to the inner critical surface, is open to criticism since corotation is unlikely to be maintained.

The evolution of close binaries with a common envelope, in which a compact component is embedded in the extended envelope of the companion, have been examined with a variety of assumptions about dissipation and angular momentum transfer between the components. Alexander et al. (1976, Astrophys. J. 204, p.879) discuss the evolution of the orbital elements including gravitational drag, for the case of a compact object orbiting through a constant velocity outflowing wind, and suggest that symbiotic stars may be experiencing orbital decay as a result of accretion by the compact component. Taam et al. (1978, Astrophys. J. 222, p.269) follow the evolution of a 16 $M_{\odot}$ supergiant as a 1 $M_{\odot}$ neutron star spirals in, through turbulent dissipation and frictional drag. Either of two outcomes results: hydrodynamic expulsion of the envelope, or coalescence of the two cores with no ejection. Mass ejection requires a more extended envelope in which the lower binding energy, and inefficient convective energy transport, both aid expulsion. Tutukov and Yungelson (IAU Symp. 83, "Mass Loss and Evolution of O-type stars") have also examined the conditions required for expulsion of the envelope. Meyer and Meyer-Hofmeister (Astron. Astrophys. in press) have studied the evolution of a 5 $M_{\odot}$ red giant and a 1 $M_{\odot}$ main sequence star through a common envelope phase, aimed particularly at accounting for the existence of cataclysmic binaries. Transfer of angular momentum between the inner, tidally dominated region of the common envelope, and the outer region, due to a form of turbulent velocity, is the crucial feature of their model.

Symbiotic stars may be related to earlier stages in the formation of cataclysmic binaries. Bath (1977, Monthly Notices Roy Astron. Soc. 178, p.203) emphasises the role of accretion in symbiotic binaries. He developes a model based on super critical accretion by either a white dwarf or a main-sequence star, with nuclear burning only able to contribute in the former case. Variability, observed in systems like Z And, is due to variable mass loss by the giant companion. Yungelson and Tutukov (Astrofizika, 1976, 12, 521) discuss the structure and evolution of symbiotic stars assuming that the hot components are generating energy primarily by nuclear burning rather than accretion. Paczyński and Rudak (Astron. Astrophys., in press) suggest a model for two classes of symbiotic star. With the accretion rate between two critical values a stably-burning hydrogen shell is the energy source, with variability produced by variable accretion. In the second class the accretion rate is below a lower critical value and hydrogen burning proceeds through a series of shell flashes.

The relative importance of nuclear burning and accretion energy in symbiotic stars has yet to be established. Many of the arguments put forward in discussion of dwarf novae and classical novae are likely to raise their hoary heads again here. If accretion is dominant then some systems should be found in which the accreting star is too massive to be a white dwarf, and is a main sequence star in which surface nuclear burning is impossible. If these systems exist, then they would be the most likely progenitors for cataclysmic binaries.

4. EFFECTS OF OPACITY AND COMPOSITION ON STELLAR EVOLUTION TRACKS IN THE H-R DIAGRAM (Arthur N. Cox)

1. Introduction

A review such as this which attempts to discuss the accuracy of stellar evolution tracks in the H-R diagram, and especially the effects of opacity, is necessarily incomplete. There are some recent results where the opacity or composition has been changed systematically. These are what will be discussed in some detail. Composition changes can be interpreted in terms of opacity changes in the most simplified case, but they obviously are not interchangable because each element has its unique pattern of absorption while opacity errors can be of any size and at any temperature and density. This review will discuss results published mostly after 1975 and before 1979.

There are many other uncertainties in the calculation of stellar evolution tracks besides opacities and compositions. Thermonuclear energy sources, especially those at high temperatures, but even the basic pp reactions, may not be known accurately and energy source representations may differ from one set of calculations to another. One example of this difficulty can be found in studies with and without the reaction $N^{14}(\alpha,\gamma)F^{18}$ by Durand, Eoll, and Schlesinger (1976, Monthly Notices Roy.

Astron. Soc. 174, p.671). Neglect of this reaction supresses the Cepheid blue loops in evolution tracks for 8 and 10 $M_\odot$ stars.

The changing composition due to nuclear burning (nucleosynthesis) is always considered in stellar evolution calculations, but just how these changes are considered in calculating the equation of state, opacity and energy generation is usually obscure. Lamb, Iben, and Howard (1976, Astrophys. J. 207, p.209) and Weaver, Zimmerman, and Woosley (1978, Astrophys. J. 225, p.1021) have shown how these composition changes develop and have followed the changes in material properties correctly by using formulas for them.

The loss of mass during evolution of stars of mass greater than 10 $M_\odot$, a new research area, produces tremendous changes in evolution tracks. The small effects of all other uncertainties are overwhelmed. Examples of this research for massive stars are Dearborn, Blake, Hainbach, and Schramm (1978, Astrophys. J. 223, p.552), Sreenivasan and Wilson (1978, Astrophys. Space Sci. 53, p.193), de Loore, de Greve, and Lamers (1977, Astron. Astrophys. 61, p.251), Chiosi, Nasi, and Sreenivasan (1978, Astron. Astrophys. 63, p.103) and Stothers and Chin (1978, Astrophys. J. 226,p.231). This latter paper shows that the blue supergiants brighterthan 3 x 10^4 $L_\odot$ at a range of T_e can be explained by either mass loss or opacity effects making it important to know the opacity accurately at temperatures near 10^6 K.

Mengel (1976, Astron. Astrophys. 48, p.83) shows the effects of mass loss for population II stars as they evolve with double shell sources off the horizontal branch to carbon ignition.

Another effect on evolution tracks for stars at 10 $M_\odot$ and above is the semi-convection and convective overshooting which can mix hydrogen into the thermonuclear burning core and prolong the star's life near the main sequence. Opacity effects in the million degree range are important for determining the thickness of this marginally convective region as well as for the outer envelope structure and the surface conditions. Maeder (1976, Astron. Astrophys. 47, p.389) has artifically inserted several versions of this overshooting in his calculations to produce evolution tracks and isochrones that match the cluster color magnitude diagrams very well. However, he has not changed the opacity to see how the arguments might be affected with different opacities. Helium abundance changes at Z = 0.1 can also produce large observed gaps according to Castellani and Tornambe (1977, Astrophys. Space Sci. 46, p.195).

In the later evolution core helium burning stages there are more features that can obscure or mitigate the importance of opacities. Blue loops seem to depend on the exact profile of the hydrogen content in the outer layers down to perhaps 10^6 K. In 1972 Robertson investigated several input parameters and found that the envelope opacities were important in order to predict Cepheid blue loops. However, the existence of the blue loops seems to depend on the red giant mass loss, nuclear reactions, and even the ratio of mixing length to scale height in convection zones, preventing any opacity measurement. The second blue loop, when models of 7 - 9 $M_\odot$ have two shell sources, has been considered by Höppner, Kähler, and Weigert (1978, Astron. Astrophys. 63, p.391).

Rotation is another parameter that is usually ignored, but it can influence stellar evolution tracks as discussed by Endal and Sofia (1976, Astrophys. J. 210, p.184 and 1978, Astrophys. J. 220, p.279). At 7 $M_\odot$ the dwarf-stage seems to be slightly less luminous and the giant stage slightly more luminous when rotation is included in several possible ways. Similar effects are seen by Mengel and Gross (1976, Astrophys. Space Sci. 41, p.407) for population II 0.85 $M_\odot$ evolution.

This review does not consider the late evolution stages for population I or II stars after the blue loops, that is, for the stages of helium shell flashes (Paczyński, 1977, Astrophys. J. 214, p.812; Iben, 1977, Astrophys. J. 217, p.788) and other core thermal oscillations during carbon burning (Iben, 1978, Astrophys. J., in press). Also, evolution track calculations for mass transfer in binary systems (see Massevich, Tutukov, and Yungleson, 1977, Astrophys. Space Sci. 40, p.115), done actually for single stars in spherical geometry anyway, will also not be discussed because the opacity and composition effects can all be seen without the complicating problems of mass addition or loss.

2. Opacity Calculations

Opacity data used for all stellar evolution calculations are based on either the Los Alamos data or the new unpublished Carson results. A tabulation of opacities for 40 mixtures of interest for stellar evolution has been given by Cox and Tabor (1976, Astrophys. J. Suppl. 31, p.271). Newer tables can be obtained from the Astrophysical Opacity Library of Huebner, Merts, Magee, and Argo (1977, LA-6760-M Los Alamos) which became available in December 1977. The library consisting of magnetic tapes for each of 20 elements has been sent out to about 10 institutions in the United States and Europe. Updates of the library will be available periodically. These Los Alamos tables, or quite frequently fits to them, in evolution programs by Demarque, Iben, Paczynski, Eggleton, Stothers, etc. or, for pulsation calculations, fits by Stellingwerf (1975, Astrophys. J. 195, p.441 and 1975, Astrophys. J. 199, p.705) are used in all calculations except a few by Carson and Stothers (1976, Astrophys. J. 204, p.472; 1977, Astrophys. J. 211, p.189; 1977, Astrophys. J. 216, p.61; 1978, Astrophys. J. 226, p.231) and by Stothers (1976, Astrophys. J. 210, p.484).

Additional mixtures have been calculated at Los Alamos. Huebner in the Proceedings of an informal conference on the status of solar neutrino research (1978, BNL 50879 1, 107. Brookhaven) discussed several mixtures. He has considered a total of 20 with compositions of C, N, O, Ne, Mg, Fe varying by 25% to see what effects there could be on the solar structure and predicted solar neutrinos. Seven more mixtures are given by Cox, King, and Hodson (1979, Astrophys. J., in press) in dwarf Cepheid studies. Finally, opacities using the same program used for the 1969 publications in the Astrophysical Journal Supplement for 16 mixtures have been used by Demarque for studies of evolution tracks with enhanced carbon and nitrogen abundances.

Detailed tables since the 1976 publication (same mentioned above) are

Older Version of Computer Program

Demarque Mix	X	Y	Z
20	0.800	0.194336	.005664
21	.500	.494336	.005664
22	.200	.794336	.005664
23	.000	.994336	.005664
24	.800	.196	.004
25	.500	.496	.004
26	.200	.796	.004
27	.000	.996	.004
28	.800	.190934	.009066
29	.500	.490934	.009066
30	.200	.790934	.009066
31	.000	.990934	.009066
32	.800	.196	.004
33	.500	.496	.004
34	.200	.796	.004
35	0.000	0.996	0.004
	New	Opacities	
Cox-Davis II	.602	.354	.044
Cox-Davis III	.700	.299	.001

Cox-Davis VI	.70	.28	.02
Cox-Hodson II	.00	.98	.02
Cox-Hodson III	.10	.88	.02
Cox-Hodson·IV	.20	.78	.02
Cox-Hodson V	.30	.68	.02
Cox-Hodson VI	.40	.58	.02
Cox-Hodson VII	.50	.48	.02
Cox-Hodson VIII	.90	.08	.02
Cox-Hodson IX	.98	.00	.02
Deupree III	.65	.34	.01
Deupree IV	.70	.29	.01
Deupree V	.75	.24	.01
Cogan II	.800	.195	.005
Cogan VI	.700	.295	.005
HE9C1	0	.900	.100 (C)
HE5C5	0	.500	.500 (C)
C505	0	0	.500 (C) . 500 (O)
C109	0	0	.100 (C) . 900 (O)

3. Intermediate Mass Population I Composition Tracks

For masses between one and ten solar masses there are several tabulations. This is the region where there is no radiation pressure induced mass loss. Also,there are no problems of semiconvection in the core.

Near the main sequence, composition (or opacity) effects have been discussed by Flannery and Ayres (1978, Astrophys. J. 221, p.175) for the α Cen binary system which consists of two equal age equal composition but non interacting systems. At 0.92 and 1.11 $M_\odot$ tracks were computed. For Y = 0.26 it appears that Z = 0.004 for these two stars if Z = 0.02 is used for the sun which serves as a calibration star. This work does not consider opacity variations separately preferring to put all the relative errors from the sun to the α Cen binary into a Z value.The evolution tracks over the first 0.8 magnitude of brightening seem satisfactory since both stars are predicted to be at their observed H-R diagram position.

Another study at constant X = 0.7 which discusses the effects of various Z values on evolution tracks is by Alcock and Paczyński (1978, Astrophys. J. 223, p.244). In addition to getting the main sequence effect of lower Z hotter T_e, these authors look at the red giant regions and the Cepheid blue loops. Lower Z gives hotter red giants, hotter blue loop tips, and higher luminosity in most cases. At 7 and 10 $M_\odot$ at Z = 0.001 and at 5, 7 and 10 $M_\odot$ at Z = 0.0004 the loops start blueward even before the evolution reaches the red giant region. Fortunately, the morphology change in the tracks needs large changes in Z, and, therefore, the opacity errors that might be present and equivalent to a Z change are not highly important in establishing the overall H-R diagram evolution picture.

A similar study by Becker, Iben, and Tuggle (1977, Astrophys. J. 218, p.633) has found the same effects plus those which occur when Y is changed. Convenient mass-luminosity relations which depend on X, Y, and Z are given for the Cepheid loops.

Harris and Deupree (1976, Astrophys. J. 209, p.402) have given tracks for 3.5, 4.5, 6.0 and 9.0 $M_\odot$ with two compositions (Z = 0.01 and 0.04) for use in isochrone construction for cluster color magnitude comparison.

Unusual opacity effect has been discussed by Lauterborn and Siquig (1976, Astron. Astrophys. 49, p.285). Earlier studies were done by Lauterborn (1976, Astron. Astrophys. 53, p.419). Evolution tracks across the Hertzsprung gap for 4, 6 and 8 $M_\odot$ have real unstable secular eigenvalues. For these stars evolving with a slight thermal imbalance, it was found that the evolution proceeded at the rate given by the most unstable eigenvalue, and it has a radial variation very similar to the associated eigenfunction. The opacity connection is that the extent of the evolution time where there are these unstable eigenvalues depends on the opacity law. At high luminosity where the opacity throughout the star is not so temperature dependent, the instability occurs over as much as half the Herzsprung gap and even includes part of the evolution along the Hayashi line.

Further studies of these instabilities using strictly static models have been made by Lauterborn and Zeuge (1978, Astron. Astrophys. 66, p.367).

Additional tracks for both population I and II compositions have been computed by Saio, Shibata, and Simoda (1977, Astrophys. Space Sci. 47, p.151) in order to determine the ages of the oldest population I cluster (NGC 188) and the oldest population II cluster (M 92). The tracks are given for 3 different masses for each population and for Z = 0.02 and 0.0002. No opacity variations were made, but the authors note that their ages are slightly longer than those obtained before by others, even though Los Alamos opacities were used by all studies.

The causes of the blue loops and whether they really are suppressed by red giant stage mass loss have been studied by Schlesinger (1977, Astrophys. J. 212, p.507) and Siquig and Sonneborn (1976, Bull. Am. Astron. Soc. 8, p.230). In the former case at 5 $M_{\odot}$ no tracks were calculated but only static models with various structures in an intermediate zone between the hydrogen burning shell and the homogeneous envelope made so by the outer convection. The intermediate zone structures and helium context give either red or blue giants. In the second work, which has appeared so far only in abstract form, the loss of mass for a 7 $M_{\odot}$ star is calculated in detail. The result is that about 20% mass loss is enough to penetrate into the evolved structure and prevent blue loops. Thus Cepheids have masses which are only slightly less than no-mass-loss evolution tracks indicate.

Carson and Stothers (1976, Astrophys. J. 204, p.461) have done the evolution tracks both with the Los Alamos and the Carson opacities. This is the only case where the opacity effects are clearly separated from abundance (such as Z) changes. The main difference between the Los Alamos and Carson opacities is that there is an opacity increase and then a decrease at temperatures above about 3×10^5 K. This bump is clearly associated with the CNO abundance because it is larger with larger Z. Thomas-Fermi potential calculations at Los Alamos which have attempted to exactly reproduce the Carson results do not find this bump even for pure carbon. The cause of the CNO bump is unknown and no details of the calculations are yet published or even available.

It should be noted that there is one other region in the ρ, T-plane where the Carson opacities differ from those computed at Los Alamos. The Carson opacity has a bump at about 40,000 K and $\rho = 10^{-6}$ to 10^{-8} g/cm^3. This small feature is not relevant for stellar evolution tracks, but it does make the pulsation instability more strong for Cepheids and puts them at a pulsation amplitude somewhat larger than observed (Vemury and Stothers, 1978, Astrophys. J. 225, p.939). Also, in the blue edge of the region for fundamental mode pulsation is bluer than for Los Alamos opacities. Both the CNO and the He bumps are surprising because they occur where the ionization of these elements is almost complete. The one remaining, bound electron is surely a hydrogen-like system which the Los Alamos program is well suited to compute.

Evolution tracks given by Carson and Stothers are for 5 and 7 $M_{\odot}$ with X = 0.73 and Z = 0.02 using both sets of opacities. They cover the evolution from the main sequence to the end of core helium burning. The Carson opacity models are fainter and bluer in the main sequence region, but they are only fainter along the common Hayashi line. Robertson in 1972 using even larger earlier Carson opacities got even lower luminosities. The Cepheid mass-luminosity relation is $L \sim M^{3.3}$ instead of the $L \sim M^{3.8}$ of Becker, Iben and Tuggle for Y = 0.25 and Z = 0.02 in their fit formula.The

Carson and Stothers blue loops occur at an earlier age because the opacities at the bottom of the red giant stage convection zone near 10^6 are larger making a deeper convection zone which promotes blue loops. This difference in evolution rate has been used with counts of blue and red giants in clusters to see if the relative lifetime for these two evolution stages can indicate the proper opacities. The theoretical evolution data is not accurate enough for the test, however.

4. High Mass Population I Composition Tracks

The tracks by various authors for models with mass above 10 $M_\odot$ are even more varied than for those of the lower masses. The problems in addition to composition and opacity differences are semiconvection and mass loss. Stothers and Chin (1976, Astrophys. J. 204, p.472) discuss the evolution tracks for these higher mass stars using the Schwarzschild rather than the Ledoux criterion for semiconvection in deep inhomogeneous composition layers. This work is based on previous studies by the two authors where no mixing and the Ledoux criterion was used. The Z values are 0.021 and 0.044 and the X values are 0.739 and 0.602. As usual, the lower Z and lower X gives higher luminosity when crossing the Hertzsprung gap. No opacity variations are included in this series of papers.

Opacity effects using Carson opacities have been studied by Stothers (1977, Astrophys. J. 210, p.434) for masses 10.9 and 15 $M_\odot$ and by Stothers and Chin (1977, Astrophys. J. 211, p.189) for 7 - 60 $M_\odot$. In the first paper, tracks just through the core hydrogen burning to helium ignition were calculated to try to match the β Cepheid variables. The larger Carson opacities give redder tracks (as does the track at a higher Z of 0.044) than those with Los Alamos opacities. Thus the β Cep variables may be in the core hydrogen burning stage instead of at the later stages (where the evolution is briefly blueward at core hydrogen exhaustion, or even after helium ignition) as deduced from using the Los Alamos opacities.

More complete evolution tracks at higher masses for the Carson opacities show that these larger opacities in the envelope make the models much larger and redder than those using Los Alamos opacities. The criterion for semiconvection does not matter in these models. Core helium burning does not start until the models enter the red super-giant region. While the existence of supergiants all across the Hertzsprung gap is explained with these new opacities, there does seem to be dis-agreement between observation and theory on the hottest T_e values. These evolution tracks give a maximum T_e of 39,000 K whereas the Los Alamos opacities give T_e up to 50,000 K.

Stothers and Chin (1977, Astrophys. J. 216, p.61) have considered the parallel problems of the evolution using Carson opacities of pure helium stars with masses between 4 and 15 $M_\odot$. Again the main sequence band is much wider and cooler above log $L/L_\odot$ = 4 than with the smaller Los Alamos opacities. The evolution for Z = 0.02 and 0.04 has been followed to the end of core carbon burning. In these models, the highest T_e at lowest mass approaches 10^5K, but it has a slope of lower T_e at higher L, different from the 1971 Paczyński slope of higher L higher T_e using Los Alamos opacities.

Further studies by Stothers and Chin (1978, Astrophys. J. 226, p.231) using the Carson opacities have been made with extensive mass loss. The recent studies of others (Sreenivasan and Wilson, 1978, Astrophys. Space Sci. 53, p.193; and Chiosi, Nasi, and Sreenivasan, 1978, Astron. Astrophys. 63, p.103) have been able to explain the appearance of many supergiants all across the Hertzsprung gap with mass loss, but the width of the main sequence still seems to require higher envelope opacities such as those of Carson. Also, at 30 and 60 $M_\odot$ both with and without mass loss, the models are red supergiants even before the core hydrogen is exhausted. Since the mass loss rates are very uncertain; and in any event the tracks do not seem very dependent on the opacity, we will not discuss these results.

The population I tracks of Lamb, Iben, and Howard (1976, Astrophys. J. 207, p.209) for 15 and 25 $M_\odot$ have been discussed in the light of semiconvection theory by Chiosi and Nasi (1978, Astrophys. Space Sci. 56, p.431). They find for 20 $M_\odot$ that the neglect of semiconvection makes the track everywhere bluer during the core helium burning stages because the different H profile outside of the H burning shell.

5. Population II Tracks

There is a very large number of papers on low mass population II composition evolution. In these cases the opacity and even the Z value are of importance. Demarque and McClure (1977, Astrophys. J. 213, p.716) show that color magnitude diagrams for 47 Tuc and NGC 2420 imply either helium or Z difference. Post horizontal branch evolution and the appearance of BL Her and W Vir variables with Z = 0.001 is presented by Gingold (1976, Astrophys. J. 204, p.116).

Taam, Kraft, and Suntzeff (1976, Astrophys. J. 201, p.204) have also evolved models from the zero age horizontal branch to see what happens for metal rich models (Z = 0.02). These are the metal rich RR Lyrae variables which comprise about 25% of the field stars but are rarely seen in even the metal rich globular clusters. Evolution times for these metal rich models are longer than for the Z = 0.001 models corresponding to the observed slow period changes for them. These metal rich models must have a mass within a few percent of 0.55 $M_\odot$ to enter the RR Lyrae variable instability strip at all, however. Changes of Z as well as X and Y (which were 0.7 and 0.28) to population II compositions would give evolution more like the normal horizontal branch evolution where a large range (like 0.2 $M_\odot$) can populate this H-R diagram region.

Higher Z (= 0.01)models have been studied by Gingold (1977, Monthly Notices Roy. Astron. Soc. 178, p.533) and there is a mass range of $\pm 10\%$ around 0.55 $M_\odot$ where RR Lyrae stars can form. The discussion centers around how much mass loss must occur if the 0.8 - 0.9 $M_\odot$ red giants which evolve to red giants in globular cluster lifetimes are the progenitors of the horizontal branch stars and RR Lyrae variables. At Z = 0.01 or 0.02 errors due to the opacities have not been considered yet.

A series of evolution tracks has been calculated for six values of Z (10^{-5}, 10^{-4}, 4×10^{-4}, 10^{-3}, 4×10^{-3}, and 10^{-2}) and four values of Y (0.1, 0.2, 0.3, 0.4) by Caloi, Castellani, and Tornambe (1978, Astron. Astrophys. Suppl. 33, p.169). Larger Y gives a larger luminosity in the horizontal branch region with masses like 0.6 $M_\odot$, and lower Z, as usual, gives bluer colors. To whatever extent one can change Z and the opacity, any errors in the opacity may be interpreted using these extensive results including core He burning semiconvection.

An even more comprehensive set of evolution tracks has been given by Sweigart and Gross (1976, Astrophys. J. Suppl. 32, p.367; 1978, Astrophys. J. Suppl. 36, p.405) for late evolution, and Mengel, Sweigart, Demarque, and Gross (1978, Astrophys. J. Suppl., in press) for early evolution.

Castellani and Tornambe (1977, Astron. Astrophys. 61, p.427) have discussed two horizontal branch problems, the relative CNO abundances and the variations in the branch location caused by opacities. They show that at Z = $10^{-4}$to 10^{-3} the opacities are affected at temperature between 10^5 and 10^7 by the CNO elements. Below 8000 K the other elements in Z give electrons to affect the H^- abundance and its absorption. Increasing the Z from 10^{-4} to 10^{-3} moves a model of 0.625 $M_\odot$with 0.5 $M_\odot$in the helium burning core along the zero age horizontal branch, the move consists of a luminosity increase and a T_e decrease due to the nuclear burning and an almost purely T_e decrease of $\Delta \log T_e$ of 0.15 due to the opacity increase. At these low Z values, the distribution of masses along the zero age horizontal branch is critically dependent on the Z and the Z_{CNO}. Errors in the hydrogen shell burning which depend on the

Z_{CNO}, and the opacity can greatly influence the interpretation of globular cluster color magnitude diagrams.

Wagner (1978, Astron. Astrophys. 62, p.9) has discussed the evolution of very metal poor stars with $Z = 10^{-4}$ for masses 0.55, 3.25, and 4.0 $M_\odot$. Among the several conclusions is the fact that these stars evolve very rapidly and might be able to increase abruptly the Z value in our galaxy in its early life.

Two recent studies have been made to investigate the helium content of RR Lyrae variables. Caputo, Castellani, and Tornambe (1978, Astron. Astrophys. 67, p.107) have used tracks from previous publications and have considered how horizontal branch evolution proceeds to either the red or blue. The Oosterhoff types of RR Lyrae period distributions have been interpretated by van Albada and Baker as: I. initial zero horizontal branch blueward evolution starting red enough to have pulsation in the fundamental mode up to almost the fundamental blue edge; and II. blueward evolution starting even bluer so that the overtone occurs until the ultimate redward evolution moves the variable to a transition line beyond which only the fundamental can occur. The immediate pre-horizontal branch evolution is redward as O^{16} burning reaches equilibrium. Depending on the pre-horizontal branch evolution either Oosterhoff group can occur. Using the post horizontal branch, tracks, and pulsation period data the authors derive a Y of 0.22 - 0.25 and a mean mass of 0.72 $M_\odot$.

Caputo, Castellani, and Wood (1978, Monthly Notices Roy. Astron. Soc. 184,p.377) have continued the horizontal branch calculations with $Z = 10^{-3}$. Using the ratio of observed horizontal branch stars to those on the first giant stage branch and the ratio of the later asymptotic giant branch stars to the number on the first giant branch, they derive Y values of 0.18 to 0.26 for 5 globular clusters. They show that indeed semiconvection on the horizontal branch is required. In both these studies no variation of Z or opacity is discussed even though there could be a dependence of the derived Y on the Z value.

5. EFFECTS OF MASS LOSS ON STELLAR EVOLUTION (D. S. P. Dearborn and D. N. Schramm)

1. Introduction

Mass loss has been observed to occur in stars of both early and late type. Ultraviolet observations beginning with Morton (1967a) and leading to such surveys as that of Snow and Morton (1976) indicate that mass loss is a general property of early type stars. Analysis of both Visible and UV data (Hutchings 1976 and references therein) show mass loss rates ranging from 10^{-7} to 10^{-5} $M_\odot$/yr. More recently, infrared observations by Barlow and Cohen (1977) of free emission coupled with an assumed velocity law have confirmed such rates for O-B stars.

Red giants and supergiants have also been observed to be losing mass. Observations of infrared excess and resonance line profiles indicate mass loss rates ranging from 10^{-7} to 10^{-4} $M_\odot$/yr (Reimers 1975, Sanner 1976, Bernat 1977 and Hagen 1978). These observations have stimulated a resurgence of interest in studying the effects of mass loss on the evolution of stars. Recent calculations on massive stars (Chiosi and Nasi 1974; DeLoore, DeGreve, and Lamers 1977; Dearborn, Blake, Hainebach, and Schramm, 1978; Stothers and Chin 1978; Czerny 1978; Dearborn and Blake 1978; Chiosi, Nasi and Streenivasan 1978) though varying somewhat in approach have achieved similar results in mapping the effects of mass loss.

In early type stars, when the mass loss time scale ($\tau_m = |M/\dot{M}|$ is much greater than the main sequence time scale, the effects on the star are relatively minor. Low rates of mass loss will cause a star to be less luminous than a mass conserving star, but overluminous for its actual mass. Higher rates of mass loss can affect the surface composition, nucleosynthetic yield, and post-main sequence evolution of a star.

Mass loss from red giants seems necessary to understand the appearance of blue horizontal branches in globular clusters (Rood 1973) and is important in determining which stars evolve to white dwarf instead of ending in a supernova. Mass loss in evolved stars may also be necessary to understand the composition anomalies sometimes observed, and the slow rotation (low angular momentum) seen in stellar remnants.

2. Mass loss mechanisms

A comprehensive theory of the mass loss mechanism in stars should predict the temperature and velocity distribution of the ejected material as well as the mass loss rate. It should also account for the observed spectral features in all wavelength regions, and explain the apparent variability observed in some stars. The mass loss mechanisms are probably different for early and late type stars, and so the theories must be discussed separately.

a) Mass loss in early type stars

No theory to date is sufficiently complete to predict all observed features in early type stars but certain points are generally agreed upon (Hearn 1978). The large terminal velocity observed in the ejected material ($V > 1000$ km s^{-1}) is probably due to radiation pressure. Also, the existence of OVI in the ejected material is not predicted by the simplest form of the radiation pressure theory and seems to require the existence of a corona. There is no consensus however on whether radiation pressure, a corona, or some unspecified mechanism initiates the mass loss.

The radiation pressure of Castor, Abbott, and Klein (1975) is the most successful theory in terms of being able to describe the expected mass loss for a star of specified mass luminosity temperature and composition. In this theory, mass loss is driven by radiation pressure on ultraviolet absorption lines of carbon, nitrogen, and oxygen. Not only is a mass loss rate predicted, but line profiles (Klein and Castor 1978) and terminal velocities (Abbott 1978).

Models using a corona, of "Warm Wind" to initiate mass loss (Thomas 1973, Cannon and Thomas 1975, Rogerson and Lamers 1975) require mechanical energy deposition in the atmosphere. Hearn (1972, 1975) proposed a mechanism in which density waves are produced by the strong radiation field in early type stars. The efficiency of this mechanism is, however, uncertain (Hearn 1975) and without constraints on the energy input, the magnitude of the resulting mass loss is unpredictable.

The existence of OVI in the outflowing material seems to indicate the existence of a corona (or at least a chromosphere, Lamers and Snow 1978). Castor, Abbot, and Klein (1976), however, point out that x-ray observations constrain this warm region to temperatures of less than 300,000K. This rules out a hot corona, but not the warm wind model of Rogerson and Lamers (1975).

The radiation pressure model of Castor et al does not exclude the existence of a corona, but their assumption of radiative energy balance does not produce one. By relaxing this assumption, a small corona may be produced and the OVI observations explained. According to Castor (1978) a small corona can exist without significantly disturbing the radiation pressure model.

Finally, all of these models are steady state theories, and so do not predict any fluctuation in the mass loss rate. Observations of Hα (Hammerschlag 1978) show variation on short time scales, possibly indicating fluctuations in the mass loss rate. Conti (1978) has pointed out that these line strength fluctuation may be caused by small variations in the mass outflow; however, Thomas (1978) argues that mass loss varies dramatically due to small scale structure under the photosphere.

b) Mass Loss in late type stars

The primary theories proposed for mass loss in red giants and supergiants are coronal models and radiation pressure on dust grains. A coronal model was developed by Fusi-Pecciand Renzini (1975) in which accoustic energy produced in the convection is deposited in circumstellar material producing a corona and mass loss. The mechanism was scaled to the solar mass loss rate, and was successful in explaining the mass loss needed to produce a blue horizontal branch. As the accoustic energy available is sensitive to the luminosity, mass loss rates increase rapidly with brightness. Coronal models may then drive sufficient mass loss to explain the observed rates.

A mass loss model should, however, be able to reproduce the observed spectral features in mass losing stars. As shown by Weymann (1963) the coronal model has difficulty. In order to avoid too much hydrogen emission, it is necessary to have a low density through the temperature transition region. It was then difficult to avoid velocities in excess of the observed terminal velocity ($v_T \sim 10$ km s^{-1}).

Radiation pressure models generally assume the radiation field acts on circumstellar dust (Kwok 1975, Goldreich and Scoville 1976). As discussed by Weymann (1978), the dust model also has difficulty in reproducing observed spectral features. The Ca, H and K emission is blue shifted suggesting that mass loss is already under way in the chromosphere. The temperature there should be too hot for grains to exist. The radiation pressure mechanism may however be rescued from this dilemma if the grain forming region is supported above the chromosphere by turbulent pressure. In this cooler circumstellar environment, grains may form and be accelerated outwards. Outflow of material to replace the accelerated gas may then cause the observed velocity at the chromosphere.

3. Evolution with mass loss

Mass loss has been parameterized in a number of ways, but most commonly:

$\dot{M}=\theta NL/C^2$ (DeLoore et al 1977)

or

$\dot{M}=\mu LR/M$ (McCray 1962).

The results obtained with these different formulations have, however, been quite similar, indicating the results are not sensitive to this choice.

As mentioned in the introduction, low rates of mass loss lead to relatively minor effects. In O or B type stars, low rates of mass loss will cause a star to be less luminous than it would have been without mass loss, but it will be overluminous for its lowered mass. The amount of mass contained within the convective core decreases at the same rate as the luminosity and so the main sequence lifetime is unaffected (i.e., low rates of mass loss will not extend the main sequence lifetime of a star). Since a slightly smaller core develops, the post main sequence evolution of a mass losing star is increased.

As the mass loss rate increases ($\tau_{\dot{m}}$ decreases), the general form of the mass loss becomes significant. Dearborn and Blake define a critical rate of mass loss at which the main sequence lifetime of a mass losing star begins to rapidly increase. Stars which begin losing mass at a rate higher than this, <u>must</u> reduce their mass loss rate, or become totally depleted.

When the time averaged mass loss rate nears the critical rate, the resulting evolution can depend on the parameterization of the mass loss rate. Different results will be obtained for an increasing mass loss rate when compared to a constant one. A star initially losing mass at a low rate will develop a large core in spite of a high mass loss rate during subsequent evolution. A constant or decreasing mass loss (as is predicted in stars losing mass at or above the critical rate) results in a smaller core on affecting the post main sequence phases of evolution, and the nucleosynthetic yield.

As shown by Dearborn and Eggleton (1977), losing 1/3 of the initial mass in stars of less than 30 $M_\odot$, or 8 to 10 $M_\odot$ in more massive stars results in ^{14}N enhancement of 3 to 5 times. Stars losing mass at a rate below the critical rate may show such ^{14}N enhancement at the end of the main sequence, or during post-main sequence evolution. Mass loss at the critical rate exposes the ^{14}N rich material early in the main sequence lifetime, and may expose deeper region in which ^{16}O has been converted to ^{14}N causing an enhancement of $\sim$10 times. Quantitative observations of single OBN stars can potentially give valuable information on how much mass has been removed. The existence of a single OBN star with ^{14}N enhancement of $\sim$10x would strongly indicate the occurence of mass loss at the critical rate.

Mass loss at the critical rate can remove the hydrogen rich envelope and expose the core. This would result in a hydrogen deficient (nitrogen enriched) star (DeLoore et al 1978), or helium star (Dearborn and Blake 1978). These models could be identified as WN stars.

Continued mass loss at rates $\dot{M}>10^{-5}$ $M_\odot$/yr (Barlow 1978) will affect the evolution of helium stars; in fact, low mass helium cores $M \lesssim 8M_\odot$ could not maintain such mass loss rates over their entire helium burning lifetime. They would necessarily evolve to lower luminosity objects with reduced mass outflows. This suggests the existence of a more quiescent low luminosity form of helium star.

Higher mass helium stars can survive mass loss rates of $\dot{M} >10^{-5}$ $M_\odot$/yr, but material once contained within the helium burning convective core will be exposed. This material will be rich in carbon and possibly oxygen. Also, as deeper regions are exposed, one sees material in which ^{14}N has been converted to ^{18}O and ^{22}Ne. Continued mass loss from helium cores then naturally results in a WN-WC sequence. Mass loss at rates below the critical rate will result in a star which attempts to become a red giant. Mass transfer in a binary system (Van beeveran and DeGreve 1978), or a very large mass loss rate ($\sim 10^{-4} M_\odot$/yr) may still allow the star to become a stripped core (or Wolf Rayet star). As expected however, the cores produced in this manner are larger than those resulting from mass loss at the critical rate.

Due to reduced core size (as compared to mass conserving stars) and the elimination of semiconvection (Falk 1978) in mass-losing stars, the C/O yield of helium burning should be increased (Dearborn 1978). After helium is depleted the C-O core contracts until temperatures are suitable for carbon burning. The post carbon evolution is so quick that mass loss should not have any further effect. The constant mass models of Arnett (1972) should adequately represent the subsequent evolution. The ratio of core mass to zero age main sequence mass will, however, be affected.

Mass loss during later stages of evolution is important to determine which stars (in the 1.4 to 8 $M_\odot$ range) lose sufficient mass to become white dwarfs. An observational search for white dwarfs in clusters (Romanishen and Angel 1978) indicates that stars of 5 $M_\odot$ and possible 7 $M_\odot$ regularly lose sufficient mass to become white dwarfs. In more massive stars, mass loss in the red giant can solve the discrepancy between pulsational mass and evolutionary mass (Cox 1978), and as mentioned above, may aid in removing the remainder of the envelope to produce a Wolf Rayet Star. Unfortunately, our understanding of the mass loss mechanism in red giants is inadequate to accurately understand its effect. Still, studies (Scalo 1978) have been made indicating the observed mass loss rates are sufficient to cause the effects stated above.

4. Summary

Observed mass loss rates ($>10^{-6}$ $M_\odot$/yr) are sufficient to affect the evolution of stars, and if continued over the main sequence life-time of an O-B star may produce single Wolf Rayet Stars. While many of the effects of mass loss have now been explored, there are still many unsolved questions.

The basic mechanism by which mass is ejected in both early and late type stars is still subject to dispute. Further observations to accurately determine mass loss rates, as well as temperature and velocity profiles in the ejected material are needed to help resolve this. Quantitative observations of ^{12}C and ^{14}N abundances in early type single stars will also be useful in determining the total amount of mass which is lost. Determining CNO abundances and isotope ratios in evolved stars may also give information on the total mass lost (Dearborn, Schramm, and Eggleton 1976).

More calculations need to be made on the effects of mass loss in binary systems (particularly those systems which become x-ray binaries) and mass loss in red supergiants. Studying the effects of mass loss in red giants is, however, difficult due to a lack of understanding the mass loss mechanism involved.

The phenomenon of mass loss is important for understanding the appearance of some stars, and their evolution. Massive stars are responsible for most of the heavy element production, mass loss can have important effects on their yield, and so affect chemical evolution of the galaxy. The interaction of the ejected material with the interstellar medium may also have interesting effects. While much has been learned on the effects of mass loss on evolution, there are more questions to answer.

References

Abbott, D: 1978, Astrophys. J. 224, p893.
Arnett, D: 1972, Explosive Nucleosynthesis, eds. D. N. Schramm and W. D. Arnett.
Barlow, M., and Cohen, M: 1977, Astrophys. J. 213, p737.
Barlow, M: 1978, IAU Symposium 83, in press.
Bernat, A: 1977, Astrophys. J. 213, p756.
Cannon, C., and Thomas, R: 1977, Astrophys. J. 211, p910.
Castinnelli, J: 1978, IAU Symposium 83, in press.
Castor, J., Abbott, D., and Klein, R: 1975, Astrophys. J. 195, p157.
Castor, J., Abbott, D., and Klein, R: 1976 Physique des mouvements dans les atmospheres stellaires.
Castor, J: 1978, IAU Symposium 83, in press.
Chiosi, C., and Nasi, E: 1974, Astron. Astrophys. 34, p355.
Chiosi, C., Nasi, E., and Sreenivasan, S: 1978, Astron. Astrophys. 63, p103.
Conti, P: 1978, Ann. Rev. Astron. Astrophys. 16, p371.
Cox, A: 1978, Sky and Telescope 55, p115.
Czerny, M: 1978, Acta. Astron. in press.
Dearborn, D., Schramm, D. N., and Eggleton, P: 1976, Astrophys. J. 203, p455.
Dearborn, D., and Eggleton, P: 1977, Astrophys. J. 213, p448.
Dearborn, D., Blake, J. B., Hainebach, K., and Schramm, D. N: 1978, Astrophys. J. 223, p552.
Dearborn, D., and Blake, J. B: 1978, Astrophys. J. in press.
Dearborn, D.: 1978,Acta. Astron. in press.
Dearborn, D., and Blake, J. B: 1978, IAU Symposium 83, in press.
DeLoore, C., DeGreve, J., and Lamers, M: 1977, Astron. Astrophys. 61, p251.
DeLoore, C., DeGreve, J. P., and Vanbeveren, D: 1978, Astron. Astrophys. 67, p373.
Falk, S: 1978, IAU Symposium 83, in press.
Fusi-Pecci, F., and Renzini, A: 1975, Astron. Astrophys. 39, p413.
Goldreich, P., and Scoville, N: 1976, Astrophys. J. 205, p144.
Hagen, W: 1978, Astrophys. J. Letters 222, L37.
Hammerschlag, G: 1978, Astron. Astrophys. 64, p399.
Hearn, A: 1972, Astron. Astrophys. 19, p417.
Hearn, A: 1973, Astron. Astrophys. 23, p97.
Hearn, A: 1975, Astron. Astrophys. 40, p355.
Hearn, A: 1978, IAU Symposium 83, in press.
Hutchings, J. B: 1976, Astrophys. J. 203, p438.

Kwok, S: 1975, Astrophys. J. 198, p585.
Lamers, H., and Snow T: 1978, Astrophys. J. 219, p504.
McCray, R: 1962, Quart. IRAS, 3 p63.
Morton, D: 1967a, Astrophys. J. 147, p1017.
Morton, D: 1967b, Astrophys. J. 150, p535.
Reimers, P: 1975, Problems in Stellar Atmospheres and Envelopes, eds. B. Baschek, W. H. Kegel and G. Traving.
Romanishen, W., and Angel, R: 1978, preprint.
Rood, R: 1973, Astrophys. J. 184, p815.
Rogerson, J., and Lamers, M: 1975, Nature 256, p190.
Sanner, F: 1976, Astrophys. J. Letters 204, L41.
Scalo, J: 1976, Astrophys. J. 206.
Snow, T., and Morton, D: 1976, Astrophys. J. Suppl. 32, p429.
Stothers, R., and Chin, C: 1978, Astrophys. J. in press.
Thomas, R: 1973, Astron. Astrophys. 27, p297.
Thomas, R: 1978, IAU Symposium 83, in press.
Vanbeveren, D., and DeGreve, J. P: 1978, Astron. Astrophys, submitted.
Weymann, R: 1963, Ann. Rev. Astron. Astrophys. 1, p97.

B. PACZYNSKI
President of the Commission

R. J. TAYLER
Vice-President of the Commission

36. THEORY OF STELLAR ATMOSPHERES (THÉORIE DES ATMOSPHÈRES STELLAIRES)

PRESIDENT: D. Mihalas
VICE PRESIDENT: G. Traving
ORGANIZING COMMITTEE: A.G. Hearn, J.T. Jefferies, K. Kodaira, L.V. Kuhi, J.M. Marlborough, B.E.J. Pagel, F. Praderie, A. Sapar, M.S. Vardya

I. INTRODUCTION, COMMISSION ACTIVITY, SYMPOSIA AND COLLOQUIA

Commission 36 has cosponsored the following colloquia and symposia: (1) IAU Colloquium No. 51, "Convection and Turbulence in Stellar Atmospheres", London, Ontario (27-30 August 1979); (2) IAU Symposium No. 80 "The HR Diagram, the 100th Anniversary of Henry Norris Russell", Washington DC (2-5 November 1977); (3) IAU Symposium No. 83 "Mass-Loss and Evolution of O-Type Stars", Qualicum Beach, Vancouver Island (5-9 June 1978).

In connection with the XVIIth General Assembly in Montreal, the commission participated in the organization of a Joint Discussion with Commissions 10, 12, 29, 35, and 44 on the topic of "Physics of the Chromosphere-Corona-Wind Complex and Mass-Loss in Stellar Atmospheres", and Joint Meetings with Commissions 29 and 45 on the topics "Stellar Abundances" and "Stellar Rotation".

Owing to limitations of space, this report will not give a comprehensive discussion of all aspects of the field, nor an exhaustive bibliography. Rather, it focuses on a few important and active areas in which significant progress has been made recently. The author of each section of the report is indicated explicitly; references are collected at the end of the report, arranged by sections.

II. RECENT PROGRESS IN THE THEORY OF STELLAR ATMOSPHERES

A. Winds from Early-Type Stars (J.P. Cassinelli)

1. Introduction

Mass-loss from O-stars, OB supergiants, Wolf-Rayet stars and central stars of planetary nebulae is revealed by broad P-Cygni lines from light ions in relatively low ion stages such as C^{+3}. These flows are characterized by large mass-loss rates (10^{-8} to 10^{-5} $M_{\odot}$/yr), and large flow speeds (600 to 3500 km/sec). The observations and the evolutionary consequences of large mass-loss rates are summarized by Conti (1978), and the basic equations of wind theory are presented by Mihalas (1978).

Infrared-continuum and ultraviolet-line surveys have been completed and provide information on the mass-loss rates, terminal velocities, and ionization-structures of the winds, but the temperature structure is still uncertain.

1) Barlow and Cohen (1976) derived mass-loss rates, $\dot{M}$, from their infrared continuum photometry. They found that $\dot{M}$ is nearly proportional to the stellar luminosity ($\dot{M} \propto L^{1.1}$) and is only weakly dependent on the effective temperature of OB supergiants and *Of* stars.

2) The catalogue of Copernicus satellite spectra by Snow and Jenkins (1977) provides a rich source of information on the wind structures. Abbott (1978a) has derived an empirical relation between terminal velocities and photospheric escape speeds: $v_{\infty} = 3\ v_{esc}$.

3) The Copernicus spectra show strong lines of anomalously high stages of ionization such as O VI in O stars, and Si IV in late B supergiants. Summaries of the anomalies, along with two different explanations, are given by Lamers and Snow (1978) and Cassinelli and Olson (1979). The discovery of these anomalies has revived widespread interest in chromospheric and coronal models for the winds, as had been proposed by Cannon and Thomas (1977) and Hearn (1975).

2. Theoretical Work

The most complete theoretical model for the winds of hot stars is that of Castor, Abbott and Klein (1975, 1976), hereinafter CAK. It is successful in explaining the observed mass-loss rates and terminal velocities. However, CAK did not address the problem of mechanical energy deposition and the accompanying production of high stages of ionization such as O VI. In the last three years two especially active areas of theoretical research have emerged: 1) Castor, Abbott and Klein have refined their original model to provide rebuttal to several early criticisms of it. 2) Primarily to explain the observed ionization anomalies, several authors have derived semi-empirical models of winds, with various assumptions concerning the temperature structure.

Line-Driven Wind Theory

In the radiation-pressure model of Lucy and Solomon (1970), as extended by CAK, the envelope is accelerated by transfer of momentum from the radiation field to particular ions by line-absorption in the ultraviolet. The observed high mass-loss rates require a large number of absorbing lines. Abbott (1977, 1978b) compiled extensive lists of lines and found that there are enough strong lines available to account for the observed mass-loss rates. Furthermore, he delimited the region in the HR diagram in which the acceleration due to line opacity in a static atmosphere exceeds gravity at large-enough optical depth to guarantee that the atmosphere must expand. This is a region in which the winds can be "initiated" solely by the radiation mechanism, and it corresponds well to the zone in the HR diagram where outflows are actually observed. While it now seems clear that winds can be produced entirely radiatively, whether they actually are or not is still under debate. Thomas (1978) criticizes the fully radiatively-driven wind models for assuming laminar, steady-state flow, and contends that the mass-loss results from subphotospheric motions in accordance with the thermodynamics of open systems.

Improvements in the predicted velocity-structure have been developed by Castor (1978). Accounting for angular redistribution of multiply-scattered line radiation, and for the velocity dependence of the ionization balance, he derived a slower velocity-rise of the flow. This agrees better with the velocity-law inferred from the infrared continuum and ultraviolet line profiles (Lamers and Morton, 1976). Klein and Castor (1978) defended the assumption that the winds have temperatures not very different from stellar effective temperatures on the basis of theoretical fits to observed Hα and He I emission lines.

The CAK theory can explain the empirical results that $\dot{M} \propto L^{1.1}$, that $v_\infty = 3\ v_{esc}$, and that the winds of supergiants are nearly independent of T_{eff} (Abbott, 1977). These successes suggest that a major part of the mass-loss mechanism has been accounted for. However, the presence of O VI in the observed spectra, and evidence of variability of the stellar winds (York, et al. 1977, Stalio and Upson 1978) indicates that the picture is far from complete.

Semi-Empirical Models

It is now possible to derive rather good estimates of the fractional abundances, and even spatial variations, of ions from the observed P-Cygni lines using the families of theoretical profiles that have been calculated by Olson (1978) and Castor and Lamers (1979). The relative abundance of the ions depends on the assumed electron temperature and the diffuse radiation field within the wind, and on the hard radiation-flux incident on it. From the observational data now available, it is not possible to derive temperatures unambiguously. Casinelli, Castor and Lamers (1978) discuss three quite different models that can explain the anomalous ionization.

a) Lamers and Morton (1976) and Lamers and Rogerson (1978) found that the overall ionization conditions in the winds of ζ Pup (O4*f*) and τ Sco (B0 V) are like that of any optically thin plasma at 2×10^5 K, and proposed a "warm wind" model.

b) Castor (1978) found that the ionization could be explained with a lower temperature, (6×10^4 K) if account is taken of the diffuse field in an optically-thick wind.

c) Cassinelli, Olson and Stalio (1978) and Cassinelli and Olson (1978) proposed a model in which the winds are cool (.8 T_{eff}), but are subjected to radiation from a small hot corona. The high ion stages are produced by the Auger mechanism following K-shell absorption of x-rays. This model explains the anomalous ionization in the O and B supergiants, and predicts that there should be a 2 kev flux that is large enough to be detected by the HEAO-B satellite, scheduled for launch in late 1978.

3. Directions for Future Work

On the observational side, better estimates of mass-loss rates ought to be forthcoming from VLA radio measurements. X-ray observations are crucial for determining the existence of coronae. Thus far only ζ Pup has been studied in great detail at all spectral wavelengths; other stars need this attention so as to avoid a distorted picture. Studies of variability, and of photospheric motions, are needed to elucidate the cause of the mechanical heating.

On the theoretical side, the semi-empirical models will need continual improvement as new data become available. Mechanisms for heating a coronal region, or a broader warm region, need to be investigated. Sufficient data are now available on the optically-thick winds of Wolf-Rayet stars (Hartmann, 1978a,b; Johnson 1978) that significant progress should be possible. A study of the effects of rotation on line-driven winds is needed to understand the flows from Oe and Be stars (Castor 1978). Much work is already underway on the effects of mass-loss on evolution (Conti 1978) and on the effects of winds on the interstellar medium (Weaver; Castor and McCray 1977).

B. Mass-Loss From Late-Type Stars (D. Reimers)

Within the last few years, our knowledge concerning circumstellar envelopes and mass-loss in red giants has increased rapidly, mainly as a result of new and better observations (new technologies) in a variety of wavelength regions. However, reliable mass-loss rates are still difficult to estimate, as the discrepant results obtained by different authors show (Sanner, Bernat, Reimers). In spite of the uncertain rates, various authors have included red-giant massloss in stellar evolution calculations (e.g. Fusi-Pecci and Renzini, Mengel, Wood and Cahn). There is still no convincing theory of mass-loss in red giants. Even the basic underlying mechanisms have not been identified with certainty.

1. Optical Observations

The P-Cygni character of the circumstellar (CS) lines in red giants has been investigated for the first time by means of high-resolution echelle spectrograms (Goldberg et al. 1975; Bernat and Lambert, 1975; Sanner, 1976) and by Fourier transform spectrograms. At high resolution, the profiles of strong lines such as the Na I and K I resonance lines show multiple components or weak satellite lines (Goldberg et al., Sanner). The amount of shell emission seems to be time-variable due to radial velocity variations of the underlying stellar Fraunhofer lines (Goldberg et al.).

Resonance-line re-emission from shells has been detected directly in α Ori and μ Cep (Bernat and Lambert, 1976; Münch et al. 1976), in particular in α Ori out to $\sim$30"=600 stellar radii-(Bernat et al. 1978). Two-dimensional images of the shell of α Ori have been constructed (Bernat et al. 1978; Lynds et al. 1977). There seem to be no gross departures from spherical symmetry on a scale an order of magnitude larger than the stellar radius.

In a study of the incidence of mass-loss in red giants, CS lines have been found in all stars cooler and more luminous than a line in the HR diagram defined by (spectral type K 5, M_V=0), (K 4, -1), (K 2, - 1.8), (G 5, -4), (G 0, -5). A simultaneous study of wind-velocities in the whole red giant region ($\sim$150 stars) revealed a loose correlation of the wind-velocity with the escape velocity at the stellar surface (Reimers 1977). In K giants like α Tau, variations of CS lines on a time-scale of months (minor changes within days) have been detected (Reimers, 1977; Kelch et al. 1978). It is not yet clear whether most of the variations indicate real changes of the flow, or changes in the degree of ionization.

Rates of mass-loss of M supergiants as determined from CS P-Cygni profiles are controversial (by a factor 100!) (Sanner, 1976; Bernat, 1977; Hagen, 1978). The main reason is that it is difficult to determine the location of the shells (inner shell radii). This difficulty can be avoided by use of CS lines visible in the spectra of close visual companions like α^2 Her and α Sco B (Reimers, 1977; Kudritzki and Reimers, 1978; van der Hucht et al. 1979). Mass-loss rates of globular-cluster stars have been derived by Cohen (1976) and Mallia and Pagel (1978).

2. Infrared Observations

Recent advances in detector technology have brought to light many new facts about circumstellar dust around highly-evolved stars (cf. review by Merrill, IAU Coll. 42). In particular systematic studies of objects in the 2-μ Sky Survey and the AFCRL/AGL Infrared Sky Survey have revealed a large number of highly-evolved objects with silicate-, Si C-, and C-dust emission (decreasing abundance-ratio O/C along the sequence M, MS, S, SC to C stars). In the most evolved M-type Miras, the 9.7-μ silicate emission even reverses to strong self-absorption (Merrill et al. 1976; Forrest, et. al. 1977). The most evolved objects are hidden in their own dust envelopes which completely absorb and re-emit the stellar flux. A number of objects have been detected first as Type II OH masers, and subsequently as infrared stars, and Mira stars with periods up to 750 days (Schultz et al. 1976).

Besides the earlier lunar-occulation technique, spatial interferometry has now been used to resolve CS dust envelopes on the sky (McCarthy, Low, Howell, 1977). In particular, the angular diameter of the silicate dust envelope of Mira itself has been detected to vary in phase with the star (McCarthy et al. 1978). The question whether dust condensation begins at the surface of the stars or only at distances of several stellar radii is not yet settled.

It has become clear that the featureless dust emission--except broad bands--does not give unambigious information about the amount of dust, the dust temperature, and the gas-to-dust ratio as long as grain opacities, condensation processes etc. are so poorly known (cf. Schmid-Burgk and Scholz, 1976). Mass-loss rates determined from dust-emission thus remain very uncertain.

3. Radio Observations

The maser lines of OH, H_2O and SiO in Mira stars and supergiants will not be discussed here (cf. review by Winnberg, IAU Coll. 42). In the present status of theory they do not yield reliable mass-loss rates.

Owing to the increased sensitivity of radio receivers at mm-wavelengths, thermal emission in lines of molecules such as CO, HCN, CS, SiO, SiS, HC_3N, C_2H_2 and CH_4 have been detected in infrared stars like IRC + 10216 (Mufson et al. 1975; Lo and Bechis 1976; Kuiper et al. 1976; Zuckerman et al. 1977). The lines are formed in cool expanding envelopes around highly evolved stars ('proto-planetary nebulae' Zuckerman 1978). Most of these are carbon-rich stars and S-type stars. These objects offer unique opportunity to obtain both fairly reliable mass-loss rates and molecular and isotopic abundances of the material ejected to the interstellar medium. The mass-loss rate of IRC + 10216 is about $2 \cdot 10^{-5}$ $M_\odot$/yr (Goldreich and Scoville 1976; Kwan and Hill 1977). The search for 21-cm HI emission in these stars has thus far been unsuccessful, probably because hydrogen is present mainly as H_2. There is a possible detection of the 21-cm line in α Ori (Zuckerman, 1978).

In the GO Ia supergiant HR 8752, radio emission has been measured between 2.7 and 90 GHz (Smolinski et al. 1977). In terms of a spherically expanding stellar wind, a mass of $0.5 \cdot 10^{-3}$ $M_\odot$ of ionized hydrogen is required. This is consistent with the [NII]-emission detected earlier, and with recent optical observations of a shell around the star (Lambert and Luck, 1978).

4. Theoretical Aspects

Accurate rates of mass-loss for various types of stars are of vital interest to stellar-evolution theory. Though we may be able to measure mass-loss rates with sufficient accuracy for a few stars by purely empirical means, it appears that the final aim, knowledge of mass-loss rates in all relevant evolutionary phases, cannot be achieved so long as the basic mass-loss mechanism(s) has not been identified and before the theory of stellar winds in late-type stars has been worked out.

For the difficulties of the various theoretical attempts [thermally (acoustically)-driven winds; wave-driven winds; radiatively-driven winds] we refer to the review by Weymann (IAU Coll. 42). Semi-theoretical estimates of mass-loss have been made by Renzini (1976) and Mullan (1978). However, the various assumptions made are questionable. One of the principal factors inhibiting progress in the theory of stellar winds for red giants is that the region where the mass-loss mechanism is operating is difficult to observe. In red giants, contrary to conditions in hot stars, the amount of mechanical energy required to produce the slow winds (a fraction of $\sim 10^{-5}$ of the stellar luminosity) is fixed by the potential energy of the escaping matter. This means that energy deposition is needed mainly quite near the star. The CS lines, however, are formed in the cool outer envelope ($r \gtrsim 5$ to 10 stellar radii). It thus appears that we cannot learn much about the mechanism from optical lines (and the same seems to apply to infrared and radio observations).

One may hope that the following observations--combined with appropriate diagnostic techniques--will help us to understand the unknown transition zone between chromospheres and CS envelopes: blue-shifted Hα emission and absorption components; FeII emission-lines and radio emission from stars like α Ori; observations of extended chromospheres and "prominence" features (satellite lines) in eclipsing binaries like ζ Aur and 31 Cyg; UV observations of M giants and supergiants (as indicated by preliminary emission-line data from IUE--cf. Commission 44). Observations of CS lines with high spectral resolution (multiple components), and of the marked time-variations of CS lines in K- and early-M giants, combined with observations of light-variations of the stars may also be helpful for identifying the mechanism.

Much remains to be done, both theoretically and observationally, before we understand mass-loss in late-type stars.

C. Stellar Chromospheres and Coronae (J. Linsky)

1. Introduction

Important advances have been made recently in this field due to the development of spectroscopic diagnostics and the growth of ultraviolet and X-ray data from space. These data lead to estimates of physical properties in chromospheres and coronae of individual stars, and to a first generation of models. In some cases crude estimates of radiative losses have been made; these pose tests of theoretical estimates of nonradiative heating rates.

2. Spectroscopic Diagnostics

The physical properties of astronomical plasmas are inferred from observed line and continuum profiles and intensities. Dupree (1978) has reviewed the status of ultraviolet and X-ray diagnostics for plasmas hotter than 5×10^4 K, and Praderie (1976) and Linsky (1977) have reviewed the literature on chromospheric diagnostics.

Basri *et al.* (1979) have shown that the incoherence fraction for redistribution in the wings of Lα is frequency dependent, but that it is feasible to derive chromospheric models using wing-intensities. Fluorescent processes that have been investigated recently include pumping of the O I resonance lines by Lβ (Haisch *et al.* 1977), of the H_2 Lyman bands by Lα (Jordan *et al.* 1978), of the fourth positive system of CO by C IV resonance lines (Bartoe *et al.* 1978) and Bowen fluorescence of O III (Raymond 1978; Margon and Cohen 1978). Stencel (1977) and Canfield and Stencel (1976) have discussed formation of emission lines in the wings of the Ca II H and K lines. Lites *et al.* (1978) analyzed the C II resonance lines, and Tripp *et al.* (1978) analyzed the Si II and Si III ultraviolet lines.

3. Theoretical Models and Empirical Tests

The goal of computing ab initio theoretical models of the outer atmospheres of stars including shock-wave or other nonradiative heating processes has not yet been attained. Cannon and Thomas (1977) suggest that an "imperfect wind tunnel" analogy, with an imposed outward velocity in the subatmosphere, may provide the proper physical framework for such calculations. Klein *et al.* (1978) have shown that low-amplitude upward-propagating waves can form shocks in A- and B-type stars, and have computed the observable effects of precusor radiation, accretion shocks, and conversion of mechanical to thermal energy.

For late-type stars, Renzini et al. (1977) and Ulmschneider et al. (1977) have computed acoustic energy-fluxes and chromospheric heating-rates. One test of this theory is the predicted mass column-density of the temperature-minimum for different stars, as reflected in the Ca II K_1-widths. Ulmschneider et al. (1978a,b) have computed the locations of temperature minima, but Cram and Ulmschneider (1978) point out inconsistencies between predictions and measured K_1 widths. A second test of the theory involves chromospheric radiative loss-rates. Linsky and Ayres (1978) and Linsky et al. (1978) point out the extrapolated total chromospheric loss-rates in lines appear to be independent of stellar gravity, whereas the computed heating-rates of Ulmschneider et al. (1977) show a large increase with decreasing gravity. Estimates by Praderie and Thomas (1976) of the mechanical energy input into the solar chromosphere have been criticized by Ulmschneider and Kalkofen (1978) and Kalkofen and Ulmschneider (1978). Indeed, even the basic hypothesis that solar-type chromospheres are heated by short-period acoustic waves has been questioned by Cram (1978).

Hearn (1975, 1977) has porposed a method for computing coronal and wind properties based on a minimum-energy principle. His equations have been modified by Haisch and Linsky (1976) and applied to dwarfs and giants by Mullan (1976a). This approach has been criticized by Vaiana and Rosner (1978), Antiochos and Underwood (1978), and Endler et al. (1978). Withbroe and Noyes (1977) have reviewed the mass-and energy-flow in the solar corona.

4. Modeling of Stellar Chromospheres

Models for chromospheres and upper photospheres have been constructed for F0-M0 dwarfs (Ayres et al. 1976, Kelch 1978, Kelch et al. 1979) and for G-K giants (Kelch et al. 1978). The analysis of the Ca II H and K line-profiles and Mg II h and k line-fluxes, may require revision if redistribution in the line wings has not been properly treated. Further, these are one-component atmospheres, an approximation just barely valid for the Sun and possibly grossly inadequate for stars like Arcturus (Heasley et al. 1978).

Bearing in mind these potential problems, the general trends found for the 18 stars now studied can be summarized as follows: (1) In most cases the temperature in the upper photosphere inferred from the K line-wings is hotter than that computed from radiative-equilibrium models, implying significant nonradiative heating. (2) With increasing nonradiative heating, as measured by the radiative loss-rate in chromospheric emission lines, the temperature minimum moves deeper into the atmosphere. (3) Active-chromosphere dwarf stars, as indicated by bright emission lines, have large chromospheric radiative-loss rates and steep chromospheric temperature-gradients. In particular, Kelch et al. (1979) have shown that Hα emission, which characterizes dMe stars, can be simply explained by the steep chromospheric temperature gradients implied by the bright K-line emission in these stars. (4) The mass column-density at the top of the chromosphere is a simple function of stellar gravity for dwarfs and giants.

Much work remains to be done to model the chromospheres of supergiants, RS CVn-type binaries, and metal-poor stars. Ultraviolet spectra from IUE yield fluxes for the Si II lines, and C I and Si I continuua which test the models based on fluxes of the Ca II, Mg II, Lα, C II-IV, Si II-IV, and N V lines, and help to extend these models into stellar transition regions. Active regions appear to play a major role in explaining the cyclic variability of the K-line in G-K dwarfs studied by Wilson (1976) and variability in dMe stars like BY Dra (Kunkel 1975). Since magnetic field-strength and K-line brightness are correlated in the Sun (Skumanich et al. 1975), strong magnetic fields are likely to be the cause of stellar plages and chromospheric activity. The fundamental question of how magnetic fields enhance nonradiative heating-rates to produce steeper chromospheric temperature-gradients and enhanced emission line-fluxes remains unanswered.

5. Location of Chromospheres and Coronae in the HR Diagram

Ca II K-line emission provides evidence for chromospheres in stars as early as γ Boo (A7 III). This is consistent with mechanical-flux calculations based on the

Lighthill-Proudman theory, but it is disconcerting that the K line ceases to be a useful chromospheric diagnostic at just this class of stars.

The identification of chromospheres and coronae in A-type and hotter stars requires ultraviolet and X-ray observations. Praderie et al. (1975) proposed a chromosphere or temperature inversion in Vega (A0 V) to explain excess emission shortward of Lα, but Snijders (1977) has shown that the observed *Copernicus* data are explained by non-LTE calculations of the C I continua without invoking a temperature inversion. The detection of soft X-ray emission from Sirius (Mewe et al. 1975) raised the question whether Sirius A (A1 V) or its white-dwarf companion is the emitter. Ultraviolet observations of Cash et al (1978a) and Brune et al (1979) appear to rule out photospheric emission from Sirius B as the source of the X-ray emission, hence the only plausible alternative is emission from a corona surrounding one of the stars (cf. Mullan 1976a; Hearn and Mewe 1976). The apparent detection of X-ray emission from Vega (Topka et al. 1977) provides further evidence for coronae in A-type stars. Cassinelli and Hartmann (1977) and others have proposed that O-type stars have coronae, and Kondo et al. (1976a) have detected Mg II emission in several B-type stars. The HEAO-B satellite should provide more complete information about coronae in hot stars.

New ultraviolet data from IUE provides powerful tools for studying chromosperes and transition regions in F-K stars. Linsky et al. (1978) have shown that: (1) the 1175-2000 Å spectra of cool stars contain emission lines indicative of chromospheres and transition regions (log T=4.3-5.4); (2) the RS CVn stars exhibit larger surface fluxes in transition lines than chromosphere lines, compared to the quiet Sun; and (3) during a large flare HR1099 shows chromospheric and transition-region emission-line fluxes comparable to those in solar flares. Prior to IUE, the outer atmosphere of Capella was studied by Dupree (1975), Vitz et al. (1976), and Haisch and Linsky (1976).

HEAO-1 soft X-ray spectra of the RS CVn stars Capella, HR 1099 and UX Ari (Cash et al. 1978b; Walter et al. 1978a,b) indicate 10^7 K coronal plasma, considerably hotter than the active solar corona. The RS CVn systems are also powerful nonthermal radio emitters (Owen et al. 1976; Owen and Spangler 1977). These systems are binaries in which one or both components are slightly evolved (Popper and Ulrich 1977), and may have a hot disk around the secondary, or dense solar-like chromospheres produced by tidal coupling (cf. Young and Koniges 1977; Naftalan and Drake 1977; Rhombs and Fix 1977).

In their survey of 22 late-type stars with IUE, Linsky and Haisch (1979) noted a sharp division of stars in two groups. The solar-type stars, i.e. F2-K4 dwarfs and G giants, show emission lines indicative of plasma at temperatures of 2.5×10^5 K and hotter. The α Orionis-type stars, i.e., the K giants and G-M supergiants, show emission lines from material no hotter than 10^4 K. It appears that either the α Orionis-type stars do not have transition regions and coronae, perhaps due to the onset of supersonic winds (Mullan 1978; Reimers 1977; Stencel 1978), or that their transition regions and coronae have emission measures far smaller than the Sun's.

Little is known about the outer atmospheres of normal M dwarfs, but dMe flare-stars exhibit bright chromospheric emission lines and both X-ray and radio emission during flares (e.g., Karpen et al. 1977; Haisch et al. 1977, 1978; Spangler and Moffett 1976; Robinson et al. 1976). Models for stellar flares have been proposed by Mullan (1976b, 1976c, 1977).

T Tauri stars exhibit extremely bright chromospheric emission lines, but coronae from these young objects have not yet been detected by X-ray or radio emission. Ulrich (1976, 1978) has reviewed the present status of models for these stars.

6. Width-Luminosity Relations

Correlations between emission-line widths and stellar luminosity exist for the Mg II and H I resonance-lines (Dupree 1976; Kondo et al. 1976b) as well as for Ca II and Hα. Lutz and Pagel (1977) have discussed the empirical dependence of Ca II widths on gravity, effective temperature, and metal abundance. Ayres (1979) has derived scaling relations for the K_1 and K_2 widths, giving their dependence on gravity and on chromospheric nonradiative heating rate; his K_1 scaling relation appears to be consistent with data for a wide range of stars (Kelch *et al.* 1979; Linsky, *et al.* 1979).

D. Model Atmospheres for Late-Type Stars (Bengt Gustafsson)

1. Contemporary models

Recent years have seen an increased interest in problems relating to late phases in stellar evolution, to nucleosynthesis, and to the composition and chemical evolution of galaxies. This has led to a greater demand for accurate methods of interpreting late-type stellar spectra. In this effort model atmospheres are necessary--they form the link between observations of stellar radiation and theories of stellar evolution.

Most recent models of late-type stellar atmospheres (Table 1) are purely "theoretical": they are constructed *ab initio* using the classical assumptions of plane-parallel stratification, hydrostatic equilibrium and LTE. The only non-radiative flux considered is convection, in the framework of the mixing-length "theory". A very important improvement has been the development of methods for handling blanketing by enormous numbers of atomic and molecular lines.

Table 1. Recent grids of blanketed model atmospheres for late-type stars

Author(s)	Ranges of parameters T_{eff}	log g	[A/H]	Method	Notes on sources of line absorption
BEGN[1,2]	3500-6000	0.0- 4.5	-3.0-+0.5	ODF	no TiO or H_2O
Johnson et al.(1978)	2500-4000	0.0- 2.0	0.0	OS	no H_2O
Kurucz (1978)[3]	5500-50000	0.0- 4.5	-2.0- 0.0	ODF	no molecules
Mould (1975,1976)	3000-4250	4.75,5.75	-2.0- 0.0	ODF	
Peytremann (1974)[4]	5000-8500	2.0 4.5	-1.0- 0.0	OS	no molecules
Querci et al. (1974) and Querci and Querci (1975)[5]	2600-4500	-1.0-+1.0	C star models	ODF	
Tsuji (1976,1978)	2200-4200	-2.0-+2.5	0.0	VAEBM	varying CNO

1) Gustafsson et al. (1975), Bell et al. (1976), Eriksson et al. (1978).
2) Colors and fluxes given and discussed by Bell and Gustafsson (1978), Gustafsson and Bell 1978) and Manduca *et al.* (1977).
3) Colors discussed by Relyea and Kurucz (1978) and Buser and Kurucz (1978).
4) Colors discussed by Peytremann (1975).
5) Fluxes discussed by Querci and Querci (1976).

A prerequisite for the computation of realistic blanketed models is an extensive compilation of spectral-line data. Thanks to the efforts of Kurucz and Peytremann (1975) and of Bell (see Bell and Gustafsson 1978) such compilations now exist and are in wide use for calculating both spectra and colors.

There are two different current methods for handling line-blanketing in the calculation of model atmospheres (cf. Table 1). These are the opacity (probability) distribution function (ODF) method, and the opacity-sampling (OS) method. The merits and drawbacks of each have been discussed by Carbon (1974), Gustafsson *et al.* (1975), Sneden *et al.* (1976) and others. An optimal scheme for computing blanketed model atmospheres has not yet been designed, especially for the upper layers where the LTE assumption fails. Methods for distributing the sample wavelength points of the OS method economically should be developed further.

Attempts to design approximate, but more easily manageable, methods to handle line-absorption have been made. One example is the so-called Voigt-analog-Elsasser band model (VAEBM) for molecular opacity used by Tsuji (1976, 1978). The merits of such methods must be studied by comparing the resulting models with more accurate ones.

2. Comparisons with Observations

Present-day models have no chromospheres or coronae--they will (at best) be model photospheres. In recent years models including chromospheric-heating mechanisms have reached some level of sophistication (Ulmschneider *et al.* 1977) but uncertainties in them are still considerable (Cram 1977b). The following discussion will therefore be confined to spectral features formed in the photospheres, i.e. continua, relatively weak spectral lines, and the wings of strong lines.

For the Sun, accurate observations and modern semi-empirical models have been compared with theoretical models (Ayres 1977, 1978, Gustafsson and Bell 1978, Kurucz 1978 and others); reasonable agreement is found. However, the theoretical models seem to show a flux excess in the ultraviolet (see below). Discrepancies in the infrared limb-darkening may be caused by the inadequacy of the mixing-length theory (Holweger 1978). Few comparisons with the observations of fine structure and dynamical phenomena in the solar atmosphere are as yet possible owing to the limitations of the standard models. However, the importance of these observations, and attempts at understanding them, for improving the theory of late-type stellar atmospheres can hardly be overestimated.

Comparisons of observed and computed low-resolution scans and colors of late-type stars have been made by several authors (see Gustafsson 1979 for references); these provide new information on the fundamental parameters of stars, and on properties of various photometric systems. Some information about the degree of realism of the models also emerged; in many respects the agreement between the fluxes of stars and models is very satisfactory, but significant discrepancies have been noticed:

Relyea and Kurucz (1978) compare theoretical and observed *uvby* colors and find deviations for late A and F stars. They suggest that these discrepancies may possibly be caused by an inadequate treatment of convection. Böhm-Vitense (1978) also suggests that an increased convective efficiency or an unknown source of continuous opacity may explain why her models for F-dwarfs are brighter in the ultraviolet than stars. For G- and K-giants Gustafsson and Bell (1978) have found a similar but stronger ultraviolet discrepancy which increases systematically with metal abundance. They tentatively ascribe it to very weak metal or molecular lines omitted in the computations; an unknown continuous opacity source, or possible departures from LTE, may also be responsible. Although this effect is pronounced in the ultraviolet flux, the effects on the atmospheric structures are quite small, except for models with T_{eff}>5000 K.

Comparisons at higher resolution of stellar spectra with synthetic ones show a reasonable overall agreement (cf., e.g., Gustafsson and Bell 1978), if a few *ad hoc* parameters describing macro-and microturbulence are introduced. However, some discrepancies for photospheric spectral lines remain:

Ramsey (1977) has observed a greater ionization of calcium than is expected from the Saha equation in late K and M stars, in qualitative agreement with the predictions by Auman and Woodrow (1975). This overionization may be produced by "hot" photoionizing radiation from deeper atmospheric layers. A similar effect for iron was found in solar-type models by Lites and Cowley (1974), and in F-dwarf models by Nissen and Gustafsson (1978). [The overionization of Fe reported by Oinas (1977) for stars later than K2 seems difficult to understand. Cohen (1978) does not find an ionization anomaly for iron in low-gravity Population II giants, although these atmospheres are much more transparent to radiation at short wavelengths.] The structural effects caused by overionization appear to be unimportant with the possible exception of stars with $T_{eff} \lesssim 3500$ K and metal-poor F and G stars. Departures from the LTE *excitation* equilibrium of metals may also affect atmospheres via blanketing effects; this is probably of importance mainly for stars with T_{eff}>5000 K.

Discrepancies also occur in the strengths of molecular bands, and at least for TiO in M stars this is important because the TiO absorption efficiently heats the outer layers in cool models (Krupp *et al.* 1978). The models predict far too strong

TiO bands for dwarfs (Mould 1975), giants and supergiants (Lengyel-Frey 1977 and Johnson et al. 1978). This might be due to uncertainties in the temperature scale or to departures from LTE for TiO and/or Ti, but the situation is not well understood at present.

Kelch et al. (1978 and papers listed therein) and Desikachary and Gray (1978) have compared observed and theoretical Ca II K profiles for more than ten late-type stars. They suggest that the upper photospheres ($\tau_{Ross} \lesssim 0.01$) of current theoretical models are too cool, typically by 100 K at $\tau_{Ross}=0.001$). Although these studies are subject to several uncertainties, this discrepancy seems rather well established. It may be a consequence of the assumption of LTE--the large surface cooling occurs because the spectral lines are assumed to be formed in absorption.

Other strong spectral lines, e.g., the vibration-rotation bands of CO (cf. Heasley et al. 1978), may be of interest for studying the structure of late-type atmospheres. For the CO lines the LTE assumption is valid in most of the photosphere, but the surface-cooling by CO in the outermost layers may be reduced by non-LTE effects (Carbon et al. 1976).

Departures from LTE are known to lead to observable discrepancies between models and spectra of late-type stars in some cases: CN 3883Å band (Mount et al. 1975), sodium lines (Kelch and Milkey 1976), oxygen triplet at 7774Å (Eriksson and Toft 1978), lithium 6708Å line (Luck 1977, de la Reza and Querci 1978). However, these departures do not appreciably affect atmospheric structures. A general study of non-LTE effects on molecular lines is given by Hinkle and Lambert (1976).

3. Models of Tomorrow

Improvements in our understanding of the physics of stellar atmospheres will emerge from the continuing confrontation of models with observations. Improvements in atomic and molecular data--identifications of spectral lines, oscillator strengths, damping parameters, dissociation energies, etc.--are needed. High-precision studies, such as the work on iron by Blackwell and collaborators (Mon. Not. Roy. Soc. 1972-78) are of great value. However, broad surveys of, e.g., weak metal lines and molecular lines, are also very important, as are studies of grain-formation and absorption in cool stellar atmospheres. For investigations where the LTE assumption is relaxed, present uncertainties in collision and other cross-sections are very troublesome.

The simultaneous solution of the statistical-equilibrium and radiative-transfer equations subject to the condition of constant flux is costly of computer time for realistic late-type models. The effort required increases rapidly with the complexity of the model atoms and the number of transitions treated; thus attempts to find general rules of how a given atom may be modeled as simply as possible without loss of physically-important information need to be made.

Departures from plane-parallel geometry can be treated with present-day techniques. Watanabe and Kodaira (1978) found sphericity effects in red supergiants to be greater than expected, owing to the coupling of temperature-sensitive molecule formation to the cooling caused by flux-dilution. Schmid-Burgk and Scholz (1977) showed that sphericity effects for a Population II F-giant might be observable.

Methods for handling radiative transfer in inhomogeneous media are now available (see, e.g., Cram 1977a and Mihalas et al. 1978 and references therein), but the construction of self-consistent model atmospheres also requires an understanding of how the inhomogeneities in late-type stellar atmospheres form and develop. Inhomogeneities and velocity fields may be connected with various phenomena: differential rotation, magnetic fields, radial and non-radial oscillations, granular and super-granular motions, mass flows, etc. However, for many of these phenomena convection is the basic generating factor, and to understand them we shall need a well-founded theory of convection in stellar atmospheres (also needed for an understanding of the energy balance). Recently, numerical simulations (Graham 1977, Nordlund 1976, 1978, Weiss 1977, Peckover and Weiss 1978) have initiated promising developments which may eventually lead to realistic predictions of convective fluxes, velocity fields and mass-flows in photospheres. Very accurate observations of line profiles (wavelength shifts and widths) in stellar spectra should be of great value in this work as will be research on the (magneto)hydrodynamics of the solar atmosphere. This work offers hope that quantitative predictions of inhomogeneities, non-thermal Doppler-shifts, departures from hydrostatic equilibrium, and non-radiative heating in stellar atmosphere might become possible within the foreseeable future. It seems likely that the model atmospheres of tomorrow will differ significantly from those of today, particularly as a result of treating phenomena not considered in classical models (cf. Cram 1978).

E. Radiative Transfer in Extended, Expanding, and Multidimensional Atmospheres (D. Mihalas)

In contrast to the previous sections of this report, the discussion in this section will concentrate primarily on developments of methodology and technique.

1. Extended Atmospheres

A number of different methods for solving the transfer equation in static spherical media exist, and this problem can now be considered to be solved. The two main approaches are: (a) ray-by-ray solutions along the tangents to a set of spherical shells within the medium, and (b) solutions of moment equations. For a broad discussion cf. Mihalas (1978, §7-6 and §11-14).

In the case that the source function is frequency-independent, a ray-by-ray solution can be used to solve directly for its depth-variation. The formulation can be made in terms of either integral-operator (Schmid-Burgk 1975) or difference-equation techniques (Mihalas and Hummer 1974); in the latter case the integration scheme devised by Auer (1976) is advantageous.

To solve the moment equations one requires closure relations among the moments, and hence information about the angular distribution of the radiation field as a function of depth. Given these closures, the moment equations may be solved stably by using a "sphericality factor" (Auer 1971). The solution may be effected by standard difference-equation techniques, both for complete and partial redistribution in the scattering process. To obtain the angle-dependent information required to determine the closure relations, a general approach is to perform a ray-by-ray formal solution assuming a given source function, and then to iterate between the formal solution and moment equations until consistency is obtained (see, e.g. Mihalas 1978 §7-6). Alternatively, one can develop approximate closure relations by a suitable generalization of the Eddington approximation to allow for the strong forward-peaking of the radiation field in the outermost layers. Such approaches have been studied by Simonneau (1976, 1978a, 1978b), Unno and Kondo (1976, 1977); and Masaki and Unno (1978); these methods appear to offer simple and accurate results in many cases.

Analyses of stellar spectra allowing for atmospheric-extension effects have been performed by Schmid-Burgk and Scholz (1976, 1977) and Watanabe and Kodaira (1978). Temperature-correction procedures for extended stellar atmospheres have been studied by Grinkevich (1976), Mihalas and Hummer (1974), Schmid-Burgk (1975) and Stebnev (1976).

2. Expanding Atmospheres

Considerable progress has recently been made in solving line-formation problems in moving media. A superb review of the physics of the problem has been given by Hummer (1976) who points out in particular the severe diagnostic difficulties implied by the breakdown of the Eddington-Barbier relation in moving atmospheres. Mathematical schemes for solving moving-medium transfer problems fall into three basic categories: (a) observer-frame, (b) comoving-frame, and (c) escape-probability (Sobolev) methods. Each approach has advantages and disadvantages, and no one scheme is well-adapted to all problems; see Mihalas (1978, Chap. 14) for a broad discussion.

Observer-frame methods are well-suited for complex velocity-fields and geometrical structures in the low-velocity regime; effective mathematical techniques are easy to formulate for both planar and spherical atmospheres; see e.g. Mihalas (1978, §14-1). A perturbation method that offers computational economies has been developed by Cram and Lopert (1976). Solutions of the transfer equation for important resonance lines (e.g. Lα, CaII H and K) have been obtained by Cannon (1976a) for multidimensional structures and velocity fields, allowing for angle-and frequency-dependent redistribution.

Comoving-frame methods can treat high-velocity flows and are well posed for studying redistribution effects, but thus far have been formulated only for monotone expansion of the atmosphere. Haisch (1976) has derived the comoving-frame equations using a tensor formalism. Methods for treating multilevel atomic models in the comoving frame have been developed by Mihalas and Kunasz (1978). Redistribution effects by atomic and electron scattering in expanding spherical media have been studied by Mihalás et al (1976b), Peraiah (1978b), and Wehrse and Peraiah (1978). Studies by Vardavas (1976), Mihalas et al (1976a), and Hamann and Kudritzki (1977) have shown that in moving media it is essential to angle-average redistribution functions in the comoving fluid frame and that redistribution functions angle-averaged in the laboratory rest frame introduce spurious effects on line-formation. The differences between results obtained from redistribution functions angle-averaged in the fluid frame and from full angle-frequency dependent redistribution fundtions are minor (Vardavas 1976). Partial redistribution effects with dipole scattering in expanding media have been treated by Peraiah (1978a).

The escape-probability (Sobolev) method has often been applied with good success to expanding flows; recent formal developments have been made by Grachov (1976, 1977, 1978). For flows in which the line-of-sight velocity is nonmonotonic (e.g. decelerating expansion or accretion) the purely local nature of the approximation is lost, and two or more separated regions of the atmosphere can be radiatively coupled. Generalizations of the method to treat this complication have been developed by Grinin (1977), Marti and Noerdlinger (1977), Rybicki and Hummer (1978), and Surdej (1977, 1978), and have been used by those authors to perform exploratory calculations in a variety of cases. Expanding and differentially-rotating atmospheres have been studied by Grinin (1976). Detailed numerical modeling of H and He II line-profiles in Of stars has been done by Klein and Castor (1978) using the Sobolev method. Karp et al. (1977) have applied escape-probability techniques to estimate the effects of spectrum lines on radiative opacities in expanding media. Karp (1978) has used Sobolev methods to assess the accuracy of observational methods of estimating velocity gradients from doppler shifts, and Duval and Karp (1978) have examined the combined effects of expansion and rotation on spectral line shapes.

3. Multidimensional Atmospheres

Relatively little work has yet been done on the difficult problem of radiative transfer in multidimensional media, and both the physical foundations and mathematical methodology relating to this problem are in a state of rapid development. A comprehensive and critical survey of the literature through 1976 has been given by Cram et al. (1976). A number of important review papers presented at the 1976 IAU General Assembly in Grenoble have been published separately (Cannon 1976a).

Radiative transfer in cylindrical media has been treated by Stenholm (1977) using Rybicki's core-saturation, wing-diffusion technique and by Heasley (1977) using moment methods with an integrating factor reminiscent of that used for spherically symmetric media (see also Schmid-Burgk 1974). Mihalas, et al. (1978) have developed an efficient method for treating radiative transport in two-dimensional moving media; the computing effort in this method scales linearly with the number of angles and frequencies, and hence it is well suited for problems in cylindrical geometry, both static (Mihalas 1979) and moving (Mihalas and Auer 1979).

BIBLIOGRAPHY

I. General References

The following books and review articles may be of interest to members of the commission:

Books

Bonnet, R. and Delache, P. (eds.) 1977. Energy Balance and Hydrodynamics of

the Solar Chromosphere and Corona, IAU Colloquium No. 36, Observatoire de Nice, Nice.
Gray, D.F. 1976. The Observation and Analysis of Stellar Photospheres, Wiley, New York.
Hauck, B. and Keenan, P.C. (eds.) 1976. Abundance Effects in Classification. IAU Symposium No. 72, Reidel, Dordrecht.
Mihalas, D. 1978. Stellar Atmospheres (2nd ed.), W.H. Freeman, San Francisco.
Schatzman, E. (ed.) 1977. Perte de Masse de Etoiles, Compte-rendu de l'école de Goutelas, Observatoire de Meudon, Paris.
Slettebak, A. (ed.) 1976. Be and Shell Stars, IAU Symposium No. 70, Reidel, Dordrecht.
Spiegel, E.A. and Zahn, J.-P. (eds.) 1977. Problems of Stellar Convection. IAU Colloquium No. 38, Springer, Berlin.
Weiss, W.W., Jenkner, H. and Wood, J.H. 1976. Physics of Ap Stars, IAU Colloquium No. 32, Universitätssternwarte Wien, Vienna.

Review Articles

Boesgaard, A.M. 1976. Pub. Astron. Soc. Pacific 88, 353.
Conti, P.S. 1978. Ann. Rev Astron. Astrophys. 16, 371.
Pecker, J.-C. and Thomas, R.N. 1976. Space Sci. Rev. 19, 217.
Smith, M.A. and Gray, D.F. 1976. Pub. Astron. Soc. Pacific 88, 809.

II. References by sections of the Report

A. Winds From Early-Type Stars

Abbott, D.C. 1977. unpublished Ph.D. Thesis, Univ. of Colorado.
Abbott, D.C. 1978. Astrophys. J. 225, 893.
Abbott, D.C. 1978. in Mass Loss and Evolution of O-Type Stars, IAU Symposium No. 83, P.S. Conti and C. de Loore, eds. Reidel, Dordrecht.
Barlow, M.J. and Cohen, M. 1977. Astrophys. J. 213, 737.
Cannon, C.J. and Thomas, R.N. 1975. Astrophys. J. 211, 910.
Cassinelli, J.P., Olson, G.L. 1979. Astrophys. J., Mar. 15, in press.
Cassinelli, J.P., Olson, G.L., and Stalio, R. 1978. Astrophys. J. 220, 573.
Castor, J.I. 1978. in Proceedings of IAU Symp No. 83.
Castor, J.I., Abbott, D.C. and Klein, R.I. 1975. Astrophys. J. 195, 157.
Castor, J.I., Abbott, D.C. and Klein, R.I. 1976. in Physique des Mouvements dans les Atmosphères Stellaires, ed. R. Cayrel and M. Steinberg, CNRS, Paris, p. 365.
Castor, J.I. and Lamers, H.J.G.L.M., 1979. Astrophys. J. Suppl., in press.
Conti, P.S. 1978. Ann. Rev. Astron. Astrophys. 16, 371.
Hartmann, L. 1978. Astrophys. J. 221, 193, and 224, 520.
Hearn, A.G. 1975. Astron. Astrophys. 40, 227.
Johnson, H.M. 1978. Astrophys. J. Suppl. 219, 773.
Klein, R.I. and Castor, J.I. 1978. Astrophys. J. 220, 902.
Lamers, H.J.G.L.M. and Morton, D.C. 1976. Astrophys. J. Suppl. 32, 715.
Lamers, H.J.G.L.M. and Rogerson, J.B. 1978. Astron. Astrophys. 66, 417.
Lamers, H.J.G.L.M. and Snow, T.P. 1978. Astrophys. J. 219, 504.
Lucy, L.B. and Solomon, P.M. 1970. Astrophys. J. 159, 879.
Mihalas, D. 1978. Stellar Atmospheres, 2nd ed. Freeman, San Francisco, pp. 511-566.
Olson, G.L. 1978. Astrophys. J. 226, 124.
Snow, T.P. and Jenkins, E.B. 1977. Astrophys. J. Suppl. 33, 269.
Stalio, R., and Upson, W. 1979. to appear in Astron. Astrophys.
Thomas, R.H. 1978. in Proceedings of IAU Symposium No. 83.
York, D.G., Vidal-Madjar, A., Laurent, C. and Bonnet, R. 1977. Astrophys. J. Letters 213, L61.
Weaver, R., McCray, R., Castor, J.I. 1977. Astrophys. J. 218, 377.

B. Mass-Loss From Late-Type Stars

Bernat, A.P. 1977. Astrophys. J. 213, 756.
Bernat, A.P., and Lambert, D.L. 1975. Astrophys. J. Letters 201, L153.
Bernat, A.P. and Lambert, D.L. 1976. Astrophys. J. 210, 395.
Bernat, A.P., Honeycutt, R.K., Kephart, J.E., Gow, C.E., Sanford, M.P. Lambert, D.L. 1978. Astrophys. J. 219, 532.
Cohen, J. 1976, Astrophys. J. Letters 203, L127.
Forrest, W.J., Gillett, F.C., Houck, J.R., McCarthy, J., Merrill, K.M., Pipher, J.L., Puetter, R.C., Russell, R.W., Soifer, B.T. and Willner, S.P. 1978. Astrophys. J. 219, 114.
Goldberg, L., Ramsey, L., Testermann, L., Carbon, D. 1975. Astrophys. J. 199, 427.
Goldberg, L. 1976. Mem. Soc. Roy. Sci. Liége 6th Ser. 9, 387.
Goldreich, P., Scoville, N. 1976. Astrophys. J. 205, 144.
Hagen, W. 1978. Astrophys. J. Letters 222, L37.
Higgs, L.A., Feldman, P.A., Smolinsky, J. 1978. Astrophys. J. Letters 220, L109.
Kelch, W.L., Linsky, J.L., Basri, G.S., Chiu, H.-Y., Chang, S.-H., Maran, S.P., Furenlid, I. 1978. Astrophys. J. 229, 962.
Kudritzki, R.P., Reimers, D. 1978. Astron. Astrophys. 70, 227.
Kuiper, E.N.R., Kuiper, T.B. H., Zuckerman, B., Kakar, R.K. 1977. Astrophys. J. 214, 394.
Kwan, J., Hill, F. 1977. Astrophys. J. 215, 781.
Lambert, D.L., Luck, R.E. 1978. Mon. Not. Roy. Astron. Soc. 184, 405.
Lo, K.Y., Bechis, K. 1976. Astrophys. J. Letters 205, L21.
Lynds, R., Harvey, J., Goldberg, L. 1978 in preparation.
Mallia, E.A., Pagel, B.E.J. 1978. Mon. Not. Roy. Astron. Soc. 184, 55P.
McCarthy, D.W., Low, F.J., Howell, R. 1977. Astrophys. J. Letters 214, L85.
McCarthy, D.W., Howell, R., Low, F.J. 1978. Astrophys. J. Letters 223, L113.
Merrill, K.M. 1977. in The Interaction of Variable Stars with Their Environment, I.A.U.Coll. No. 42, (ed. R. Kippenhahn, J. Rabe, W. Strohmeier), Remeis Sternwarte, Bamberg, p. 446.
Merrill, K.M., Russell, R.W. Soifer, B.T. 1976. Astrophys. J. 207, 763.
Münch, G., Roester, S., Trauger, J. 1976. Carnegie Yearbook 1975, p. 283.
Mufson, S.L., Lyon, J., Marionni, P.A. 1975. Astrophys. J. Letters 201, L85.
Mullan, D.J. 1978. Astrophys. J. 226, 151.
Reimers, D. 1975. Mem. Soc. Roy. Sci. Liége 6th Ser. 8, 369.
Reimers, D. 1977a, Astron. Astrophys. 57, 395.
Reimers, D. 1977b. Astron. Astrophys. 61, 217.
Renzini, A., 1976. Lectures at the 7th Advanced Course Saas-Fee "Advanced Stages in Stellar Evolution".
Sanner, F. 1976. Astrophys. J. Suppl. 32, 115.
Schmid-Burgk, J., Scholz, M. 1976. Astron. Astrophys. 51, 209.
Schultz, B.V., Kreysa, E., Sherwood, W.A. 1976. Astron. Astrophys. 50, 171.
Van der Hucht, K.A., Bernat, A.P., Kondo, Y. 1978. preprint.
Weymann, R. 1977. in The Interaction of Variable Stars with Their Environment, I.A.U. Coll. No. 42, (ed. R. Kippenhahn, J. Rabe, W. Strohmeier), Remeis Sternwarte, Bamberg, p. 577.
Winnberg, A. 1977. in The Interaction of Variable Stars with Their Environment, I.A.U. Coll. No. 42, (ed. R. Kippenhahn, J. Rabe, W. Strohmeier), Remeis Sternwarte, Bamberg, p. 495.
Zuckerman, B. 1978. in Planetary Nebulae, I.A.U. Symp. No. 76, (ed. Y. Terzian) Reidel, Dordrecht, p. 305.
Zuckerman, B., Gilra, D.P., Turner, B.E., Morris, M., Palmer, P. 1976. Astrophys. J. Letters 205, L15.
Zuckerman, B., Palmer, P., Gilra, D.P., Turner, B.E., Morris, M. 1978. Astrophys. J. Letters 220, L53.

C. Stellar Chromospheres and Coronae

Antiochos, S.K. and Underwood, J.H. 1978. Astron. Astrophys. 68, L19.
Ayres, T.R. 1979. Astrophys. J. in press.
Ayres, T.R., Linsky, J.L., Rodgers, A.W. and Kurucz, R.L. 1976. Astrophys. J. 210, 199.
Bartoe, J.-D.F., Brueckner, G.E., Sandlin, G.D., van Hoosier, M.E., and Jordan, C. 1978. Astrophys. J. Letters in press.
Basri, G.S., Linsky, J.L., Bartoe, J.-D.F., Brueckner, G., and van Hoosier, M.E. 1979. submitted to Astrophys. J.
Brune, W.H., Feldman, P.D., and Mount, G.H. 1979. Astrophys. J. Letters in press.
Canfield, R.C. and Stencel, R.E. 1976. Astrophys. J. 209, 618.
Cannon, C.J. and Thomas, R.N. 1977. Astrophys. J. 211, 910.
Cash, W., Bowyer, S., and Lampton, M. 1978a. Astrophys. J. Letters 221, L87.
Cash, W., Bowyer, S., Charles, P.A., Lampton, M., Garmire, G., and Reigler, G. 1978b. Astrophys. J. Letters 223, L21.
Cassinelli, J.P. and Hartmann, L. 1977. Astrophys. J. 212, 488.
Cram, L.E. 1978. Astron. Astrophys. 59, 151.
Cram, L.E. and Ulmschneider, P. 1978. Astron. Astrophys. 62, 239.
Dupree, A.K. 1975, Astrophys. J. Letters 200, L27.
Dupree, A.K. 1976. in Physique des Mouvements dans Les Atmospheres Stellaires, ed. R. Cayrel and M. Steinberg (Coll. Int. du CNRS No. 250).
Dupree, A.K. 1978. Advances in Atomic and Molecular Physics Vol. 14, ed. D.R. Bates and B. Bederson.
Endler, F., Hammer, R., and Ulmschneider, P. 1978. Astron. Astrophys. in press.
Haisch, B.M. and Linsky, J.L. 1976. Astrophys. J. Letters 205, L39.
Haisch, B.M., Linsky, J.L., Lampton, M., Paresce, F., Margon, B., and Stern, R. 1977. Astrophys. J. Letters 213, L119.
Haisch, B.M., Linsky, J.L., Weinstein, A., and Shine, R.A. 1977. Astrophys. J. 214, 785.
Haisch, B.M. Linsky, J.L., Slee, D.B., Hearn, D.R., Walker, A.R., Rydgren, A.E., and Nicholson, G.D. 1978. Astrophys. J. Letters 225, L35.
Hearn, A.G. 1975. Astron. Astrophys. 40, 355.
Hearn, A.G. 1977. Solar Phys. 51, 159.
Heasley, J.N., Ridgeway, S.T., Carbon, D.F., Milkey, R.W., and Hall, D.N.B. 1978. Astrophys. J. 219, 970.
Jordan, C., Brueckner, G.E., Bartoe, J.-D.F., Sandlin, G.D., and van Hoosier, M.E. 1978. Nature in press.
Kalkofen W. and Ulmschneider, P. 1978. Astrophys. J. in press.
Karpen, J.T., et al. 1977. Astrophys. J. 216, 479.
Kelch, W.L. 1978. Astrophys. J. 222, 931.
Kelch, W.L., Linsky, J.L., Basri, G.S., Chiu, H.Y., Chang, S.H., Maran, S.P. and Furenlid, I. 1978. Astrophys. J. 220, 962.
Kelch, W.L., Linsky, J.L. and Worden, S.P. 1979. Astrophys. J. in press.
Klein, R.I., Stein, R.F., and Kalkofen, W. 1978. Astrophys. J. 220, 1024.
Kondo, Y., Modisette, J.L, Dufour, R.J., and Whaley, R.S. 1976a. Astrophys. J. 206, 163.
Kondo, Y., Morgan, T.H., and Modisette, J.L. 1976b. Astrophys. J. 209, 489.
Kunkel, W.E. 1975. in Variable Stars and Stellar Evolution, ed. V.E. Sherwood and L. Plaut, Reidel, Dordrecht, p. 42.
Linsky, J.L. 1977. in The Solar Output and Its Variation, ed. O.R. White, Colorado Univ. Press, Boulder, Chap. 7.
Linsky, J.L. and Ayres, T.R. 1978. Astrophys. J. 220, 619.
Linsky, J.L. et al. 1978. Nature 275, 389.
Linsky, J.L. and Haisch, B.M. 1979. submitted to Astrophys. J. Letters.
Linsky, J.L., Worden, S.P., McClintock, W., and Robertson, R.M. 1979. submitted to Astrophys. J.
Lites, B.W., Shine, R.A., and Chipman, E.G. 1978. Astrophys. J. 222, 333.
Lutz, T.E. and Pagel, B.E.J. 1977. Bull. Amer. Astron. Soc. 9, 634.

Margon, B. and Cohen, J.G. 1978. Astrophys. J. Letters 222, L33.
Mewe, R., Heise, J., Gronenschild, E.H., Brinkman, A.C., Schrijver, J., and den Boggende, A.J.F. 1975. Astrophys. J. Letters 202, L67.
Mullan, D.J. 1976a. Astrophys. J. 209, 171.
Mullan, D.J. 1976b. Astrophys. J. 207, 289.
Mullan, D.J. 1976c. Astrophys. J. 210, 702.
Mullan, D.J. 1977. Astrophys. J. 212, 171.
Mullan, D.J. 1978. Astrophys. J. in press.
Naftalan, S.A. and Drake, S.A. 1977. Astrophys. J. 216, 508.
Owen, F.N., Jones, T.W., and Gibson, D.M. 1976. Astrophys. J. Letters 210, L27.
Owen, F.N. and Spangler, S.R. 1977. Astrophys. J. Letters 217, L41.
Popper, D.M. and Ulrich, R.K. 1977. Astrophys. J. Letters 212, L131.
Praderie, F. 1976. Mem. Soc. Astron. Ital.
Praderie, F. and Thomas, R.N. 1976. Solar Phys. 50, 333.
Praderie, F., Simonneau, E., and Snow, T.P. 1975 Astrophys. Space Sci. 38, 337.
Raymond, J.C. 1978. Astrophys. J. 224, 259.
Reimers, D. 1977. Astron. Astrophys. 57, 395.
Renzini, A., Cacciari, C., Ulmschneider, P., and Schmitz, F. 1977. Astron. Astrophys. 61, 39.
Rhombs, C.G. and Fix, J.D. 1977. Astrophys. J. 216, 503.
Robinson, R.D., Slee, O.B., and Little, A.G. 1976. Astrophys. J. Letters 203, L91.
Skumanich, A., Smythe, C. and Frazier, E.N. 1975. Astrophys. J. 200, 747.
Snijders, M.A.J. 1977. Astrophys. J. Letters 214, L35.
Spangler, S.R. and Moffett, T.J. 1976. Astrophys. J. 203, 497.
Stencel, R.E. 1977. Astrophys. J. 215, 176.
Stencel, R.E. 1978. Astrophys. J. Letters 223, L37.
Topka, K., Fabricant, D., Harnden, F.R., and Gorenstein, P. 1977. Bull. Amer. Astron. Soc. 9, 560.
Tripp, D.A., Athay, R.G., and Peterson, V.L. 1978. Astrophys. J. 220, 314.
Ulmschneider, P. and Kalkofen, W. 1978. Astron. Astrophys. 69, 407.
Ulmschneider, P., Schmitz, F., Renzini, A., Cacciari, C., Kalkofen, W. and Kurucz, R.L. 1977. Astron Astrophys. 61, 515.
Ulmschneider, P., Schmitz, F., Renzini, A., Cacciari, C., Kalkofen, W., and Kurucz, R. 1978a. Astron. Astrophys. in press.
Ulmschneider, P., Schmitz, F., and Hammer, R. 1978b. Astron. Astrophys. in press.
Ulrich, R.K. 1976. Astrophys. J. 210, 377.
Ulrich, R.K. 1978. UCLA Preprint No. 37.
Vaiana, G.S. and Rosner, R. 1978. Ann. Rev. Astron. Astrophys. 16, 393.
Vitz, R.C., Weiser, H., Moos, H.W., Weinstein, A., and Warden, E. 1976. Astrophys. J. Letters 205, L35.
Walter, F., Charles, P., Bowyer, S. 1978a. Nature 274, 569.
Walter, F. 1978b. Astrophys. J. Letters 225, L119.
Wilson, O.C. 1976. in Basic Mechanisms of Solar Activity, eds. V. Bumba and J. Kleczek, Reidel, Dordrecht, p. 447.
Withbroe, G.L. and Noyes, R.W. 1977. Ann. Rev. Astron. Astrophys. 15, 363.
Young, A. and Koniges, A. 1977. Astrophys. J. 211, 836.

D. Model Atmospheres for Late-Type Stars

Auman, J.R., Woodrow, J.E.J. 1975. Astrophys. J. 197, 163.
Ayres, T.R. 1977. Astrophys. J. 213, 296.
Ayres, T.R. 1978. Solar Phys. 57, 19.
Bell, R.A., Gustafsson, B. 1978. Astron. Astrophys. Suppl., in press.
Bell, R.A., Eriksson, K., Gustafsson, B., Dordlund. A. 1976. Astron. Astrophys. Suppl. 23, 37.
Böhm-Vitense, E. 1978. Astrophys. J. 223, 509.
Buser, R., Kurucz, R.L. 1978. submitted to Astron. Astrophys.
Carbon, D.F. 1974. Astrophys. J. 187, 135.

Carbon, D.F., Milkey, R.W., Heasley, J.N. 1976. Astrophys. J. 207, 253.
Cohen, J.G. 1978. Astrophys. J. 223, 487.
Cram, L.E. 1977a. Astron. Astrophys. 56, 401.
Cram, L.E. 1977b. Astron. Astrophys. 59, 151.
Cram, L.E. 1978. Astron. Astrophys. 67, 301.
Desikachary, K., Gray, D.F. 1978, preprint.
de la Reza, R., Querci, F. 1978. Astron. Astrophys. 67, 7.
Eriksson, K., Toft, S.C. 1978. Astron. Astrophys., in press.
Eriksson, K., Bell, R.A., Gustafsson, B., Nordlund, Å. 1978. in preparation.
Graham, E. 1977. in Problems in Stellar Convection, IAU Coll. No. 38, eds. E. A. Spiegel and J.P. Zahn. Springer, Berlin p. 151.
Gustafsson, B. 1979. in Stars and Star Systems, ed. B. Westerlund, Reidel, Dordrecht.
Gustafsson, B., Bell, R.A. 1978. Astron. Astrophys., in press.
Gustafsson, B., Bell, R.A., Eriksson, K., Nordlund, Å. 1975. Astron. Astrophys. 42, 407.
Heasley, J.N., Ridgeway, S.T., Carbon, D.F., Milkey, R.W., Hall, D.N.B. 1978. Astrophys. J. 219, 970.
Hinkle, K.H., Lambert, D.L. 1975. Mon. Not. Roy. Astron. Soc. 170, 447.
Holweger, H. 1978. Introductory report, 22nd Liège Symposium, Liège.
Johnson, H.R., Krupp, B.M., Bernat, A.P. 1978. preprint.
Kelch, W.L. 1978. Astrophys. J. 222, 931.
Kelch, W.L., Milkey, R.W. 1976. Astrophys. J. 208, 428.
Kelch, W.L., Linsky, J.L., Basri, G.S., Chiu, H.Y., Chang, S.H., Maran, S.P., Furenlid, I. 1978. Astrophys. J. 220, 962.
Krupp, B.M., Collins, J.G., Johnson, H.R. 1978. Astrophys. J. 219, 963.
Kurucz, R.L. 1978. Astrophys. J. Suppl., in press.
Kurucz, R.L., Peytremann, E. 1975. Smithsonian Astrophys. Obs. Spec. Report No. 362.
Lengyel-Frey, D.A. 1977. Ph.D. dissertation, University of Maryland.
Lites, B.W., Cowley, C.R. 1974. Astron. Astrophys. 31, 361.
Luck, R.E. 1977. Astrophys. J. 218, 752.
Manduca, A., Bell, R.A., Gustafsson, B. 1977. Astron. Astrophys. 61, 809.
Mihalas, D., Auer, L.H., Mihalas, B.R. 1978. Astrophys. J. 220, 1001.
Mould, J.R. 1975. Astron. Astrophys. 38, 283.
Mould, J.R. 1976. Astron. Astrophys. 48, 443.
Mount, G.H., Ayres, T.R., Linsky, J.L. 1975. Astrophys. J. 200, 383.
Nissen, P.E., Gustafsson, B. 1978. in Astronomical Papers Dedicated to Bengt Strömgren, eds. A. Reiz and T. Andersen, Copenhagen Univ. Obs., p. 95.
Oinas, V. 1977. Astron. Astrophys. 61, 17.
Peckover, R.S., Weiss, N.O. 1978. Mon Not. Roy. Astron. Soc. 182, 189.
Peytremann, E. 1974. Astron. Astrophys. 33, 203; Astron. Astrophys. Suppl. 18, 81.
Peytremann, E. 1975. Astron. Astrophys. 38, 417.
Querci, F., Querci, M. 1975. Astron. Astrophys. 39, 113.
Querci, M., Querci, F. 1976. Astron. Astrophys. 49, 443.
Querci, F., Querci, M., Tsuji, T. 1974. Astron. Astrophys. 31, 265.
Ramsey, L.W. 1977. Astrophys. J. 215, 827.
Relyea, L.J., Kurucz, R.L. 1978. Astrophys. J. Suppl. 37, 45.
Schmid-Burgk, J., Scholz, M. 1977. Mon. Not. Roy. Astron. Soc. 179, 563.
Sneden, C., Johnson, H.R., Krupp, B.M. 1976. Astrophys. J. 204, 281.
Tsuji, T. 1976. Publ. Astron. Soc. Japan 28, 543.
Tsuji, T. 1978. Astron. Astrophys. 62, 29.
Ulmschneider, P., Schmitz, F., Renzini, A., Cacciari, C., Kalkofen, W., Kurucz, R. 1977. Astron. Astrophys. 61, 515.
Watanabe, T., Kodaira, K. 1978. Publ. Astron. Soc. Japan 30, 21 and preprint.
Weiss, N.O. 1977. in Problems in Stellar Convection, IAU Coll. No. 38, eds. E.A. Spiegel and J.P. Zahn, Springer, Berlin, p. 142.

E. Radiative Transfer in Extended, Expanding, and Multidimensional Atmospheres

Auer, L.H. 1971. J. Quant. Spectrosc. Rad. Transf. 11, 573.
Auer, L.H. 1976. J. Quant. Spectrosc. Rad. Transf. 16, 931.
Cannon, C.J. 1976a. Astron. Astrophys. 52, 337.
Cannon, C.J. (ed.) 1976b. Interpretation of Atmospheric Structure in the Presence of Inhomogeneities. Report of Commission 12 session at IAU General Assembly in Grenoble. Univ. of Sydney Printing Office, Sydney.
Cram, L.E. and Lopert, P.B. 1976. J. Quant. Spectrosc. Rad. Transf. 16, 347.
Cram, L.E., Durrant, C.J., and Kneer, F. 1976. Report from the Workshop on Pluridimensional Radiative Transfer, Osservatorio Astronomico di Capodimonte, Naples.
Duval, P. and Karp, A.H. 1978. Astrophys. J. 222, 220.
Grachov, S.I. 1976. Vestnik Leningrad Univ. No. 1, 128.
Grachov, S.I. 1977. Vestnik Leningrad Univ. No. 19, 114.
Grachov, S.I. 1978. Vestnik Leningrad Univ. No. 1, 129.
Grinin, V.P. 1976. Izv. Krymskoj Astrofiz. Obs. 54, 176.
Grinin, V.P. 1977. Izv. Krymskoj Astrofiz. Obs. 56, 3.
Grinkevich, F.G. 1976. Trudy. Kazan. Gorod. Astr. Obs. 41, 31 and 41, 36.
Haisch, B.M. 1976. Astrophys. J. 205, 520.
Hamann, W.R. and Kudritzki, R.P. 1977. Astron. Astrophys. 54, 525.
Heasley, J.N. 1977. J. Quant. Spectrosc. Rad. Transf. 18, 541.
Hummer, D.G. 1976, in Be and Shell Stars, IAU Symposium No. 70, Dordrecht, Reidel, p. 281.
Karp, A.H. 1978. Astrophys. J. 222, 578.
Karp. A.H., Lasher, G., Chan, K.L. and Salpeter, E.E. 1977. Astrophys. J. 214, 161.
Klein, R.I. and Castor, J.I. 1978. Astrophys. J. 229, 902.
Marti, F. and Noerdlinger, P.D. 1977. Astrophys. J. 215, 247.
Masaki, I. and Unno, W. 1978. Pub. Astron. Soc. Japan 30, 547.
Mihalas, D. 1978. Stellar Atmospheres (2nd ed.), W.H. Freeman, San Francisco.
Mihalas, D. 1979. in preparation.
Mihalas, D. and Auer, L.H. 1979. in preparation.
Mihalas, D. and Hummer, D.G. 1974. Astrophys. J. Suppl. 28, 343.
Mihalas, D. and Kunasz, P.B. 1978. Astrophys. J. 219, 635.
Mihalas, D. Shine, R., Kunasz, P.B., and Hummer, D.G. 1976a. Astrophys. J. 205, 492.
Mihalas, D., Kunasz, P.B., and Hummer, D.G. 1976b. Astrophys. J. 210, 419.
Mihalas, D., Auer, L.H., and Mihalas, B.R. 1978. Astrophys. J. 220, 1001.
Peraiah, A. 1978a. Kodikanal Obs. Bull. Series A. 2, 107.
Peraiah, A. 1978b. Astrophys. Space Sci. in press.
Rybicki, G.B. and Hummer, D.G. 1978. Astrophys. J. 219, 654.
Schmid-Burgk, J. 1974. J. Quant. Spectrosc. Rad. Transf. 14, 979.
Schmid-Burgk, J. 1975. Astron. Astrophys. 40, 249.
Schmid-Burgk, J. and Scholz, M. 1976. Astron. Astrophys. 40, 249.
Schmid-Burgk, J. and Scholz, M. 1977. Mon. Not. Roy. Astron. Soc. 179, 563.
Simonneau, E. 1976. J. Quant. Spectrosc. Rad. Transf. 16, 741.
Simonneau, E. 1978a. J. Quant. Spectrosc. Rad. Transf. 19, 439.
Simonneau, E. 1978b. J. Quant Spectrosc. Rad. Transf. 20, 49.
Stebnev, V.I. 1976. Trudy Kazan. Gonod. Astr. Obs. 41, 24.
Stenholm, L.G. 1977. Astron. Astrophys. 54, 577.
Surdej, J. 1977. Astron. Astrophys. 60, 303.
Surdej, J. 1978. Astron. Astrophys. 66, 45.
Unno, W. and Kondo, M. 1976. Pub. Astron. Soc. Japan 28, 347.
Unno, W. and Kondo, M. 1977. Pub. Astron. Soc. Japan 29, 693.
Vardavas, I.M. 1976. J. Quant. Spectrosc. Rad. Transf. 16, 781.
Watanabe, T. and Kodaira, K. 1978. Pub. Astron. Soc. Japan 30, 21.
Wehrse, R. and Peraiah, A. 1978. Astron. Astrophys., in press.

D. MIHALAS, President of the Commission

42. CLOSE BINARY STARS (ETOILES DOUBLES SERREES)

PRESIDENT: G. Larsson-Leander
VICE-PRESIDENT: B. Warner
ORGANIZING COMMITTEE: A.H. Batten, A.M. Cherepashchuk, M.G. Fracastoro, K. Gyldenkerne, T.J. Herczeg, M. Kitamura, Y. Kondo, M. Plavec, S.D. Sinvhal, J. Smak, E.P.J. van den Heuvel, J.A.J. Whelan

1. Introduction

Studies of close binary stars were persued more vigorously than ever: about 3000 research papers and notes pertaining to the field were published during the triennium 1976 - 1978. Many major advances and spectacular discoveries were made, mostly due to increased observational efficiency and precision, especially in the X-ray, radio, and ultraviolet domains. The high activity is also shown by the number of conferences devoted to close-binary problems.

The proceedings of our 1975 Cambridge Symposium, IAU No. 73, appeared in print in 1976: *Structure and Evolution of Close Binary Systems*, eds. P. Eggleton, S. Mitton, and J. Whelan; Reidel Publ. Co. Likewise, IAU Symposium No. 70, dedicated to the memory of P.W. Merrill and D.B. McLaughlin, was printed by Reidel in 1976 (*Be and Shell Stars*, ed. A. Slettebak). At the Grenoble Meeting decision was taken to co-sponsor a symposium on "Mass-Loss and Evolution of O-type Stars". It was held, as IAU Symposium No. 83, at Qualicum Beach, Vancouver Island (Canada), in June 1978; many papers and much of the discussion dealt with close binaries. As will be known to all readers, the Commission is sponsoring IAU Symposium No. 88, "Close Binary Stars: Observation and Interpretation", to be held in Toronto, August 1979, with M. Plavec as Chairman of the Scientific Organizing Committee. Further, the Commission offered co-sponsorship for a symposium on "Fundamental Problems in the Theory of Stellar Evolution", planned for Kyoto, July 1980. IAU Colloquia of special interest were: No. 29, *Multiple Periodic Variable Stars*, ed. W.S. Fitch, published 1976 in two volumes by the Hungarian Academy of Sciences (Vol. 1, with the invited papers, was also issued by Reidel); No. 33, "Observational Parameters and Dynamical Evolution of Multiple Stars", published 1977 as Vol. 3 of *Rev. Mexicana Astron. Astrophys.*; No. 42, "The Interaction of Variable Stars with their Environment" appeared in book form 1977 (eds. R. Kippenhahn, J. Rahe, and W. Strohmeier, *Veröff. Remeis-Sternw. Bamberg*, No. 121) just a few months after the colloquium; No. 46, "Changing Trends in Variable Star Research", held at Hamilton, New Zealand, November 1978. The conference on "X-ray Binaries" at Goddard Space Flight Center in October 1975 was edited by E. Boldt and Y. Kondo and published early 1976 as *NASA SP*-389. Reidel printed 1977 *Novae and Related Stars* (ed. M. Friedjung), the proceedings of the September 1976 Paris conference. The 1974 Moscow conference on evolutionary problems was followed up by a second conference in Warsaw, June 1977 (*Non-stationary Evolution of Close Binaries*, ed. A.N. Zytkow, Polish Scientific Publ.,1978).

Two important monographs were published, *viz.*, Z. Kopal, *Dynamics of Close Binary Systems*, Reidel Publ. Co., 1978, and J. Sahade and F.B. Wood, *Interacting Binary Stars*, Pergamon Press, 1978. Of utmost importance for close-binary workers is the appearance 1978 of *Seventh Catalogue of the Orbital Elements of Spectroscopic Binary Systems*, by A.H. Batten, J.M. Fletcher and F.J. Mann (*Publ. Dominion Astrophys. Obs.* 15, No. 5). Also, we should mention that a new edition of the *Finding List for Observers of Eclipsing Variables* will soon be published by F.B. Wood, R.H. Koch, J.P. Oliver, and D.R. Florkowski. The *Bibliography and Program Notes on Close Binaries* continued to be issued from Lund Observatory, with G. Larsson-Leander as editor. Nos. 27 - 31 appeared during the triennium; regional contributors were B. Cester (Italy), M. de Groot and R.F. Sisteró (Southern Hemisphere), J. Grygar (Central and Eastern Europe), D.S. Hall and

G.W. Henry (USA, except the West Coast), M. Kitamura (Japan), H. Mauder (Germany and *IAU Circ.*), C.D. Scarfe (Canada, US West Coast, Mexico), A. Shulberg (USSR), S.D. Sinvhal and C.D. Kandpal (India and Indonesia), F. van't Veer (Western Europe).

The present Report is based on published literature and information from Commission members. Members of the Organizing Committee kindly consented to write specific Sections, and for reviewing radio observations I got the expert help of Dr. D.M. Gibson. The names of these contributors are indicated at proper places. In addition, on my request Dr. A.M. Cherepashchuk wrote an extensive summary of the work done in the USSR, excerpts of which were communicated to the contributors. However, to keep the Report at the prescribed length, make the presentation as homogeneous as possible, and take into account the most recent literature, the contributions were more or less heavily edited. Much material was especially added to Tables 1 - 4, in an attempt to make them as complete as possible. Lack of space made it necessary to omit names of authors in the tables. As it was realized that many details on work in progress could not be included in the Report, much of this information was given in No. 31 of the *Bibliography*. I am deeply thankful to all contributing authors and Commission members who have helped me so generously.

Key to the references:

AA = *Acta Astron.*
AAp = *Astron. Astrophys.*
AAp Sup = *Astron. Astrophys., Suppl. Ser.*
AASin = *Acta Astron. Sinica*
Af = *Astrofizika*
Af Issl = *Astrofiz. Issled. Izv. Spets. Astrofiz. Obs.*
AJ = *Astron. J.*
AmAf = *Astrometr. Astrofiz., Kiev*
AN = *Astron. Nachr.*
Ann Rev AAp = *Annu. Rev. Astron. Astrophys.*
Ann Tokyo = *Ann. Tokyo Astron. Obs., Second Ser.*
ApJ = *Astrophys. J.*
ApJ Sup = *Astrophys. J., Suppl. Ser.*
ApL = *Astrophys. Lett.*
ApSpSc = *Astrophys. Space Sci.*
ATs = *Astron. Tsirk.*
AZh = *Astron. Zh. Akad. Nauk SSSR*
BAAS = *Bull. American Astron. Soc.*
BAC = *Bull. Astron. Inst. Czechoslovakia*
Bamberg Ver = *Veröff. Remeis-Sternw. Bamberg*
BASI = *Bull. Astron. Soc. India*
Byull Abastuman = *Byull. Abastuman. Astrofiz. Obs.*
Com Ankara = *Commun. Astron. Dep. Univ. Ankara*
ComAp = *Comments Astrophys.*
Contr Cluj-Napoca = *Contrib. Astron. Obs. Univ. "Babes-Bolyai", Cluj-Napoca*
IAUC = *IAU Circ.*
IBVS = *Inf. Bull. Variable Stars*
Izv AN AzSSR = *Izv. Akad. Nauk Azerb. SSR, Ser. fiz.-tekn. i mat. n.*
Izv Ehngelg = *Izv. Astron. Ehngel'gardt. Obs., Kazan*
Izv Krym = *Izv. Krymskoj Astrofiz. Obs.*
JAAVSO = *J. American Assoc. Variable Star Obs.*
JRASC = *J. R. Astron. Soc. Canada*
Kodaikanal Bull = *Kodaikanal Obs. Bull.*
Liège Coll = *Colloque Int. Astrophys. Liège*
MittAG = *Mitt. Astron. Ges.*
MN = *Mon. Not. R. Astron. Soc.*
MRAS = *Mem. R. Astron. Soc.*
MSAI = *Mem. Soc. Astron. Italiana*
Obs = *Observatory*
PASP = *Publ. Astron. Soc. Pacific*
PerZv = *Perem. Zvezdy, Byull.*
PerZv Pril = *Perem. Zvezdy, Prilozhenie*
Pis AZh = *Pis'ma v Astron. Zh.*
Publ DAO = *Publ. Dominion Astrophys. Obs.*
Publ Tartu = *Publ. Tartu Astrofiz. Obs.*
QJRAS = *Q. J. R. Astron. Soc.*
Rep Ege = *Sci. Rep. Fac. Sci. Ege Univ., Izmir*
Rep Lund = *Rep. Obs. Lund*
Soob Zelenchuk = *Soobshch. Spets. Astrofiz. Obs.*
Stud Babes-Bolyai = *Stud. Univ. Babes-Bolyai, Ser. Math.-phys.*
Tokyo Bull = *Tokyo Astron Bull., Second Ser.*
Trudy Alma-Ata = *Trudy Astrofiz. Inst. Alma-Ata*
Trudy Kazan = *Trudy Kazan. Gorod. Astron. Obs.*
Ts Lvov = *Tsirk. Astron. Obs., L'vov.*
Ts Shemak = *Tsirk. Shemakhinsk. Astrofiz. Obs.*
Villanova Contr = *Villanova Univ. Obs. Contrib.*
Vistas = *Vistas Astron.*

2. Observational Techniques
(B. Warner)

A wide range of techniques are currently used for the study of close binary stars. The more conventional photometric and spectrographic observations, polarimetry, X-ray and radio observations are discussed in Section 4. Extraterrestrial observations, including those made with the IUE, are summarized in Section 10. In addition we mention here the ultraviolet satellite observations of β Lyr (Kondo *et al.*, *ApSpSc* 41,121), V1500 Cyg (Jenkins *et al.*, *ApJ* 212,198; Wu and Kester, *AAp* 58,331), W UMa (Rucinski, *AA* 26,227), and V Pup (York *et al.*, *ApJ* 210,143).

High-resolution spectroscopy of HR Del (Gallagher and Anderson, *ApJ* 203,625) used an Echelle system, in V1500 Cyg photoelectric scans gave good results for interstellar lines (Tomkin *et al.*, *AAp* 48,319). Infrared spectroscopy (2 - 3 μ) detected coronal lines in V1500 Cyg (Grasdalen and Joyce, *Nature* 259,187); 1 - 20 μ spectra of this nova were obtained by Ennis *et al.* (*ApJ* 214,478). An application of the surface-brightness technique to V1500 Cyg (Barnes, *MN* 177,53P) provided a new way of estimating distances to novae.

Rapid photometry was widely used for studies of dwarf and other novae (Nevo and Sadeh, *MN* 177,167, 182,595; Robinson and Nather, *PASP* 89,572; Szkody, *ApJ* 207,190; Warner and Brickhill, *MN* 182,777), quasi-periodic variations were found in VW Hyi (Haefner *et al.*, *AAp* 61,L37), RU Peg (Patterson *et al.*, *ApJ* 214,144), and WZ Sge (Robinson *et al.*, *ApJ* 219,168). Rapid flickering was detected in AM Her (Bailey *et al.*, *MN* 180,35P) and in the symbiotic variable CH Cyg at outburst (Slovak and Africano, *MN* 185,591). Optical pulsations were detected in HZ Her (Middleditch and Nelson, *ApJ* 208,567) and from 4U 1626-67 (Ilovaisky *et al.*, *AAp* 70,L19). Optical bursts were reported from MXB 1735-44 (McClintock *et al.*, *Nature* 274,567). Rapid variations in line profiles were seen in V1500 Cyg (Campbell, *ApJ* 207,L41), in EZ CMa (Vojkhanskaya *et al.*, *Af* 12,180), and in β Lyr (Sanyal, *ApJ* 210,853). Pulse-resolved spectrophotometry was obtained for DQ Her by Margon *et al.* (*ApJ* 208,L35), and possible pulsations in the spectral lines were reported (Margon *et al.*, *PASP* 89,300).

We note the increasing importance of speckle interferometry (Blazit *et al.*, *ApJ* 214,L79; McAlister, *ApJ* 215,159, 225,932; Morgan *et al.*, *MN* 183,701) for spectroscopic binaries. Detailed studies by McAlister include η Ori (*PASP* 88,957), 12 Per (*ApJ* 223,526), and 51 Tau (*ApJ* 212,459). Lunar occultations continue to furnish information on new and old systems (*e.g.*, Africano *et al.*, *AJ* 82,631, 83,1100). Long-focus astrometry by van de Kamp yielded parallaxes and orbital data for ε Aur (*AJ* 83,975) and VV Cep (*AJ* 82,750). New interferometric techniques (Currie *et al.*, *BAAS* 9,625) seem very promising.

3. Methods of Analyzing Light Curves

The methods of analyzing light changes in the frequency domain were further developed, primarily by Kopal and his collaborators in the long series of papers entitled "Fourier Analysis of the Light Curves of Eclipsing Variables" in *ApSpSc* (Vol. 34 - 57, 19 papers). Other contributions were given by Jurkevich *et al.* (*ApSpSc* 44,63), Rovithis-Livaniou (*ApSpSc* 51,77, 52,271), and by several investigators dealing with specific systems.

As outlined by Kopal (*ApSpSc* 34,431), one of the most suitable methods of transforming from the time to the frequency domain is to consider light loss to be a function of even powers of the sine of the phase angle and to evaluate certain integrals A_{2m}, called moments, from which the elements are determined. The relations between the A_{2m} and the geometrical elements, although simple for uniformly bright stars, were found complicated for limb-darkened stars. A simpler and more direct method to obtain these relations was indicated by Smith (*ApSpSc* 40,315). Kopal (*ApSpSc* 46,87) generalized the procedure, introducing the complete Fourier transform and considering also odd and non-linear moments. In Paper XI of the series, Kopal (*ApSpSc* 50,225) defined the light loss as the cross correlation of two apertures, representing the eclipsing and eclipsed disks. This new approach is shown to have considerable advantages compared to the geometrical one. The fractional light loss of arbitrarily limb-darkened stars, equal to the associated alpha-functions α_l^0, may be described as Hankel transforms of products of two Bessel functions, with orders depending on the physical characteristics of the components, while the geometry of the system enters only through their arguments. Kopal (*ApSpSc* 51,439) gave explicit forms for the α_l^0 for different types of eclipses together with the corresponding moments A_{2m}, and alternative forms of the α_l^0 were given by Demircan (*ApSpSc* 52,189). Kopal and Demircan (*ApSpSc* 55,241) described the practical procedures for deriving the geometrical elements for any type of eclipse, any proximity of components, and any limb-darkening. Demircan (*ApSpSc* 56,389) gave new general expressions for the moments A_{2m}, allowing an automated method of analysis, and discussed (*ApSpSc* 56,453) which moments should be chosen for best determinacy. Photometric perturbations were considered by Edalati and Budding (*ApSpSc* 57,181) and by Kopal (*ApSpSc* 57,439). All the above papers assume circular orbits. The generalization to excentric orbits was given by Kopal and Al-Naimiy in Paper XIX (*ApSpSc* 57,479). Several more papers of the series are in press. A book on the Fourier methods has already been completed by Kopal in manuscript form. It should be available as preprint to interested workers at the end of 1978, and will hopefully appear from the printer during 1979.

A computer program, based on Kopal's analytical method for solution in the time domain, was worked out by Söderhjelm (*Rep Lund* 10). Botsula (*Izv Ehngelg* 39,99) and the Peking Research Group (*AASin* 18,68) discussed the convergence of successive approximations in Kopal's method. Lanzano (*ApSpSc* 42,425, 45,419,483) discussed the J-integrals and derived recursion formulae useful for numerical computations.

Lavrov (*PerZv Pril* 2,349) published a program suited to the rectifiable Russell-Merrill method, and coefficients for polynomials approximating the $\alpha(p,k)$ functions were tabulated by Lavrov and Sokolova (*BAC* 27,301). Horak (*BAC* 26,257) presented a modification of his iterative minimization method. Ureche (*Contr Cluj-Napoca*, p. 13; *Stud Babes-Bolyai*, in press) discussed the influence of the quadratic and a possible sine term in the limb-darkening law, and he extended his studies of the ellipsoid-ellipsoid approximation to the case of elliptic orbits (*BAC*, in press). Budding reports that a 16-parameter curve-fitting program was put in operation in Manchester.

Binnendijk in a lucid paper (*Vistas* 21,359) described a new method for computing light curves and deriving orbital parameters for contact binaries, assuming Roche potentials. A slightly more theoretical treatment was presented by Sapar (*Publ Tartu* 44,124). Eaton (*BAAS* 9,624) wrote a program for application on RS CVn systems, assuming unevenly distributed spots on one component. Eclipses by an elliptical torus were treated by Wilson (*AJ* 80,719).

Among the various existing programs for light-curve synthesis, especially those of D.B. Wood and Wilson and Devinney have been used extensively by many workers. For the benefit of users, Wood issued a number of *WINK Status Reports*, with improvements and amendments. In the Wilson-Devinney program, Wilson added an optional stellar-atmosphere provision, the capability of treating non-synchronous rotation, and the generation of radial-velocity curves. Accuracy-related improvements were described by Wilson and Biermann (*AAp* 48,349). The procedures are now being generalized to include effects of orbital excentricity, with proper accounting for the variable tidal field on the figures of the components. Wilson further intends to incorporate the effects of radiation pressure and to develop programs for simultaneous solution of light and radial-velocity curves.

Comparisons of various methods on specific systems have been made in a number of cases. Binnendijk, in his paper referred to above (*Vistas* 21,359), gave parameters for V566 Oph, RZ Tau, and AW UMa. Results of other methods for these and other stars were reduced to the same set of parameters, thus making direct comparisons possible. The Russell-Merrill, Kitamura, Wood, and Wilson-Devinney procedures were used on CW Cas by Burchi *et al.* (*ApSpSc* 47,35) and on RT Per by Mancuso *et al.* (*ApSpSc* 47,277). For RZ Cas, Chambliss (*PASP* 88,22) obtained with the Wood model elements which differ significantly from those found with the Russell-Merrill and Kopal-Jurkevich methods. Chambliss and Leung (preprint) comparing the Russell-Merrill, Kopal-Jurkevich, and Wilson-Devinney approaches on SX Aur found the third procedure to yield the most consistent solution.

4. Observational Data

A. Photometric Observations and Solutions (T. Herczeg)

Tables 1 and 2 list publications reporting new photoelectric observations and orbital solutions. Photoelectric work continued to increase; among observatories that recently became very active in this field, we mention Athens, Biruni, Ege, Helwan, and Villanova. Quite frequently the number of observations are so high that their detailed publication becomes impractical, and the observations are deposited with one of the established repositories or made available upon request to the observer. Also, there are cases where publication is postponed to some later time, and only the observed light curves or other results are published presently. Table 1 contains papers that are obviously based on a "considerable amount" of new photoelectric data, thus, in contrast to previous practice, not only papers presenting individual observations. Satellite UV-photometry is included.

Novae at outburst are not regarded as belonging to the field of this Commission. The papers on V1500 Cyg quoted in Table 1 deal more or less with the periodic fluctuations in the light curve; if interpreted as an orbital phenomenon similar to those found in cataclysmic variables, they would suggest a very strong change of orbital period (*cf.* Semeniuk *et al.*, *AA* 27,301).

Photographic studies are not listed in Table 1. Important ones are the optical histories of X-ray binaries: AM Her (Feigelson *et al.*, *ApJ* 222,263), X Per (Gottlieb *et al.*, *ApJ* 202,L13), V818 Sco (Wright *et al.*, *ApJ* 200,171), and HD 245770 (Stier and Liller, *ApJ* 206,257), all derived from Harvard plates.

B. Spectroscopic and Spectrophotometric Investigations (A. Batten)

Table 3 lists spectroscopic and spectrophotometric studies, including determinations of orbital elements (references indicated by an asterisk). To save space, papers already cited in the *Seventh Catalogue of the Orbital Elements of Spectroscopic Binary Systems* (Batten *et al.*, *Publ DAO* 15,121) are omitted.

Important fundamental work on detached main-sequence systems was continued by Popper and the Copenhagen group (especially Andersen). Popper extended his studies to include Algol-type systems, and reports the detection of the secondary D-lines in the spectra of a number of them, which should lead to better knowledge of their masses. A noteworthy individual achievement was the detection of the secondary spectrum of Algol itself by Tomkin and Lambert (*ApJ* 222,L119). Tomkin (*ApJ* 221,608) also detected the secondary spectrum of δ Lib. Early-type systems attracted considerable interest. Hutchings continues his work on mass loss and stellar winds. Conti and his associates published a number of critical re-investigations of systems containing O-type stars. Popper (*ApJ* 220,L11) indicated that masses of some early-type systems - especially that of V382 Cyg, one of the most massive known - should be revised downwards. Wolf-Rayet binaries were the subject of considerable discussion at IAU Symposium 83; among others, Niemela reported on her continued work. A new study of θ Mus by Moffat and Seggewiss (*AAp* 54,607) should be particularly noted. Van Blerkom (*ApJ* 225,175) found supporting evidence that HD 45166 is binary, containing a low-mass star in Wolf-Rayet clothing.

Table 1. Photoelectric Observations

RT And IBVS 1409, AB And Com Ankara 77, CN And IBVS 1087, ζ And Izv Krym 56,16, λ And AJ 83,176, DX Aqr IBVS 1199, V337 Aql PerZv 20,375, V822 Aql ATs 956, V889 Aql AA 28,221, RW Ara AAp Sup 26,227, SS Ari ATs 888, UX Ari AA 27,281, AJ 83,176, ApJ 225,919, SX Aur PerZv Pril 2,51, IBVS 1278, WW Aur BAAS 7,463, CQ Aur RicA 8,555, IU Aur AA 28,63, KR Aur Pis AZh 3,510, ε Aur Publ Tartu 45,284, TZ Boo IBVS 1086, 1487, AAp Sup 33,63, XY Boo AJ 82,648, AC Boo AAp Sup 29,57, 44i Boo IBVS 1353, 1374, 1497, AAp Sup 32,361, SZ Cam PerZv 20,473, AS Cam IBVS 1090, ApSpSc 38,87, TX Cnc Obs 98,205, WY Cnc IBVS 1484, BAC 27,335, RS CVn BAAS 10,418, VZ CVn AAp 56,75, Rep Ege 225, AM CVn AA 25,371, BAAS 10,419, R CMa AJ 82,51, UW CMa IBVS 1235, ApJ 220,582, CW CMa AJ 81,1134, HH Car AAp Sup 22,263, RZ Cas PASP 87,909, 88,22, TV Cas AJ 82,740, BAC 28,41, AO Cas Publ Tartu 43,103, BM Cas IBVS 1117, DO Cas AAp Sup 30,223, IBVS 1413, MN Cas AJ 81,665, SV Cen IBVS 1162, SZ Cen AAp 55,401, BH Cen ApJ 211,844, KT Cen AAp Sup 22,263, U Cep ATs 882, BAAS 9,352, ApJ Sup 31,1, ApJ 220,251, 222,635, VV Cep Tokyo Bull 247, 248, PASP 89,621, IBVS 1286, 1385, XY Cep BASI 4,83, CQ Cep Pis AZh 2,505, CW Cep AAp Sup 25,151, EM Cep JRASC 69,307, ER Cep MN 184,33, TX Cet AAp 70,355, XY Cet ApSpSc 38,79, RS Cha AAp Sup 33,87, RZ Cha AAp 44,343, RW Com BAAS 8,521, CC Com PASP 88,777, ATs 918, Y Cyg RicA 8,543, SS Cyg Pis AZh 2,116, CG Cyg IBVS 1458, BAAS 9,623, MR Cyg AASin 18,68, MY Cyg AJ 80,976, V367 Cyg Publ Tartu 43,114, V388 Cyg AAp Sup 27,435, V444 Cyg PerZv 20,1, ApJ 221,193, V470 Cyg PASP 87,923, V1073 Cyg IBVS 1116, V1329 Cyg PerZv 20,345, V1341 Cyg = Cyg X-2 Pis AZh 4,402, AZh 53,511, Soob Zelenchuk 15,5, V1357 Cyg = Cyg X-1 MN 173,15P, 177,63P, 182,315, ApJ 205,855, BAAS 9,645, MittAG 43,225, PerZv 19,305, Nature 263,393, V1500 Cyg AZh 54,583,592,600,620, AA 27,301, PASP 89,37,44, ApJ 211,L71, 225,954, AAp 55,171, 58,331, 31 Cyg AAp Sup 21,189, 32 Cyg AAp Sup 21,189, TW Dra AAp Sup 32,57, BS Dra IBVS 1079, 1100, BY Dra ApJ 214,140, CM Dra ApJ 218,444, 71 Dra IBVS 1104, RU Eri BASI 4,83, CW Eri MittAG 38,231, U Gem ApJ 206,790, YY Gem Tokyo Bull 240, AJ 83,618, X Gru AAp Sup 29,51, AK Her ApSpSc 52,387, AM Her ApJ 211,859, 212,L113, 215,166, 219,597, 225,542, 226,397, PASP 90,61, MN 183,73P, AAp 68,L1, Pis AZh 4,183, Nature 271,335, DQ Her ApJ 196,L113, 207,195, HZ Her = Her X-1 PerZv 19,305, Pis AZh 1 No7,27, 4,356, ApJ 208,567,L31, 4 Her AA 27,265, 88 Her BAC 29,278, TU Hor AAp 61,161, TT Hya BASI 5,105, VZ Hya ApSpSc 35,249, AI Hya IBVS 1118, AAp Sup 33,103, HS Hya AAp 42,303, χ^2 Hya AAp Sup 31,307, RT Lac ApJ Sup 31,93, AJ 82,998, AR Lac PerZv 19,461, PASP 88,762, Ts Shemak 37,9, XY Leo Obs 98,205, AJ 83,1452, AM Leo MittAG 43,186, T LMi PASJ 29,289, RR Lyn Ann Tokyo 16,1, SW Lyn BAC 28,120, β Lyr JAAVSO 4,18,80, 5,16, AAp Sup 30,231, ApSpSc 41,121, BAC 27,215, MN 174,217, TY Men IBVS 1013, 1149, AAp 47,315, UX Men AAp 48,49, AO Mon ApSpSc 40,3, AR Mon ApJ 208,142, AU Mon IBVS 1348, V616 Mon AAp 48,141, 50,445, ApJ 211,L15, IBVS 1173, Pis AZh 2,112,402, TU Mus AAp 45,107, θ Mus AAp 54,607, GM Nor AAp Sup 22,263, U Oph ApJ 214,423, BAAS 10,410, V502 Oph PerZv Pril 2,161, V566 Oph AN 298,117, PASP 88,473, 89,47, IBVS 1457, VV Ori BAAS 8,306, BM Ori AA 26,91, η Ori IBVS 1418, θ^1 Ori A IBVS 1238, 1274, PASP 89,530, σ Ori E Nature 262,116, ApJ 205,L87, MW Pav ApSpSc 46,155, U Peg IBVS 1010, X Per MN 176,217,225, AAp 61,47, ApJ 218,504, IBVS 1359, BAAS 9,349, RT Per ApSpSc 47,277, RW Per AA 28,207, ST Per AA 26,15, AG Per ApSpSc 57,17, BAAS 10,410, DM Per AAp Sup 25,291, IW Per BAAS 9,626, IZ Per BAAS 9,288, LX Per IBVS 1178, β Per AJ 81,57, 82,67, MN 172,235, 173,271, 180,461, 184,523, PASP 87,745, AE Phe AAp Sup 24,399, AI Phe IBVS 1419, ζ Phe AAp Sup 23,261, RR Pic AAp 41,15, SZ Psc AJ 81,250, IBVS 1297, UV Psc IBVS 1381, V Pup AA 28,63, KX Pup AAp Sup 22,263, NO Pup AAp 50,79, TY Pyx AA 28,231, IBVS 1489, U Sge PASP 88,688, UU Sge ApJ 223,252, WZ Sge ApJ 219,168, V1647 Sgr AAp 58,121, V523 Sco PASP 87,901, V818 Sco = Sco X-1 Soob Zelenchuk 15,5, AZh 53,511, AAp Sup 28,119, MN 176,91, σ Sco IBVS 1432, RT Scl AAp Sup 28,389, RY Sct Bamberg Ver 121,386, RZ Sct PASP 88,262, CV Ser ApJ 199,432, Af 11,49, Obs 97,76, AA 28,55, RW Tau PASP 89,533, CD Tau ApSpSc 40,15, AJ 81,855, V471 Tau AAp 46,197, ApSpSc 57,219, BAAS 10,410, V711 Tau AJ 82,47, 83,176, ApJ 225,919, IBVS 1394, 1411, Obs 98,207, X Tri AAp Sup 23,439, Byull Abastuman 48,13, RW Tri IBVS 1251, AQ Tuc AAp Sup 34,207, W UMa AA 26,227, MSAI 48,83, Obs 98,205, VV UMa AAp Sup 27,285,

Table 1. (Continued)

XY UMa *AA* 27,273, AN UMa *BAAS* 8,346, AW UMa *AA* 25,417, *IBVS* 1176, *BAAS* 10,410, RT UMi *AAp* 29,333, RU UMi *AA* 27,187, AO Vel *IBVS* 1111, CV Vel *AAp* 58,131, EO Vel *AAp Sup* 22,263, γ^2 Vel *AAp* 57,151, AH Vir *BAC* 28,157, *Obs* 98,205, BF Vir *AA* 26,253, DL Vir *AAp* 61,107, ER Vul *IBVS* 1481, HR 3827A *IBVS* 1440, HR 4665 *IBVS* 1432, HR 5110 *IBVS* 1459, HD 5980 *PASP* 90,101, HD 13970 *AAp* 70,105, HD 20301 *IBVS* 1317, *AAp Sup* 29,313, HD 38735 *AAp Sup* 29,313, HD 47732 *ApJ* 222,574, HD 60168A *AAp Sup* 29,313, HD 71581 *AAp Sup* 29,313, HD 77581 = Vel X-1 *AAp* 55,473, *AAp Sup* 27,433, *MN* 183,813, HD 86441 *IBVS* 1265, HD 93206 *IBVS* 1265, HD 128141 *PASP* 87,877, HD 152235 *IBVS* 1265, HD 153919 = 3U 1700-37 *AAp* 52,139, 54,543,683, 58,L5, HD 159176 *BAAS* 7,533, HD 162724 *ApJ* 201,792, HD 173198 *IBVS* 1041, 1306, HD 193793 *MN* 185,467, HD 199497 *IBVS* 1214, HD 224085 *PASP* 89,280, HDE 271213 *ApJ* 207,329, HDE 271227 *ApJ* 207,329, BD -3°5357 *PASP* 89,616, BD -7°3007 *PASP* 87,716, CD -33°12119 *PASP* 89,720, CD -42°14462 *ApJ* 214,471, LB 3459 *MN* 183,523, LS 55°-8 *PASP* 90,191, PG 1413+01 *ApJ* 224,892, NGC 2264 W92 *PASP* 89,874, Sk 160 = SMC X-1 *AAp* 54,307, *AAp Sup* 29,339, Wray 977 = 3U 1223-62 *ApJ* 203,689, *AAp* 49,321, 54,733, *MittAG* 43,227, Cir X-1 *MN* 183,335, Cyg X-3 *ApJ* 207,78, LMC X-4 *AAp* 59,L9, 4U 1626-67 *ApJ* 225,1001.

Two objects continuing to attract much attention are β Lyr and VV Cep. Hack *et al.* (*ApJ Sup* 34,565) presented evidence for the secondary spectrum in the far UV that would set an upper limit of 3:1 (secondary:primary) for the mass-ratio. On the other hand, Dadaev (*Izv Pulkovo* 193,41, 194,18) claims to have found the secondary spectrum in the photographic region and gives the mass-ratio 1.1:1. Dadaev (*Izv Pulkovo* 195,98, and subsequent papers) is now studying the emission-line spectrum and relating it to the pattern of gas streams. Batten, Ringuelet, and Sahade continue their studies of emission lines. The other system, VV Cep, underwent eclipse during the triennium. Wright now has high-dispersion spectrograms covering a complete orbital cycle. He published orbital elements and a discussion of gas streams (*JRASC* 71,152), and continued observations through eclipse. For a third famous star, ε Aur, IUE observations indicated the presence of a hot companion (Hack and Selvelli, *Nature* 276,376).

Binaries containing intrinsic variables were studied by several groups. Abt and Levy (*PASP* 90,188) drew attention to the probable duplicity of the cepheid Y Oph; and the dwarf cepheid RS Gru was shown to have a companion by Balona and Martin (*MN* 184,1). Rodonò (*AAp* 66,175) studied flare-active systems. The RS CVn group of binaries naturally attracted much attention, because of the discovery that they are radio sources. The study of optical counterparts of X-ray binaries continued. New identifications were made and attempted, and orbital elements of known binaries were refined (Conti, *AAp* 63,225; Hutchings, *ApJ* 226,264).

Orbital elements of several multiple systems were determined, especially by Fekel, who is continuing his work. Geyer and Patkos report that ADS 12019 is quadruple (component A is the eclipsing variable BH Dra). Morbey *et al.* (*PASP* 89,851) showed that one component of ADS 11060 is a close binary. The visual pair has now been followed through periastron and is the spectroscopic binary with the most excentric orbit known (e=0.96). Radial velocities of other visual binaries have been measured, especially δ Equ (Popper and Dworetzky, *PASP* 90,71; Scarfe *et al.*, in press).

C. Polarimetric Studies (A.M. Cherepashchuk)

"Ordinary" close binaries show variable linear polarization due to scattering from circumstellar matter, and possibly to some degree also from other mechanisms. Pfeiffer and Koch (*PASP* 89,147), reviewing existing data (in many cases obtained by the authors themselves) found that, with the exception of U Oph, unevolved binaries do not show intrinsic polarization. Phase-locked polarization was detected in AO Cas (Pfeiffer, *AJ* 81,1000; Rudy and Kemp, *ApJ* 207,L125), u Her (Rudy and Kemp, *ApJ* 216,767), HD 47129 (Rudy and Herman, *PASP* 90,163), and in Algol, U Sge and V444 Cyg

Table 2. Photometric Solutions

RT And *AAp Sup* 32,351, FK Aql *Trudy Kazan* 41,3, V805 Aql *AAp Sup* 32,351, σ Aql *AAp Sup* 33,91, RW Ara *AAp Sup* 26,227, V535 Ara *ApJ* 222,917, RX Ari *BAC* 26,257, RZ Aur *Trudy Kazan* 41,3, TT Aur *AAp Sup* 33,91, WW Aur *BAAS* 7,463, *AAp Sup* 33,91, AR Aur *AAp Sup* 33,91, IU Aur *AA* 28,63, LY Aur *AAp* 62,291, ZZ Boo *AAp Sup* 32,351, AC Boo *ApSpSc* 29,57, *AAp* 63,193, 44i Boo *AAp Sup* 32,361, SZ Cam *ApSpSc* 46,407, AS Cam *ApSpSc* 38,87, AY Cam *ApSpSc* 46,261, S Cnc *Trudy Kazan* 39,42, *ApSpSc* 47,375, RZ Cnc *ApJ* 208,142, *AAp* 61,469,809, TX Cnc *Obs* 98,205, RS CVn *BAAS* 10,418, VZ CVn *AAp* 56,75, *Rep Ege* 225, R CMa *AJ* 82,51, UW CMa *ApJ* 220,582, 222,924, *MN* 185,485, CW CMa *AJ* 81,1134, HH Car *AAp Sup* 22,263, RZ Cas *PASP* 88,22, TV Cas *ATs* 881, *BAC* 28,41, YZ Cas *ApSpSc* 56,389, 57,71, AO Cas *ApJ* 223,202, CW Cas *ApSpSc* 47,35, DO Cas *AAp Sup* 30,223, MN Cas *AJ* 82,290, V523 Cas *ATs* 873, RR Cen *MittAG* 38,225, SV Cen *PASP* 88,244, *MN* 176,625, SZ Cen *AAp* 55,401, BH Cen *ApJ* 211,844, KT Cen *AAp Sup* 22,263, U Cep *BAAS* 9,352, *AAp* 61,469, VW Cep *Publ Tartu* 43,130,176, WX Cep *AAp Sup* 32,351, XY Cep *BASI* 4,83, ZZ Cep *AAp Sup* 33,91, AH Cep *AAp Sup* 33,91, CW Cep *AAp Sup* 25,151, 33,91, EI Cep *Ann Tokyo* 16,37, *AAp Sup* 32,351, EK Cep *ATs* 971, ER Cep *MN* 184,33, SS Cet *Trudy Kazan* 41,3, TX Cet *AAp* 70,355, XY Cet *ApSpSc* 38,79, 56,293, RS Cha *BAC* 26,257, *AAp* 44,459, RZ Cha *AAp* 44,343, CC Com *PASP* 88,777, U CrB *AAp* 61,469, RV Crt *IBVS* 1272, SW Cyg *Trudy Kazan* 41,3, GG Cyg *BAAS* 9,623, MR Cyg *AASin* 18,68, *AAp* 66,161, *AAp Sup* 33,91, *Villanova Contr* 1, MY Cyg *BAC* 27,125, *AJ* 80,976, V367 Cyg *Publ Tartu* 43,114, V382 Cyg *AAp Sup* 33,91, V388 Cyg *AAp Sup* 27,435, V444 Cyg *ApJ* 221,193, V453 Cyg *BAAS* 9,352, *AAp Sup* 33,91, V729 Cyg *ApJ* 224,565, V1073 Cyg *ApJ* 222,917, V1357 Cyg = Cyg X-1 *MittAG* 43,225, *ApJ* 226,264, RR Dra *Trudy Kazan* 41,3, TW Dra *AAp Sup* 32,57, CM Dra *ApJ* 218,444, 71 Dra *IBVS* 1104, RU Eri *BASI* 4,83, AS Eri *AAp* 62,291, CW Eri *MittAG* 38,231, RW Gem *ApSpSc* 46,261, YY Gem *Tokyo Bull* 240, *AJ* 83,618, X Gru *AAp Sup* 29,51, RX Her *AAp Sup* 33,91, TX Her *AAp Sup* 32,351, AK Her *ApSpSc* 47,79, 52,387, *BAAS* 10,411, HZ Her = Her X-1 *ApJ* 208,567, V338 Her *BAC* 26,257, u Her *AAp* 61,469, 66,161, VZ Hya *ApSpSc* 35,249, *AAp Sup* 32,351, AI Hya *AAp* 66,377, HS Hya *AAp* 42,303, χ^2 Hya *AAp* 67,15, RT Lac *ApJ Sup* 31,93, *AJ* 82,998, AR Lac *PASP* 88,762, *ApSpSc* 52,213, *Ts Shemak* 51-52,20, CM Lac *BAAS* 8,305, *AAp Sup* 32,351, UV Leo *BAC* 26,257, *AAp Sup* 32,351, XY Leo *Obs* 98,205, *AJ* 83,1452, T LMi *PASJ* 29,289, RS Lep *PerZv Pril* 3,149, RR Lyn *Ann Tokyo* 16,1, SW Lyn *BAC* 28,120, β Lyr *ApSpSc* 41,121, UX Men *AAp* 48,49, RW Mon *Trudy Kazan* 41,3, TU Mon *AAp* 61,469, AO Mon *ApSpSc* 40,3, AR Mon *ApJ* 208,142, *AAp* 61,809, IM Mon *AAp Sup* 33,91, V616 Mon *AAp* 50,445, TU Mus *AAp* 45,107, GM Nor *AAp Sup* 22,263, U Oph *ApJ* 214,423, *AAp Sup* 33,91, *BAAS* 10,410, WZ Oph *AAp Sup* 32,351, V566 Oph *Vistas* 21,359, *PASP* 88,473, 89,366, V1010 Oph *ApJ* 211,853, VV Ori *AAp Sup* 33,91, θ^1 Ori A *IBVS* 1129, MW Pav *ApSpSc* 46,155, AW Peg *AAp* 62,291, RT Per *ApSpSc* 47,277, 58,3, RW Per *AA* 28,207, RY Per *Trudy Kazan* 41,3, ST Per *AA* 26,15, AG Per *ApSpSc* 57,17, *BAAS* 10,410, DM Per *AAp Sup* 25,291, IZ Per *BAAS* 9,288, β Per *PASP* 87,745, *AJ* 81,57, 82,67, *ApSpSc* 47,459, 51,265, ζ Phe *AAp* 46,205, SZ Psc *AJ* 81,250, V Pup *AAp* 61,275,469, *AA* 28,63, AU Pup *ApJ* 222,917, KX Pup *AAp Sup* 22,263, NO Pup *AAp* 50,79, U Sge *PASP* 88,688, *ApSpSc* 47,361, 56,219, *AAp* 61,469, UU Sge *ApJ* 223,252, WZ Sge *ApJ* 219,168, V356 Sgr *AAp* 61,469, V1647 Sgr *AAp* 58,121, V523 Sco *PASP* 87,901, V701 Sco *AAp* 61,137, μ^1 Sco *AAp* 61,469, RZ Sct *PASP* 88,262, RW Tau *Izv Ehngelg* 41-42,196, *ApSpSc* 47,361, RZ Tau *Vistas* 21,359, CD Tau *AJ* 81,855, *ApSpSc* 40,15, V471 Tau *AAp* 46,197, λ Tau *AAp* 62,291, X Tri *AAp Sup* 23,439, *Byull Abastuman* 48,13, AQ Tuc *AAp Sup* 34,207, W UMa *Obs* 98,205, TX UMa *Villanova Contr* 1, *ApSpSc* 47,375, *AAp* 61,469, VV UMa *AAp Sup* 27,285, AW UMa *Vistas* 21,359, *ApSpSc* 56,219, *BAAS* 10,410, RT UMi *AAp Sup* 29,333, RU UMi *AA* 27,187, CV Vel *AAp* 58,131, EO Vel *AAp Sup* 22,263, AH Vir *Obs* 98,205, BF Vir *AA* 26,253, DL Vir *AAp* 61,107, Z Vul *AAp* 61,469, RS Vul *AAp* 61,469, BV845 *PASP* 90,97, HD 47732 *ApJ* 222,574, HD 77581 = Vel X-1 *Pis AZh* 1 No12,13, *ApJ* 202,L131, 226,264, *AAp* 54,167, HD 101799 *PASP* 88,936, HD 153919 = 3U 1700-37 *ApJ* 226,264, HD 159176 *BAAS* 7,533, HD 162724 *ApJ* 201,792, HD 173198 *IBVS* 1306, LB 3459 *MN* 183,523, PG 1413+01 *ApJ* 224,892, Sk 160 = SMC X-1 *ApJ* 204,551.

Table 3. Spectrographic Observations

ζ And *Izv Krym* 56,16, λ And *BAAS* 8,353, *PASP* 88,137, *Nature* 275,389, ν And *ApJ Sup* 36,241*, ο And *ApSpSc* 46,379, V599 Aql *Izv Krym* 58,56, V603 Aql *BAAS* 9,627, UX Ari *IBVS* 1075, *ApJ* 216,503, 225,919, *AJ* 83,795, *MN* 182,77, κ Ari *AAp Sup* 27,35, 35 Ari *ApJ Sup* 36,241*, EO Aur *ApJ* 220,L11, KR Aur *Pis AZh* 3,271,510, α Aur *Nature* 275,389, ε Aur *ApSpSc* 49,179, *Nature* 276,376, ζ Aur *ApSpSc* 36,273, TZ Boo *AAp Sup* 33,63, ZZ Boo *AAp Sup* 32,347*, α Cam *ApJ* 223,908, S Cnc *Obs* 96,9, RS CVn *AA* 26,301, *ApJ* 216,503, *MN* 182,77, TX CVn *AAp Sup* 34,211, UW CMa *ApJ* 208,760, EZ CMa *Af* 12,180, SX Cas *ApJ* 209,821, ξ Cas *ApJ Sup* 36,241*, ο Cas *ApJ Sup* 36,241*, β Cen *MN* 172,639, U Cep *ApJ* 209,821, 212,446, VV Cep *AAp* 46,317, 64,253, EM Cep *Izv Krym* 56,11, α CrB *ApSpSc* 46,379, SS Cyg *Af Issl* 8,3, *Nature* 275,385, V367 Cyg *ApSpSc* 53,345, V382 Cyg *ApJ* 220,L11, V389 Cyg *AAp* 68,437*, V453 Cyg *ApJ* 220,L11, V478 Cyg *ApJ* 220,L11, V1329 Cyg *Bamberg Ver* 121,383, V1341 Cyg = Cyg X-2 *Pis AZh* 2,534, *ApJ* 207,L171, V1357 Cyg = Cyg X-1 *ApJ* 213,815, *BAAS* 9,645, *Nature* 275,400, V1500 Cyg *PASP* 89,37, *ApJ* 207,L41, 211,184, 217,775, 224,899, *AAp* 56,181, 9 Cyg *AJ* 83,615*, 57 Cyg *ApJ Sup* 36,241*, AI Dra *AA* 28,41*, BY Dra *PASP* 89,69, CM Dra *AA* 28,167, α Equ *MN* 184,265*, δ Equ *JRASC* 71,408*, *PASP* 90,71*, YY Gem *PASP* 88,451, RS Gru *MN* 184,1, Z Her *MN* 182,77, AD Her *PASP* 90,312*, AM Her *ApJ* 212,L117, 218,L121, *PASP* 89,374, HZ Her = Her X-1 *AAp* 59,441, *ApJ* 209,547, 222,L33, *PASP* 89,285, *Nature* 275,400, V600 Her *AAp* 51,1, ι Her *ApJ Sup* 36,241, 88 Her *IBVS* 1496, TU Hor *AAp* 61,161, RT Lac *IBVS* 1130, AR Lac *PerZv* 19,377, 20,207, *Ts Shemak* 35,8,11, 36,11, *ApJ* 216,503,508, *MN* 182,77, 5 Lac *PASP* 90,184*, T LMi *PASJ* 29,289, FX Lib *AAp Sup* 34,51, δ Lib *ApJ* 221,608*, RR Lyn *Ann Tokyo* 16,1, β Lyr *Izv Pulkovo* 193,41, 194,18, *AZh* 52,710, *Ts Lvov* 51,20, *ApSpSc* 38,353, *ApJ* 206,777, 208,468, 210,853, *AAp* 50,335, *Af Issl* 9,3, *PASP* 88,899, *ATs* 966, *Izv Krym* 58,75, η Lyr *ApJ Sup* 36,241*, VV Mon *AAp Sup* 33,323*, V616 Mon *ApJ* 206,260, 211,872, 217,181, *AAp* 56,311, *MN* 180,657, U Oph *ApJ* 220,L11, Y Oph *PASP* 90,188*, κ Oph *ApJ Sup* 36,241*, BM Ori *PerZv* 20,361, δ Ori *Izv Krym* 54,128, η Ori A *ApSpSc* 46,379, θ^1 Ori A *IBVS* 1211*, ξ Ori *ApJ Sup* 36,241*, σ Ori E *ApJ* 205,L87, ψ Ori *ApJ Sup* 36,241*, 22 Ori *ApJ Sup* 36,241*, 64 Ori *BAAS* 9,623*, AG Peg *BAAS* 10,410, κ Peg *PASP* 89,857, X Per *PASP* 88,754, *MN* 176,233, 181,685, *Izv Krym* 57,45, *IBVS* 1461, LX Per *MN* 182,77, β Per *MN* 176,5P, *IBVS* 1312, *ApJ* 222,L119*, 20 Per *PASP* 90,297, AE Phe *AA* 28,49*, RR Pic *BAAS* 9,627, SZ Psc *MN* 182,77, V Pup *ApJ* 210,143, VV Pup *ApJ* 225,201, U Sge *ApJ* 209,821, 41 Sex A *AJ* 83,303*, V818 Sco = Sco X-1 *ApJ* 206,L49, *BAAS* 9,627, RY Sct *Bamberg Ver* 121,386, W Ser *IBVS* 1482, V711 Tau *AJ* 81,771*, 83,795, *ApJ* 225,919, *BAAS* 9,624, *Nature* 275,389, δ Tau *AJ* 82,176*, θ^1 Tau *AJ* 82,176, τ Tau *ApJ Sup* 36,241*, 51 Tau *BAAS* 9,600*, RR Tel *MN* 182,57P, RW Tri *BAAS* 9,556, α Tri *AA* 28,235*, γ^2 Vel *MN* 176,29P, *ApJ Sup* 36,217, α Vir *ApSpSc* 46,379, 1 Vul *ApJ Sup* 36,241*, 15 Vul *Obs* 97,2, HR 1034 *ApJ Sup* 36,241*, HR 1970 *Obs* 98,122*, HR 2013 *Obs* 96,54, HR 2172 *AAp Sup* 27,31, HR 2317 *Obs* 98,232*, HR 3080 *ApJ* 203,435, HR 4665 *BAAS* 10,419, HR 5049 *AAp* 58,93, HR 6388 *Obs* 98,14*, HR 6902 *AJ* 83,615*, HR 7083 *Obs* 97,235*, HD 698 *PASP* 90,179*, HD 5303 *IBVS* 1342, HD 28033 *AJ* 83,1114*, HD 30738 *AJ* 83,1114*, HD 47732 *ApJ* 222,570*, HD 50896 *ApJ Sup* 36,217, HD 67084 *PASP* 90,204*, HD 67198 *PASP* 90,204, HD 77581 = Vel X-1 *AAp Sup* 30,195*, *MittAG* 43,219, HD 92740 *ApJ Sup* 36,217, *AAp* 70,69, HD 135774-5 *AJ* 83,615*, HD 137569 *BAAS* 9,351*, HD 147508 *Obs* 98,47*, HD 150958 *ApJ* 225,165, HD 152270 *MittAG* 43,148, HD 153919 = 3U 1700-37 *AAp* 47,19, 56,433, 64,399*, *AAp Sup* 32,375, *ApJ* 217,L35, *Nature* 275,400, HD 155989 *Obs* 98,158*, HD 165052 *ApJ* 224,558*, HD 167771 *ApJ* 224,558*, HD 183030 *Obs* 98,118*, HD 187399 *Af* 12,623, HD 188753 *Obs* 98,122*, HD 190918 *JRASC* 71,407*, HD 204934 *Obs* 98,122*, HD 214419 *Kodaikanal Bull* 2,89, HDE 245770 *AAp* 56,311, *ApJ* 215,568, 223,530, BD +27°4642 *Izv Krym* 57,31, BD +24°692 *AJ* 83,1114*, BD -3°5357 *PASP* 89,616, BD -7°3007 *PASP* 89,716*, ADS 14893 B *ApJ* 205,194*, Feige 24 *ApJ* 223,260*, Gliese 815 *AJ* 83,1445*, Wray 977 = 3U 1223-62 *MittAG* 43,227*, G 61-29 *AA* 25,227, CRL 2104 *MN* 185,47, CRL 2179 *MN* 185,47, ω Cen V78 *BAAS* 10,411, NGC 1851 UV5 *AAp* 59,L23, SMC X-2 *ApJ* 223,L79, SMC X-3 *ApJ* 223,L79, LMC X-4 *AAp* 59,L9, *ApJ* 225,548*, 4U 1538-52 *ApJ* 225,L63, 3U 1728-24 *ApJ* 211,866.

(Rudy and Kemp, *ApJ* 221,200). These observations allow determinations of orbital inclinations (Rudy and Kemp, *ApJ* 221,200; McLean, *Obs* 98,205). Tapia (*ApJ* 212,L125) made the exciting discovery that the radiation from AM Her exhibited high linear and circular polarization in the *V* and *B* wave-length bands, indicating a magnetic field of about 2×10^8 gauss for the white-dwarf component. Further observations were published by Michalsky *et al.* (*ApJ* 216,L35), Stockman *et al.* (*ApJ* 217,815), Bailey *et al.* (*MN* 183,73P), and Priedhorsky *et al.* (*ApJ* 225,542). Similar polarization properties were found for AN UMa (Krzeminski and Serkowski, *ApJ* 216,L45; Downes and Urbanski, *PASP* 90,458) and for VV Pup (Tapia, *IAUC* 3054). In a search for other "polars", Vojkhanskaya *et al.* (*Pis AZh* 4,272) picked out MV Lyr as a promising candidate. X-ray binaries with harder spectra than AM Her show only weak intrinsic polarization in the optical region. For Cyg X-1 Michalsky and Swedlund (*ApJ* 212,221) reported slight variations of the circular polarization synchronous with the orbital period, 5.6 d, but this periodicity was not apparent in 1977 (Michalsky *et al.*, *ApJ* 225,599). For the linear polarization Kemp *et al.* found a period of 39.2 d from UV-data (*Nature* 270,227) and, from an extensive set of observations, orbital synchronism at long wave-lenghts (*ApJ* 220,L123). Long-term variations are probably present.

D. X-ray Observations (Y. Kondo)

X-ray satellites that have been operational during the triennium and used for observations of close binaries are listed below (with launch dates and experimenters):

Copernicus	1972	University College London
Ariel-5 (UK-5)	1974	Mullard Space Sci. Lab., Univ. of Leicester, Imperial College, Goddard Space Flt. Ctr., Appleton Lab.
SAS-3	1975	M.I.T.
OSO-8	1975	Columbia Univ., Goddard Space Flt. Ctr., Lockhead, Univ. of Wisconsin
HEAO-1	1977	Goddard Space Flt. Ctr., C.I.T., Univ. of Calif. Berkeley, Naval Res. Lab., Ctr. for Astrophysics, M.I.T., Univ. of Calif. San Diego
HEAO-2	1978	Goddard Space Flt. Ctr., Ctr for Astrophysics, M.I.T., Columbia Univ.

In addition, rocket experiments were flown to observe selected X-ray binaries, *e.g.* by Boldt and co-workers. Balloon-borne experiments also produced interesting new results, *e.g.*, Trümper and co-workers.

Some 40 galactic X-ray sources have now been identified with optical counterparts, the great majority being close binaries. An informative review on positions and identifications was given by Bradt *et al.* at the COSPAR/IAU Symposium on X-ray Astronomy, held at Innsbruck 1978. Reference may also be made to the series of letters to *Nature* (269,112, 270,586, 271,225, 272,701, 273,364) by the SAS-3 team. As the number of optical identifications increases, it is possible to segregate different classes of objects. One classification scheme is the following:

A. Sources comprising compact object (white dwarf, neutron star, black hole) and early-type star (often supergiants). Examples: SMC X-1, Vel X-1, Cen X-3, Cyg X-1, V861 Sco, 4U 1700-37.
B. Compact object with late-type companion: HZ Her, AM Her, Sco X-1, Cyg X-3.
C. Dwarf novae: SS Cyg, U Gem, EX Hya.
D. RS CVn systems: V711 Tau, UX Ari, RS CVn.
E. Other transient X-ray sources - novae, X-ray novae, etc.: V616 Mon, A 0535+26.
F. Other binaries identified as X-ray sources: Algol, Sirius, Capella, γ Cas.
G. X-ray bursters: Several bursters have been identified with stellar objects, making close-binary models most probable. Many bursters appear associated with globular clusters.

There are additional optically identified objects that can be fitted into some of these categories; the ones above are merely examples. As more optical identifications are made, additional classes of sources may be found. Of course, a variety of other classification schemes are possible.

Publications on X-ray emitting binaries are very numerous, and only some few items can be mentioned within the limited space available. New binary orbits from pulse-phase data were determined for Cen X-3 (Fabbiano and Schreier, *ApJ* 214,235), Vel X-1 (Rappaport *et al.*, *ApJ* 206,L103; Charles *et al.*, *MN* 183,813), SMC X-1 (Primini *et al.*, *ApJ* 210,L71, 217,543), 4U 0115+63 (Rappaport *et al.*, *ApJ* 224,L1), 3U 1223-62 (White *et al.*, *MN* 184,67P), and 3U 1538-52 (Davison *et al.*, *MN* 181,73P). Trümper and his co-workers (*ApJ* 219,L105) discovered in the spectrum of Her X-1 a line feature at about 58 keV, interpreted as cyclotron emission. Iron lines are reported by many groups in the spectra of several X-ray binaries. Observations of the γ-ray flux from Cyg X-3 were discussed by Vladimirsky *et al.* (*Izv Krym* 58,44).

E. Radio Observations (D.M. Gibson)

Radio surveys of binaries from one or more of the classes below were reported by Altenhoff *et al.* (*AAp* 46,11), Feldman (*BAAS* 10,418), Gibson (Thesis, UVa), Owen and Gibson (*AJ* Dec. 1978), Spangler *et al.* (*AJ* 82,989), Woodsworth and Hughes (*AAp* 58,105), and Wright and Allen (*MN* 184,893).

RS CVn and related binaries. Detected RS CVn systems include UX Ari, RT Lac, AR Lac, SZ Psc, TY Pyx (probable), V711 Tau (HR 1099), HD 224085 (Owen and Gibson, *AJ* Dec. 1978), and HR 5110 (Feldman, pri. com.). Emission from the related long-period systems λ And (Bath and Wallerstein, *PASP* 88,759), and σ Gem, σ CrB, HD 216489, and the short-period system UV Psc (Spangler *et al.*, *AJ* 82,989) was also reported. Algol and b Per can probably be regarded as similar objects. The above sources are highly variable - from about 10^{16} to 10^{19} erg s^{-1} Hz^{-1} (Gibson *et al.*, *AJ* Dec. 1978; Feldman *et al.*, *AJ* Dec. 1978; Gibson *et al.*, *PASP* Nov. 1978) - as befits their compact size (few $R_\odot$; Clark *et al.*, *ApJ* 206,L107; Owen and Spangler, *ApJ* 217,L41). The gyrosynchrotron emission (Owen *et al.*, *ApJ* 210,L27) arises just beyond the X-ray coronae (Gibson *et al.*, *AJ* Dec. 1978), and thus their spectra can be affected by either self-absorption (Owen *et al.*, *ApJ* 210,L27) or thermal absorption (Spangler, *AJ* 82,169). The circular polarization can be quite large (74%; Mutel and Weisberg, *AJ* Dec. 1978) and has been observed to be rapidly variable (Brown and Crane, *AJ* Dec. 1978). The emission is associated with "super solar-flare" phenomena intrinsic to one of the late-type (subgiant) components (Weiler *et al.*, *ApJ* 225,919), *not* with mass exchange. However, binarity - particularly faster rotation, larger differential rotation, and smaller surface gravity relative to single stars (Hall and Shore, *BAAS* 10,418) - seems to enhance their radio emissivity relative to their single counterparts (Gibson, *BAAS* 9,600).

X-ray/radio binaries. Continued observations of Cyg X-1 (Hjellming and Gibson, in prep.), including some correlated with X-ray bursts (Hjellming, *X-ray Binaries*, NASA SP-389, p. 495; Braes and Miley, *Nature* 264,731), now indicate the emission is non-thermal, but the spectrum is probably affected by the O-star's wind (Johnson *et al.*, in prep.). Hjellming and Helfand (in prep.) and Hjellming (*ApJ* 221,225) reconfirmed the identifications of Cyg X-2 and GX 17+2 with G-type stars. Further observations of Cyg X-3 (McEllin, *MN* 175,5P; Mason *et al.*, *ApJ* 207,78; Ledden *et al.*, *Nature* 252,669) were useful in determining conditions in the emission region both during outbursts (Seaquist, *ApJ* 207,88; Shields and Wheeler, *ApL* 17,69; Woodsworth and Hughes, *ApJ* 208,863) and quiscent periods (Seaquist and Gregory, *ApL* 18,65). Several new types of X-ray/radio binaries were detected, including the highly unusual counterpart to Cir X-1 (Whelan *et al.*, *MN* 181,259; Thomas *et al.*, *MN* 185,29P). Non-thermal radio counterparts to the transient sources A 0620-00 (Owen *et al.*, *ApJ* 203,L15; Little *et al.*, *Nature* 261,113) and one near the galactic centre (Davies *et al.*, *Nature* 261,476) were also detected. The controversy over the nature of the globular-cluster transients (*cf.*, Smith, *Nature* 261,453) or whether they have even been detected at radio frequencies (Johnson, *ApJ* 208,706) remains unresolved.

Novae. While radio emission from V1500 Cyg was apparently weak and short-lived (Pynzar *et al.*, *ATs* 893; Altunin, *Pis AZh* 2,299; Hjellming, *Bamberg Ver* 121,279), emission from FH Ser and HR Del is still being monitored (Hjellming and Vandenberg, in prep.). A thick inhomogeneous shell model for FH Ser was developed by Seaquist and Palimaka (*ApJ* 217,781).

Wolf-Rayet binaries and symbiotic stars. Variable thermal emission from WR-binaries was detected (Florkowski and Gottesman, *IBVS* 1101), with γ^2 Vel (Seaquist, *ApJ* 203,L35) and HD 193793 (Florkowski and Gottesman, *MN* 179,105) receiving continuing monitoring, the latter in conjunction with a recent IR-outburst. Variable thermal emission associated with recent outbusts was monitored for the symbiotic objects V1016 Cyg (FitzGerald *et al.*, *BAAS* 8,372), AG Peg (Gregory *et al.*, *ApJ* 211,429), and HM Sge (Feldman, *JRASC* 71,386).

Monitoring of the thermal-type, compact stellar-wind binaries CC Cas and β Lyr continued (Gibson and Newell, in prep.). Monitoring of α Sco resulted in detection of both A and B components (Gibson, in prep.). The latter's spectrum, while non-thermal, is varying, which suggests it is either a coherent emitter or has an unusual particle-acceleration mechanism. Observations of the binary pulsar, PSR 1913+16, continued; a recent progress report was given by Fowler and Taylor (*BAAS* 10,447).

5. Physical Data

A. Absolute Dimensions (T. Herczeg)

New determinations of absolute dimensions are listed in Table 4. A novel feature here is the fairly extensive rediscussion of earlier light-curves by applying modern programs of analysis and then using existing spectrographic data to obtain revised absolute dimensions. A considerable part of this work was done by Cester and his collaborators with Wood's WINK program. Leung, Schneider, and Wilson, in several papers, systematically rediscussed the "W UMa-like" light-curves of early, massive contact (or near-contact) systems. Here the Russell-Merrill method is clearly inadequate; they used the Wilson-Devinney code and found in several cases various degrees of over-contact between the components.

B. Period Changes (T. Herczeg)

Some few theoretical investigations continued the work toward establishing a relationship between period changes and the modes and rates of mass transfer and mass loss (Amnuel and Guseinov, *AAp* 54,23; Chau, *ApJ* 219,1038; Mukhametkalieva and Omarov, *Trudy Alma-Ata* 28,24; Piotrowski and Ziolkowski, *AA* 28,295). Several detailed evolutionary studies touched upon period changes in connection with the time scales involved. In general, however, we are still rather far from being able to use observed period changes as checks on evolution theories. One reason may be the apparently erratic behaviour of the periods, short-term fluctuations exhibited by a surprising number of systems. Also, the time scale is in most cases perhaps too long for the only 30 - 40 years of more systematic and accurate observations, with the possible exception of a few objects in rapid evolution, such as β Lyr and SV Cen. This, of course, emphasizes the importance of collecting as many well-defined minimum epochs as possible, for the benefit of future studies.

At IAU Colloquium 42, Kreiner (*Bamberg Ver* 121,393) reviewed period changes of W UMa stars, and Smak (*Bamberg Ver* 121,365) discussed secular mass-loss and period changes of cataclysmic variables. Of the many individual systems observed and discussed during the triennium, only a few particularly interesting ones will be mentioned. In the case of XY UMa, Geyer (*ApSpSc* 48,137) demonstrated the occurrence of spurious changes due to light-curve anomalies. Hall *et al.*, *AJ* 81,1138) discussed the "migrating wave" in the light curve of AR Lac and its relation to observed period changes; they concluded that although a correlation exists, the period changes are almost certainly real. Other studies of AR Lac are by Babaev (*Izv AN AzSSR* 4,3), Chambliss (*PASP* 88,762), and Theokas (*ApSpSc* 52,213). A detailed study of U Oph by

Table 4. Absolute Dimensions

RT And *AAp Sup* 32,351, V805 Aql *AAp Sup* 32,351, σ Aql *AAp Sup* 33,91, TT Aur *AAp Sup* 33,91, WW Aur *BAAS* 7,463, *AAp Sup* 33,91, AR Aur *AAp Sup* 33,91, IU Aur *AAp* 59,9, LY Aur *AAp* 62,291, ε Aur *AAp* 69,23, ZZ Boo *AAp Sup* 32,347, 33,91, WX Cep *AAp Sup* 32,351, AH Cep *AAp Sup* 33,91, CW Cep *AAp Sup* 33,91, EI Cep *AAp Sup* 32,351, SZ Cam *ApSpSc* 36,329, AS Cam *ApSpSc* 38,87, RZ Cnc *ApJ* 208,142, *AAp* 61,469, UX CVn *AAp* 70,451, UW CMa *PASP* 89,668, *ApJ* 220,582, 222,924, *MN* 185,485, RZ Cas *PASP* 88,22, TV Cas *ATs* 881, *BAC* 28,41, AO Cas *ApJ* 223,202, AZ Cas *PASP* 89,882, V523 Cas *ATs* 873, SV Cen *PASP* 88,244, *MN* 176,625, SZ Cen *AAp* 45,203, 55,401, BH Cen *ApJ* 211,844, U Cep *AAp* 61,469, VV Cep *JRASC* 71,152, CW Cep *AAp Sup* 25,151, EI Cep *Ann Tokyo* 16,37, EK Cep *ATs* 971, ER Cep *MN* 184,33, ξ Cep *PASP* 88,944, XY Cet *ApSpSc* 38,79, 56,293, 13 Cet *PASP* 88,50, RS Cha *AAp* 44,445, RZ Cha *AAp* 44,343,349, CC Com *PASP* 89,684, U CrB *AAp* 61,469, MR Cyg *AAp* 66,161, *AAp Sup* 33,91, MY Cyg *BAC* 27,125, V382 Cyg *AAp Sup* 33,91, V444 Cyg *ApJ* 221,193, V453 Cyg *AAp Sup* 33,91, V729 Cyg *ApJ* 224,565, V1073 Cyg *ApJ* 222,917, AI Dra *AA* 28,41, CM Dra *ApJ* 218,444, δ Equ *PASP* 90,71, AS Eri *AAp* 62,291, CW Eri *MittAG* 38,231, U Gem *AA* 26,277, YY Gem *AJ* 83,618, RX Her *AAp Sup* 33,91, TX Her *AAp Sup* 32,351, AD Her *PASP* 90,312, HZ Her = Her X-1 *ApJ* 208,567, 215,L21, V600 Her *AAp* 51,1, u Her *AAp* 61,469, 66,161, VZ Hya *ApSpSc* 35,249, *AAp Sup* 32,351, HS Hya *AAp* 42,303, χ^2 Hya *AAp* 44,445, 67,15, AR Lac *Ts Shemak* 44,3, *PASP* 88,762, CM Lac *AAp Sup* 32,351, UV Leo *AAp Sup* 32,351, XY Leo *AJ* 83,1452, T LMi *PASJ* 29,289, δ Lib *ApJ* 221,608, RR Lyn *Ann Tokyo* 16,1, SW Lyn *BAC* 28,120, β Lyr *AZh* 52,710, *Izv Pulkovo* 195,98, *ApSpSc* 41,121, UX Men *AAp* 48,49, TU Mon *AAp* 61,469, VV Mon *AAp Sup* 33,323, AO Mon *ApSpSc* 40,3, AR Mon *ApJ* 208,142, IM Mon *AAp Sup* 33,91, V616 Mon *AAp* 50,445, TU Mus *AAp* 45,107, U Oph *AAp Sup* 33,91, WZ Oph *AAp Sup* 32,351, V1010 Oph *ApJ* 211,853, VV Ori *AAp Sup* 33,91, BM Ori *ApJ* 205,462, AG Peg *ApJ* 201,404, AW Peg *AAp* 62,291, X Per *AAp* 54,817, RT Per *ApSpSc* 47,277, β Per *PASP* 87,745, *AJ* 80,836, *ApJ* 222,L119, φ Per *ApSpSc* 39,495, b Per *ApJ* 208,152, ζ Phe *AAp* 46,205, SZ Psc *AJ* 81,250, V Pup *AAp* 61,275,469, U Sge *PASP* 88,688, *AAp* 61,469, UU Sge *ApJ* 223,252, WZ Sge *ApJ* 219,168, V356 Sgr *AAp* 61,469, V523 Sco *PASP* 87,901, V818 Sco = Sco X-1 *ApJ* 207,907, μ^1 Sco *AAp* 61,469, RY Sct *PASP* 88,456, RZ Sct *PASP* 88,262, CV Ser *Af* 11,49, CD Tau *ApSpSc* 40,15, *AJ* 81,855, V471 Tau *AAp* 46,197, V711 Tau *AJ* 81,771, λ Tau *AAp* 62,291, X Tri *Byull Abastuman* 48,13, TX UMa *AAp* 61,469, VV UMa *AAp Sup* 27,285, CV Vel *AAp* 44,355, 58,131, BF Vir *AA* 26,253, DL Vir *AAp* 61,107, Z Vul *AAp* 61,469, RS Vul *AAp* 61,469, HD 698 *PASP* 90,179, HD 47732 *ApJ* 222,574, HD 77581 = Vel X-1 *ApJ* 202,L131, 206,L103, *Pis AZh* 1 No12,13, HD 93205 *ApJ* 207,502, HD 101799 *PASP* 88,936, HD 159176 *BAAS* 7,533, HD 160861 *AAp Sup* 23,277, HD 162724 *ApJ* 201,792, HD 165052 *ApJ* 224,558, HD 167771 *ApJ* 224,558, HD 173198 *IBVS* 1306, HDE 228766 *ApJ* 218,431, BD -3°5357 *PASP* 89,616, Gliese 815 *AJ* 83,1445, PG 1413+01 *ApJ* 224,892, Sk 160 = SMC X-1 *MN* 174,29, *ApJ* 210,L71, 217,186,543, LMC X-4 *ApJ* 225,548, 4U 1538-52 *ApJ* 225,L63.

Koch and Koegler (*ApJ* 214,423) illustrated the complicated nature of this system and the great difficulties in its interpretation. For SV Cam, Hilditch reports supporting evidence for a third body with $P \approx 64$ y, while for the white-dwarf binary V471 Tau, Oliver and Rucinski (*IBVS* 1444) brought strong arguments against light-time interpretation. A somewhat similar case may be UX UMa, discussed by Quigley and Africano (*PASP* 90,445); here some doubt was already cast on the triple-system interpretation by Kukarkin (*MN* 180,5P). The system is of great interest, because the period changes are markedly different from those of any other star of this class.

C. Apsidal Motion (T. Herczeg)

Several important studies concentrated on the somewhat neglected topic of relativistic periastron advance. A programme of systematic observations of five eccentric systems with relatively long periods, EK Cyg, α CrB, V1143 Cyg, DI Her, and RR Lyn, is being carried out by Koch (*AJ* 82,653). DI Her was also studied by Martynov and

Khalioullin; they report a marked discrepancy between observed and theoretically predicted rates of relativistic apsidal motion. Koch's value is apparently much closer to theoretical prediction; however, he does not consider the observed magnitude of the effect statistically significant, as yet. V1143 Cyg is also observed by Ebbighausen. Among the cases of "classical" apsidal motion, Y Cyg was discussed by O'Connell (*RicA* 8,543), including results of the Commission campaign, V477 Cyg and DR Vul were re-examined by Todoran (*AA* 27,59). Apsidal motion was detected for the systems KX Pup and GM Nor (Söderhjelm, *AAp Sup* 22,263) and for V1647 Sgr (Clausen *et al.*, *AAp* 58,121). Several X-ray binaries were considered (Tanzi and Treves, *AAp* 60,431), and for X Per, apsidal motion was discussed as possible explanation of the 581-d periodicity (Henrichs and van den Heuvel, *AAp* 54,817). Theoretical apsidal-motion coefficients were calculated by Vila (*ApJ* 213,464) for highly evolved objects, such as helium main-sequence and pre-white-dwarf stars.

D. Proximity Effects and Limb Darkening (M. Kitamura)

Considerable work was directed towards the theoretical study of tides and related problems. Seguin (*ApJ* 207,848) investigated the effects of tidal distortion on the stability of components against development of turbulence and convection. Horedt (*AAp* 44,461) studied synchronization of rotation, and Press *et al.*, *ApJ* 202,L135) circularization of orbits as well, due to tidally induced turbulence. In a comprehensive study, Zahn (*AAp* 41,329) dealt with the dynamical tide produced by non-adiabatic oscillations of components in the rotating gravitational field. For stars with convective envelopes, Zahn (*AAp* 57,383) found that the most efficient tidal-friction mechanism is turbulent viscosity retarding the equilibrium tide, while for radiative envelopes the action of radiative damping on the dynamical tide is most efficient. Effects of tidal distortion on velocity curves and ellipsoidal light-variations were considered by Wilson and Sofia (*ApJ* 203,182) for extreme mass-ratios. Gravity-darkening of W UMa stars was discussed by Ivanov (*PerZv* 20,99). Eaton (*BAAS* 9,433) noted that not all stars with radiative envelopes obey the von Zeipel law.

Perrenod (*ApJ* 206,876) and Efremov *et al.* (*PerZv* 19,407), among others, discussed reflection effects in X-ray binaries. With an application to the bolometric effect in W UMa stars, Pustylnik (*AA* 27,251) studied non-radial energy transport. Kirbiyik and Smith (*MN* 176,103) modeled circulation currents in irradiated atmospheres; Budding and Ardabili (*ApSpSc* 59,19) gave a generalized approach to evaluation of the reflection effect.

Lavrov derived linear limb-darkening coefficients for nine systems (*ATs* 971) and non-linear coefficients for four systems (*ATs* 990). Discussing the limb-darkening law, Rubashevsky (*AmAf* 27,94) found a two-parameter representation as best approximation. Theoretical limb-darkening coefficients were computed by Al-Naimiy (*ApSpSc* 53,181) for wide ranges of T, g, and λ, and by Manduca *et al.* (*AAp* 61,809), who give *UBV* and *uvby* coefficients for late-type giants.

E. Atmospheric Abundances (M. Kitamura)

Determinations of chemical abundances for double-line binaries were made by several investigators. Heacox (*BAAS* 8,522) reported LTE analyses for components of the Hg-Mn type, including 49 Dra and 66 Eri. The system δ Del, consisting of two nearly identical δ Sct variables, was analysed in detail by Reimers (*AAp* 53,377), who concluded that both components are hybrids between Am and δ Sct stars. Kitamura *et al.* (*Ann Tokyo* 16,22) found intensity variations for characteristic lines of WW Aur during eclipses, indicating that the distribution of metallic-line features is non-uniform over the surface of either component of this Am binary. κ Ari, also with two Am components, was analysed by Mitton (*ApA Sup* 27,35). Naftilan (*ApJ* 206,785) found mild metal deficiency for the secondary of U Sge, abundances relative to iron resemble those of Am stars. Contrary to this, Parthasarathy *et al.* (preprint) report quite normal abundances in U Cep B. More data on abundances in Algol-type secondaries are expected from this Texas group and from Plavec. In AR Lac, Naftilan and Drake (*ApJ* 216,508) found the secondary show high chromospheric activity (strong H and K emission) and be under-abundant in most metals, while the primary has solar-type abundances. Fine-analysis of the F-component of ε Aur (Castelli, *AAp* 69,23) gave normal abundances. From spectrophotometry and model-atmosphere calculations, Leushin *et al.* (*Af Issl* 9,3) found for the visible component of β Lyr the value 1.55 for the He/H number-ratio. The Copenhagen group derived He abundances for a number of systems, by combining *uvby* photometry with stellar-model calculations (*e.g.*, Jørgensen and Gyldenkerne, *AAp* 44,343). Model atmospheres, illuminated from a companion star, were computed by Muthsam (*AA* 28,281).

F. Circumstellar Matter (M. Kitamura)

Observational evidence for circumstellar matter: gaseous disks or rings, streams between components, expanding shells, stellar-wind outflow, etc., continued to increase. In addition to all kind of optical observations, important contributions came from X-ray, satellite UV, and radio data. Thus, for instance, observations with the Copernicus satellite confirmed the existance of a gas stream flowing from Algol B in the direction of A (Cugier and Chen, *ApSpSc* 52,169), detected mass-flow in UW CMa (McCluskey and Kondo, *ApJ* 208,760), a stationary H II-region around V Pup (York *et al.*, *ApJ* 210,143), and variations in the stellar wind from δ Ori A (Snow and Hayes, *ApJ* 226,897). Among optical results we note studies of gas expanding from ζ Aur and 31 Cyg (Saito and Kawabata, *ApSpSc* 45,63), evidence for an expanding shell around δ Ori (Galkina, *Izv Krym* 54,128), and IR-photometry indicating that a thin shell of graphite grains condensed around the WR binary HD 193793 (Williams *et al.*, *MN* 185,467).

Wu (*ApSpSc* 36,407) proposed that double-component emission lines in RW Tau and other binaries primarily originate from gas accumulated at the Lagrangian triangular points. Profiles of lines from disks or rings around one component are computed by Kříž (*BAC* 30, in press), using a refined theory. Nariai (*PASJ* 28,587) studied circulation currents in the common envelope of a contact binary, taking energy transfer into account. Stability against mass overflow in contact binaries was discussed by Nariai and Sugimoto (*PASJ* 28,593); they showed that a binary filling the outer Roche-lobe should be unstable against mass loss through L_2. Particle motions from L_2 were calculated by Nariai (*PASJ* 29,263). The interaction with the environment of gas outflowing from young W UMa-stars was studied by Van't Veer (*Bamberg Ver* 121,388). Particle mechanics was used by Angeletti (*ApSpSc* 44,23) and by Piotrowski and Ziolkowski (*AA* 28,295) to elucidate gas flow in semi-detached binaries. The vertical structure of the flow was studied hydrodynamically by Lubow and Shu (*ApJ* 207,L53). Opaque rings or disks, pictured to surround one component of unusual binaries, such as β Lyr, were approximated by Wilson (*AJ* 80,719) with a torus having elliptical meridian-sections. This geometry for the eclipsing body was used with success to explain the light curve of V356 Sgr (Wilson and Caldwell, *ApJ* 221,917). Apsidal motion of rings was treated by Castle (*PASP* 89,862).

The properties of accretion disks, formed around components with small radii, were discussed in a large number of studies, only some of which can be mentioned. Warner (*Obs* 96,49) reviewed relevant observations. Theoretical studies of gas flow leading to formation of disks were made by Flannery (*ApJ* 201,661), Lin and Pringle (*IAU Symp* 73,237), Shu (*IAU Symp* 73,253). Models were discussed by Paczynski (*ApJ* 216,822, *AA* 28,91,241), Paczynski and Jaroszynski (*AA* 28,111), and Tylenda (*AA* 27,235); theoretical *UBV* colours were computed by Schwarzenberg-Czerny and Rozyczka (*AA* 27,429). Among papers treating structure and stability of disks, we mention Lin (*Obs* 98,208), Livio and Shaviv (*AAp* 55,95), Paczynski (*AA* 28,253), Papaloizou and Pringle (*MN* 181,441), Piran (*ApJ* 221,652), Pringle (*MN* 177,65), Kato (*MN* 185,629), Shakura *et al.* (*AAp* 62,179), Shulikowskij (*Pis AZh* 4,134), Stewart (*AAp* 42,95, 49,39), Stoeger (*MN* 182,647), Weber (*BAAS* 9,632). Accretion onto stars was discussed by Inoue (*PASJ* 28,293), Kippenhahn and Thomas (*AAp* 63,265), Neo *et al.* (*PASJ* 29,249), Nomoto *et al.* (preprint), Ulrich and Burger (*ApJ* 206,509). Of the many papers dealing with disk accretion onto magnetic neutron stars and black holes, we quote a few: Aarons and Lea (*ApJ* 207,914), Bisnovatyi-Kogan and Blinnikov (*AAp* 59,111), Ghosh *et al.* (*ApJ* 217,578), Lamb *et al.* (*ApJ* 224,969, 225,582), Liang (*ApJ* 218,243), Mytrophanov and Tsygan (*AAp* 70,133), Petterson (*ApJ* 214,550, 216,827, 226,253), Shakura and Sunyaev (*MN* 175,613).

6. Structure and Models of Close Binaries

A. Early-Type Systems

The structure of the extended atmospheres of Wolf-Rayet stars was studied using eclipses of especially V444 Cyg (Khaliullin and Cherepashchuk, *Af* 11,593; Limber and Corbin, *BAAS* 9,305; Hartmann, *ApJ* 221,193). Hartmann found that the stellar wind from the WN5 component expands at a nearly constant velocity deep in the atmosphere, in contrast to theories of O-star winds driven by radiation pressure in the lines. An extensive set of UV observations, including three WC+O binaries, were obtained with the TD-1 satellite and analysed in terms of model atmospheres (Willis and Wilson, *MN* 182,559). Ultraviolet variability was discussed by Burton *et al.* (*MN* 183,605). Spectrophotometry of γ^2 Vel by Moffat (*AAp* 57,151) revealed variations with time-scales 13 - 19 d of certain emission features. Schild and Liller (*ApJ* 199,432) found no eclipses in CV Ser. Cherepashchuk (*Pis AZh* 2,356) and Prilutskii and Usov (*AZh* 53,6) drew attention to WR-binaries as potential X-ray sources.

Among the O-type binaries, especially UW CMa has been the subject of several studies (Hutchings, *PASP* 89,668; Eaton, *ApJ* 220,582; Leung and Schneider, *ApJ* 222,924; Parthasarathy, *MN* 185,485). Eaton, using OAO light-curves, found a surprisingly sharp limb for the Of primary, while the expanding low-density region above is uniform. All three photometric studies agree in that the primary fills or over-flows its Roche-lobe. Leung and Schneider found UW CMa a system in over-contact with mass-ratio m_2/m_1 less than unity, contrary to Hutchings' spectrographic data; the masses suggested are 46 and 34 $M_\odot$. For V729 Cyg, Leung and Schneider (*ApJ* 224,565) found the same degree of over-contact (about 30%) and about 60 $M_\odot$ for the primary mass, making it one of the most massive stars known. Bohannan and Conti (*ApJ* 204,797) suggest this system is on the way becoming a WR-binary. For AO Cas, a third evolved system, only small over-contact was found and masses 25 and 29 $M_\odot$ (Schneider and Leung, *ApJ* 223,202). Another system of somewhat later type but thought to be very massive, *viz.* RY Sct, was discussed preliminarily by Ciatti *et al.* (*Bamberg Ver* 121,386). IU Aur, for which Mayer (*BAC* 27,308) found the eclipse depths increasing, was studied by Eaton (*AA* 28,63). The depths of minima continue to increase, and the only realistic explanation is a change of orbital inclination due to perturbations from the third body. Nakamura *et al.* (preprint) computed an evolutionary sequence for SV Cen. The rate of mass transfer for this contact system is believed to be 4×10^{-4} $M_\odot$/y, which may be 50 times higher than in β Lyr. Numerous studies were made of β Lyr itself, some of which have been mentioned earlier in this Report. Huang and Brown completed their investigations with three papers (*ApJ* 208,780, 218,461, 222,627). The emission spectrum in satellite UV was interpreted by Hack *et al.* (*ApJ Sup* 34,565) and Kondo *et al.* (*ApJ* 208,468) in terms of a "super-corona" concentrated towards the invisible component but enveloping the whole system. While the possibility that the secondary component is a black hole must still be seriously considered, Jameson and King (*AAp* 63,285) from radio observations argued strongly in favour of a B-star. β Lyr was believed to be a unique object for its UV spectrum, but Plavec and Koch (*IBVS* 1482) discovered quite similar spectra in W Ser and some other early-type systems.

We end with a few notes on various kinds of other early-type objects. Several β Cep stars are confirmed binaries; for a recent review, see Lesh and Aizenman *Ann Rev AAp* 16,215. For one of these binaries, α Vir, the pulsation has decreased so that it is now almost undetectable (Lomb, *MN* 185,325). LB 3459, a foreground star in the LMC field, has been shown (Kilkenny *et al.* (*MN* 183,523) to be a short-period eclipsing system of two O-type sub-dwarfs. UU Sge, central star of a faint planetary nebula (Bond, *PASP* 88,192) was confirmed to be an eclipsing binary (Miller *et al.*, *IAUC* 2974; Tsessevich, *IBVS* 1320). Méndez and Niemela report velocity variations for the central stars of the planetaries NGC 1360 and 2346, indicating that also these objects are close binaries.

B. Algols and Related Systems (M. Plavec)

Significant progress was accomplished in accumulation of important observational data on semi-detached binaries. The concept of "R CMa systems" (both components grossly undermassive for their luminosities), although almost certainly quite fallacious, continued to stimulate observers to efforts to eliminate the few remaining members of this spurious group. Okazaki (*PASJ* 29,289) found that T LMi is an ordinary Algol-system with a normal main-sequence primary. Very important is RZ Sct, studied by Wilcken *et al.* (*PASP* 88,262); here the primary is a B2 II giant. Most important for confrontation with evolutionary calculations are systems where masses of both components can be determined directly from the radial-velocity curves. Most new results come from Popper, using the Varo image-tube at Lick to observe the D-lines of the secondaries; he will soon have complete orbits for the secondaries in 12 new systems. Peculiar is the secondary of the pre-main-sequence system BM Ori in the Trapezium (Popper and Plavec, *ApJ* 205,462): it appears to be the first observed star to be greatly flattened by differential rotation. While the velocity curves of the secondaries now become increasingly more accessible, some of the primaries are found to give very poor curves. It appears that a circumstellar shell and/or a stream distort the line profiles more or less seriously. Very distorted are S Cnc, AW Peg, TT Hya, RY Per, moderate cases are RY Gem, AD Her, AU Mon, XX Pup, while TW And, TW Dra, RS Vul probably have good photospheric spectra (after Popper). The eccentric orbit of AD Her reported by Batten and Fletcher (*PASP* 90,312) may be a test case. On the photometric side, Walter reports indications of gas streams and polar-region hot-spots in RW Ara (*AAp Sup* 26,227), X Gru (*AAp Sup* 29,51), TW Dra (*AAp Sup* 32,57), and XZ Sgr. He also found long-period variations of light curves for TV Cas and U Cep and interprets them as precession (*AAp* 42,135, 69,437).

Ultraviolet excesses in U Cep, U Sge, SX Cas, and several RS CVn systems were studied by Rhombs and Fix (*ApJ* 209,821, 212,446, 216,503) with spectrum scanner. They suggest the excess is due to free-free and bound-free emission associated with the hotter component in the Algols; however, in the RS CVn stars, the excess seems to be associated with the cooler sub-giant. Rucinski observed the spectrum of Algol with the Copernicus satellite and found that the primary rotates as a solid body and in synchronism with orbital motion. Prilutskij *et al.* (*Pis AZh* 2,294) explained the X-ray emission of Algol in terms of an emitting hot spot formed at the surface of the primary by an impinging gas jet.

The idea that at least some Be-stars may be interacting binaries was further followed mainly at Ondřejov. Harmanec *et al.* announced the binary nature of 4 Her (*BAC* 27,47), found a period of 87 d in 88 Her (*BAC* 29,278), and 39 d in HD 218393. Periodicity and duplicity of o And remain controversial (Fracassini *et al.*, *ApSpSc* 49,145; Harmanec *et al.*, *IBVS* 1296). Binary nature of the periodic shell star HR 2142 was confirmed by Peters.

Sub-giant systems with H and K emission, more often called RS CVn stars, received broad attention. They appear to evolve from normal main-sequence systems with components of nearly equal mass. Only low rates of mass loss and/or mass transfer are probably required. The many instabilities appear to develop as the components cross the Hertzsprung gap (Popper and Ulrich, *ApJ* 212,L131). Weiler (*MN* 182,77) suggests, from a survey of H, K, and Hα emissions, an extensive chromospheric activity on the later-type component. Naftilan and Drake (*ApJ* 216,508) add to it a disk or shell surrounding the secondary in AR Lac. Related to the RS CVn stars are several interesting objects. HD 224085 was shown by Rucinsky (*PASP* 89,280) to be a strongly spotted, cool but probably young object. CM Dra is the eclipsing binary with the smallest and faintest known red-dwarf components; it is probably a population II system (Lacy, *ApJ* 218,444; Rucinski, *AA* 28,167). Dworetsky *et al.* (*MN* 181,13P) discovered a 9-d eclipsing binary, BD $-3^{o}5357$, containing a red sub-giant and a hot OB sub-dwarf; the evolution of this system poses a problem. Mengel *et al.* (*ApJ* 204,488) suggested that in general the sdB stars are produced by mass transfer and mass loss in close binary systems. Great potential significance for our understanding of the helium-rich stars has the discovery (Hesser *et al.*, *Nature* 262,116; Groote and Hunger, *AAp* 56,129) that one of the classical objects of this kind, σ Ori E, is most likely an interacting binary with ring.

C. W Ursae Majoris Systems (J.A.J. Whelan)

Further modifications were made to the basic Lucy (*ApJ* 151,1123) common convective envelope model of W UMa contact systems. Lucy (*ApJ* 205,208), Flannery (*ApJ* 205,217), Robertson and Eggleton (*MN* 179,359) discussed a model in which the component stars are not in thermal equilibrium. The system oscillates about a state of marginal contact and spends substantial part of the cycle out of contact with components of very unequal temperatures. Since few, if any such systems are known at the characteristic periods of the models, efforts were made to reduce the out-of-contact part of the cycle. Robertson and Eggleton were able to construct models, with low total angular momentum, which spend only 5% of the cycle out of contact. The paradoxical possibility arises that as cycling models are constructed which approach 100% of the cycle in contact, their average properties might formally approach those of the "equilibrium" contact models, which are known to exist only with equal-mass components at the chemical compositions considered. A crucial factor in the cycling-model calculations is the detailed way in which the energy transfer is switched off as the degree of contact is decreased. Not only is some physical picture of energy transfer necessary, but also a very short time-step must be used, since the relative changes of contact depth occur very rapidly and the stellar reaction-times to changing rates of energy and mass flows are very short. The time-step required for accurate computation is probably much shorter than that used so far, and thus the conclusions of Lucy, Flannery, and Robertson and Eggleton must possibly be re-appraised. To cope with short time-steps in a feasible way, Hazlehurst and Refsdal introduced the "response-function method". The stellar response functions (Hazlehurst *et al.*, *AAp* 58,47) are calculated with the necessary "time resolution" to permit very rapid changes to be followed. Further detailed work is in progress along these lines (Hazlehurst, *Obs* 98,204).

A new, "contact discontinuity" model was introduced by Shu *et al.* (*ApJ* 209,536, 214,798, 216,517, 221,926). A discontinuity in the run of physical variables is maintained by fluid flow; the discontinuity provides an extra degree of freedom, which enables models with surface properties in agreement with observations to be constructed. Hazlehurst and Refsdal (*AAp* 62,L9) showed that in its first specific form the contact discontinuity would be smoothed out on a very short timescale, thus invalidating Shu *et al.*'s original model. Shu *et al.* (*ApJ* in press) have re-cast somewhat their model; they argue that a contact discontinuity is necessary for the existence of a dynamically stable configuration with unequal-mass components and that fluid flow may maintain a contact discontinuity. Various kinds of observational tests of the theories were collected and discussed by Lucy and Wilson (preprint). The conclusion is that the contact-discontinuity models are in substantial disagreement with observational data, but as several disagreements may no longer apply when the theory has reached its final form, it should not yet be regarded as contradicted.

Several attempts were made to attack the difficult problem of the details of the energy transfer between components of a contact binary (Moses, *MN* 176,161; Moses and Smith, *IAU Symp* 73,333; Nariai, *PASJ* 28,587; Ivanov, *Af* 12,475; Webbink, *ApJ* 215,851; Shu *et al.*, *ApJ* in press). There is not yet agreement about appropriate length-scales, dominant forces, etc. The large-scale circulation model (Webbink), driven by rotation and involving mass motions along the entropy gradients which generate or absorb thermal energy, ties up with the absence of thermal equilibrium in the cycling models. However, Shu *et al.* argue the mechanism is insufficient.

The question of which component of a W-type W UMa system is hotter was discussed in detail by Wilson and Biermann (*AAp* 48,349). It was concluded, for TX Cnc, that nearly radiative gravity-darkening, with a continuous temperature variation over the components, could account for the light curve as well, or better, than convective gravity-darkening and a hotter secondary. Further work on this subject is in progress by Hilditch (*Obs* 98,205). Rucinski (*AA* 26,227) showed how UV-observations are relevant to the gravity-darkening problem. Woodward and Wilson (*ApSpSc* 52,387) found that AK Her is in geometrical contact but with $\Delta T \approx 1000$ K between the components. Wilson (*ApJ* 224,885) discussed A-type systems selected for having accurately known parameters, and showed that the radii are larger than those of zero-age main-sequence stars.

D. Novae and Dwarf Novae (B. Warner)

The continued increase in observational data on cataclysmic variables, and their similarities to many X-ray objects, has generated much interest and attempts at understanding their structure and mechanisms. The three principal reviews on dwarf novae are those by Warner (*IAU Symp* 73,85), dealing with observations, Bath (*IAU Symp* 73,173), on theories, and Robinson (*Ann Rev AAp* 14,119). Classical novae were reviewed by Gallagher and Starrfield (*Ann Rev AAp* 16,171), and two recent semi-popular reviews concern explosions of novae (Bath, *QJRAS* 19,442) and the problem of novae *versus* dwarf novae (Mallama and Trimble, *QJRAS* 19,430).

Theoretical studies of gas flow between components were quoted in Section 5F; among these, Flannery (*ApJ* 201,661) confirmed the empirical hot-spot model. Nelson and Olson (*ApJ* 207,195) reported changes in DQ Her that indicate variations in the rate of mass flow; this was considered further by Nelson (*ApJ* 209,165). Works on accretion-disk structure were also mentioned in Section 5F; here we add the study of U Gem by Madej and Paczynski (*Bamberg Ver* 121,313). The structure of dwarf novae at minimum light was elucidated by discussion of multi-colour photometry by Szkody (*ApJ* 207,824, 217,140) and refinement of the U Gem model by Smak (*AA* 26,277). Masses of cataclysmic variables were deduced by Robinson (*ApJ* 203,485) and Ritter (*MN* 175,279). Interpretations of the rapid oscillations in DQ Her (Katz, *ApJ* 200,298; Patterson *et al.*, *ApJ* 224,570) and in dwarf novae during outburst were assisted by study of *r*-modes (Papaloizou and Pringle, *MN* 182,423).

Dwarf-nova outbursts and their relationship to binary properties were discussed by Sparks and Starrfield (*Liège Coll* 19,407) and Bath (*MSAI* 45,793). Photometric observations of short and long eruptions of VW Hyi (Haefner *et al.*, *AAp* in press) provided additional clues. However, there is still no consensus on the nature of the outbursts. Dynamical instabilities of the secondary were further studied by Papaloizou and Bath (*MN* 172,339). The most convincing application of the suggestion of dynamical instabilities was made by Webbink (*Nature* 262,271) to the recurrent nova T CrB.

Much work was performed in studies of classical novae; most of this concerns mechanism of ejection and properties of shells, but there are also results relevant to the binary nature. V1500 Cyg provided a wealth of information. Rapid variations in emission-line profiles indicated rapid changes in the central star (Rush and Thompson, *ApJ* 211,184). Rapid flickering and the presence of a 0.14-d (or 0.28-d) periodic light-variation (Kemp *et al.*, *ApJ* 211,L71; Kupo and Leibowitz, *AAp* 56,181; Young *et al.*, *PASP* 89,37; Patterson, *ApJ* 225,954) provided clues to binary nature. Interpretation of infrared and other observations (Ennis *et al.*, *ApJ* 214,478; Boyarchuk *et al.*, *AZh* 54,477; Vrba *et al.*, *ApJ* 211,480) gave an ejected mass of about $7\times10^{-5}\ M_\odot$. A search-light model was proposed by Hutchings and McCall (*ApJ* 217,775) and further elaborated by Hutchings *et al.* (*ApJ* 224,899). Models for ejection from novae (Bath and Shaviv, *MN* 175,305; Friedjung, *MSAI* 45,757; Gallagher and Ney, *ApJ* 204,L35; Clayton and Wickramasinghe, *ApSpSc* 42,463) were deduced partly from infrared observations; continouos-ejection models are in general favoured. This type of model was also proposed for the slow nova RR Tel by Thackeray (*MRAS* 83,1).

Finally we note that the discovery of magnetic close binaries, AM Her, AN UMa, and VV Pup (Section 4C), has closed the gap between conventional cataclysmic variables and X-ray binaries of the Sco X-1 type. In addition to references given already in Section 4C, we mention Chanmugam and Wagner (*ApJ* 222,641), Cowley and Crampton (*ApJ* 212,L121), Fabian *et al.* (*MN* 179,9P), Greenstein *et al.* (*ApJ* 218,L121), Liebert *et al.* (*ApJ* 225,201), and Stockman (*ApJ* 218,L57). The properties of the AM Her variables were recently used by Papaloizou and Pringle (*AAp* 70,L65) for explaining the different "hump"-periods of VW Hyi.

X-ray Binaries (E.P.J. van den Heuvel)

For recent reviews of the field, see (1) *Proceedings 8th Texas Conference on Relativistic Astrophysics* (ed. G. Papagiannis), 1977, *Ann. New York Acad. Sci.* 302 (especially the papers by Lamb, Rappaport and Joss, Cowley, Flannery, Schreier, Gursky, Ziolkowski, Sanford, Giacconi, Lewin, Ostriker, and van den Heuvel; each paper contains an extensive list of up-to-date references), (2) *Physics of Neutron Stars and Black Holes, Proc. 46th Enrico Fermi School of Physics* (eds. R. Giacconi and R. Ruffini), North Holland Publ. Co., Amsterdam, 1978. In this Report we only summarize the highlights of the last three years.

Massive X-ray Binaries. The most important discoveries concerning the massive X-ray binaries were probably those enumerated and described below.

1. A large number of slow X-ray pulsators with periods in the order of several minutes were recognized. These appear to be more common than the short-period pulsators, such as Her X-1 (1.24 s), Cen X-3 (4.84 s), and SMC X-1 (0.71 s). Only two more short-period pulsators were discovered: 3U 1626-67 (P = 7.7 s), probably in a low-mass close binary (Joss *et al.*, *ApJ* 221,645), and 4U 0115+63 (P = 3.6 s), in an eccentric massive system (P_{orb} = 24.3 d; Rappaport *et al.*, *ApJ* 224,L1). Presently 11 slow pulsators are known, most of them associated with massive early-type companions, either super-giants (*e.g.* 4U 0900-40, P = 383 s, P_{orb} = 8.96 d) or B-type emission stars (3U 0352+30 = X Per, P = 835 s). In one case, 3U 1728-24 (P = 138 s), the companion seems to be an M-giant (Davidsen *et al.*, *ApJ* 211,866).

2. An important characteristic of all X-ray pulsators is that they spin up rapidly; the timescales are in some cases (3U 1728-24) as short as decades, in others (Cen X-3, SMC X-1) of the order of 10^3 to 10^4 y, in Her X-1 about 3×10^5 y (see papers by Schreier and Rappaport and Joss in ref. 1). These large spin-up rates are a clear manifestation of the accretion process, which supplies the compact star with mass and angular momentum. The combination of X-ray luminosity (yielding accretion rate) with spin-up rate (proportional to the mass-transfer rate divided by stellar moment of inertia) provided unambiguous proof that the X-ray pulsators are neutron stars and not white dwarfs (Rappaport and Joss, *Nature* 266,683).

3. The discovery of the X-ray source OAO 1653-40 and its positional agreement with V861 Sco (P = 7.848 d) provided a second black-hole candidate; the mass function suggests a mass of at least 5 $M_\odot$ for the compact star (Polidan *et al.*, *IAUC* 3234). The third black-hole candidate, Cir X-1, was identified with a very faint red star, at least at a distance of 10 kpc (Whelan *et al.*, *MN* 181,259).

4. The discovery of an orbital period of 1.41 d in LMC X-4 by Chevalier and Ilovaisky (*AAp* 59,L9) is the first case of a massive binary where binary character was discovered photometrically before it was found in the X-rays.

5. The derivation of good optical and X-ray orbits proved to be more complicated than was hoped originally (see Bahcall, *Ann Rev AAp* 16,241). Much accurate optical photometric and spectroscopic information is required - together with accurate X-ray data - in order to derive accurate masses of compact stars. So far a complete analysis has been carried out only for the system 4U 0900-40 = HD 77581, by van Paradijs *et al.*

(*AAp Sup* 30,195), yielding a neutron-star mass of 1.74±0.25 $M_{\odot}$. A second system, where such an analysis seems possible, is SMC X-1 (Hutchings *et al.*, *ApJ* 217,186); here, however, due to the presence of an extra lightsource in the system near conjunction (*cf.* van Paradijs and Zuiderwijk, *AAp* 61,L19), the errors are considerably larger.

6. The Her X-1 system was studied extensively by Boynton and co-workers and by Nelson and Middleditch (see their contributions in ref. 2). In a beautiful analysis Boynton *et al.* found unambiguous evidence for the presence of a large precessing disk in the system, which causes the 35-d X-ray and optical variations.

7. The rather low rate of mass transfer by Roche-lobe overflow in the Her X-1 system inspired a detailed study of the effects of hydrodynamics and of angular-momentum transfer on the details of the mass-transfer process. Savonije (*AAp* 62,317) showed that especially the effects of hydrodynamics may extend the duration of the phase of very slow mass-transfer ($\dot{M} < 10^{-8}$ $M_{\odot}$/y) by Roche overflow to $\geq 10^4$ y in massive systems, provided the massive star is in the phase of hydrogen burning. Therefore, in the short-period massive binaries such as Cen X-3 and SMC X-1, Roche-lobe overflow, probably mixed with wind, may be the main source of mass accretion (see also: Conti, *AAp* 63,225; Ziolkowski, ref. 1).

Low-Mass X-ray Sources, X-ray Bursters, Galactic-Bulge Sources and Globular-Cluster X-ray Sources. After the discovery of the first X-ray burst-source, 3U 1820-30 (in the globular cluster NGC 6624), with the ANS satellite (Grindlay *et al.*, *ApJ* 205,L127), the SAS-3 team discovered more than 30 new burst-sources. Not more than five of them are in globular clusters. They show a galactic distribution probably characteristic for an old galactic-bulge population (see review by Lewin in ref. 1). At least four bursters have been optically identified, all of them with blue intrinsically faint stars. The spectrum of one of these (4U 1735-44), obtained by McClintock *et al.* (*ApJ* 223,L75) is almost identical to that of Sco X-1, suggesting that we are dealing with a low-mass binary system. The time-integrated burst luminosities of all bursters are such that the bursts can be satisfactorily explained by helium-burning flashes on the surface of an accreting neutron star (Joss, *ApJ* 225,L123). The steady luminosities of the sources, which are about 100 times larger, would, in this model, be due to the accretion process itself. Using all observational evidence, Joss and Rappaport (*AAp* 71, in press) argue that many of the galactic-bulge sources (including bursters and globular-cluster sources) are short-period binaries, consisting of a low-mass normal star ($M \lesssim 0.5$ $M_{\odot}$) together with a compact star. The analysis of various capture processes in globular clusters suggests that compact binaries of this type can be formed in sufficient quantities by means of two or three-body captures (Hills and Day, *ApL* 17,87; Finzi, *AAp* 62,149; *cf.* Heggie, *ComAp* 7,43).

7. Statistical Investigations

The frequency of binaries and multiples in various stellar samples was studied by several authors mainly by means of new radial velocities. Abt and Levy found among solar-type stars (F3-G2 IV or V) 46% binaries and 11% multiples (*ApJ Sup* 30,273) and for B2-B5 IV or V (including Be) respectively 36% and 13% (*ApJ Sup* 36,241). For late B-stars (B7-B9) Wolff (*ApJ* 222,556) obtained a binary frequency of about 45%. In a sample of 18 O-type stars, Bohannan and Garmany (*ApJ* 223,908) found only two which might be binaries, while several previously considered to be binaries proved to have constant velocities. For OB-stars with enhanced N or C lines, Bolton and Rogers (*ApJ* 222,234) estimated the binary frequency to 50-100%. Binary frequency in IC 4665 was brought down to about 50% by Crampton *et al.* (*ApJ* 204,502): previously reported very high frequency is suggested spurious. Among other open clusters searched for binaries we have NGC 6475 (Gieseking, *AAp* 60,9) and NGC 2516 (Gieseking, *AAp Sup* 32,17). The use of colour-magnitude diagrams of open clusters for obtaining binary frequencies and mass ratios was discussed by Dabrowski and Beardsley (*PASP* 89,225), Jaschek (*AAp* 50,185), and by Trimble and Ostriker (*AAp* 63,433), who especially point to the influence of stellar rotation. With reference to globular-cluster X-ray sources, Trimble searched for close binaries in NGC 6809 (*BAAS* 8,443) and discussed the incidence of close binaries in globular clusters (*MN* 178,335). The relation of mass-ratio *versus* period for eclipsing binaries with $P < 5$ d was studied by Sinvhal and Srivastava (*ApSpSc* 54,239). Trimble (*Obs* 98,163) discussed the mass-ratio distribution of spectroscopic binaries and found peaks at $M_2/M_1 = 0.2$ to 0.3 and 0.9 to 1.0. Van't Veer (*AAp* 70,91) studied the distribution of mass-ratios for W UMa systems.

8. Origin and Evolution of Close Binaries

There are three main theoretical avenues for the formation of binary and multiple stars: (1) collapse and fragmentation of an interstellar cloud into protostars, some of which forming gravitationally bound systems, (2) dynamical evolution of small clusters may as end products have a binary or multiple star, (3) fission may produce very close binaries. Exept perhaps for the very close pairs, the first mechanism is a most promising one, and much work was done in this field (*e.g.*, Larson, *IAU Symp* 75,255; Mouschovias, *ApJ* 211,147; Horedt, *ApSpSc* 54,253; Bodenheimer, *ApJ* 224,488; Norman and Wilson, *ApJ* 224,497). Numerical studies of cluster dynamics have generally assumed that all stars were single at the outset. If binaries already exist in the cluster, much of the *N*-body computations may not be applicable. This is especially so, if a sufficient number of strongly bound pairs exist (Dokuchaev and Ozernoy, *AZh* 55,27; Heggie, *Obs* 98,206). In globular clusters such pairs are likely to form by tidal captures, and they may also be present primordially. Fabian *et al.* (*MN* 172,15P) and Sutantyo (*AAp* 44,227) suggested that cluster X-ray sources are close binaries formed from tidal capture of compact objects by ordinary stars. This mechanism was worked out in detail by Press and Teukolsky (*ApJ* 213,183). Work on the modern versions of the fission theory (Lebovitz, *Liège Coll* 19,47) has been limited. Gingold and Monaghan (*MN* 184,481) studied fission of damped, differentially-rotating polytrops, and Lucy (*AJ* 82,1013) made a successful numerical demonstration of the fission hypothesis, applied to optically thick protostars.

The consequences of mass transfer on the evolution of close binaries were reviewed by Thomas (*Ann Rev AAp* 15,127), to which we refer for work mainly prior to 1977. Evolutionary sequences for massive binaries with conservative mass-exchange were pursued further by De Grève and de Loore. They found (*ApSpSc* 50,75) that the sequences bifurcate at an initial primary mass of about 14 $M_{\odot}$: larger masses are progenitors to neutron stars, lower masses end as white dwarfs. In a comparison with observations (De Grève *et al.*, *ApSpSc* 53,105), it is concluded that conservative mass-exchange, with low initial mass-ratio ($q_i < 0.5$), leads to acceptable first-order models of W-R and massive X-ray binaries. Evolutionary sequences, taking into account mass loss by stellar winds, were calculated by Vanbeveren *et al.* (*AAp* in press) for initial masses 30+50, 40+20, and 60+30 $M_{\odot}$. The restrictive condition $q_i < 0.5$ for the W-R systems disappears in these more refined computations. De Grève (preprint) studied similarly a still more massive system, 100+60 $M_{\odot}$, which after lobe overflow turns into a helium star with a massive O-type companion.

Very serious for the theory of Algol systems were the papers by Ulrich and Burger (*ApJ* 206,509), Flannery and Ulrich (*ApJ* 212,533), Kippenhahn and Meyer-Hofmeister (*AAp* 54,539), and Neo *et al.* (*PASJ* 29,249). All these authors agree that at the phase of rapid mass-transfer, the originally less massive component is uncapable of accomodating the high rate of mass influx, begins to swell and a contact system forms. This is illustrated, for instance, by the model sequence computed by Packet and De Grève (*AAp* in press) for a 10+3 $M_{\odot}$ system, similar to that assumed by Ziolkowski (*ApJ* 204,512) for constructing an evolution model for β Lyr. Instead of a system matching β Lyr, Packet and De Grève obtain a contact binary. The evolution subsequent to established contact is less clear, but it is probably rapid mass-loss from the system.

For a low-mass binary (1.50+0.75 $M_{\odot}$), Webbink (*ApJ Sup* 32,583) followed evolution from an original detached state to contact. Evolution, when already in contact (Webbink, *ApJ* 211,881; Rahunen and Vilhu, *AAp* 56,99) and a possible end-point of contact-binary evolution, namely coalescence (Webbink, *ApJ* 209,829) were also considered. The possibility of an absence of thermal equilibrium during the contact phase ties up with the cycling models of W UMa systems (Section 6C).

Besides other evolutionary studies, many papers discussed changes of orbital elements due to regular, slow evolution or to catastrophic mass-loss from symmetric or asymmetric supernova explosions. Lack of space prohibits us to enter these fields.

9. Report of the Coordinated Programs Committee
(K. Gyldenkerne)

At the Grenoble IAU Meeting it was recommended to carry out coordinated observing campaigns for RX Cas and VV Cep. For RX Cas, the coordinator, D.Ya. Martynov, reports successful *UBV*-photometry in the USSR: by Zaitseva at the Pulkovo Crimean Station, by Kalv in Tallinn, and by Kumsiashvili at Abastumani; nothing has been reported concerning spectroscopy. From the photometric series, confirmation was obtained for the lengthening of the period. For VV Cep, A. Galatola acted as coordinator and got photometric contributions from observers in USA and Japan, but probably the system has also been observed elsewhere. Wright, Faraggiana and others obtained spectrographic observations; Faraggiana and observers in USA also made IUE spectroscopy. It is suggested to have a discussion on the VV Cep problems at the Montreal Meeting.

Concerning X-ray binaries, Y. Kondo reports on three coordinated campaigns in 1977. The systems observed and the coordinators are: HDE 226868 (V1357 Cyg) = Cyg X-1 (T. Bolton), HZ Her = Her X-1 (J. Nelson), and X Per = 4U 0352+30 (C. de Loore). About a dozen groups or individual observers participated in each campaign. The X-ray observations of Cyg X-1 and Her X-1 were performed with SAS-3; the X Per ground-based observations were timed simultaneous with OSO-8 X-ray observations.

10. Report of the Committee for Extra-Terrestrial Observations
(Y. Kondo)

The Committee continued its usual task of monitoring space experiments that are of interest to the Commission members. In coordination with the Coordinated Programs Committee and IAU Commission 44, we have been involved in various campaigns to observe X-ray binaries; a summary of these activities is given in Section 9. For information on satellites that have been active for X-ray observations of close binaries and for some observational results, reference is made to Section 4D. We add here that the satellite HEAO-B was successfully launched on 1978 November 13. Those interested in observations coordinated with HEAO X-ray observations may contact Dr. J. Swank at Code 661, Goddard Space Flt. Ctr., Greenbelt, MD 20771, USA.

During the three-year period since 1976, the following two satellites were used to observe close binaries in the ultraviolet (launch dates and organizations/experimenters are indicated):

Copernicus	1972	Princeton University
IUE	1978	NASA (Goddard Space Flt. Ctr.), ESA, British Science Res. Council.

In addition, a balloon-borne ultraviolet stellar spectrometer (BUSS) was flown four times to obtain spectra in the mid-ultraviolet. Among binaries observed are β Lyr, ϵ Aur, υ Sgr, and μ Sgr (de Jager, Kondo and co-workers).

A large number of close binaries are being investigated with the two satellites. Copernicus was effective in observing at high resolutions (0.2 and 0.05 Å) in the spectral region 1000 - 1400 Å. IUE has the capability to observe at slightly lower resolutions (0.1 to 0.2 Å in the far-ultraviolet and 0.2 to 0.3 Å in the mid-ultraviolet), alternatively, it can observe much fainter objects at a resolution of 6 to 7 Å in the range 1200 - 3100 Å.

The results already published from Copernicus data include papers on β Lyr, β Per, UW CMa, and HD 153919 (4U 1700-37). R. Polidan prepared the following list of close binaries observed with Copernicus since 1976 January 1: λ And, UX Ari, 54 Cam, R CMa, AO Cas, AR Lac, HK Lac, 17 Lep, δ Lib, θ Mus, X Per, β Per, V Pup, μ Sgr, υ Sgr, V861 Sco, V 711 Tau, λ Tau, ε UMi, γ^2 Vel, HR 2142, 4665, 5110, 7275, 8703, HD 72754, 77581, 153919; and a list of observed, possibly interacting binaries: ζ And, π Aqr, α Aur, α CMa, τ CMa, υ Cyg, 66 Oph, δ Ori, η Ori, θ Ori, ι Ori, λ Ori, π^5 Ori, σ Ori, ε Per, o Per, ψ Per, δ Pic, α Sco, β^1 Sco, μ Sco, π Sco, α Vir, HR 4830, 7084, HD 50896. B. Perry compiled the following list of examples of mostly non-X-ray eclipsing binaries observed with the IUE by guest observers, Hack: ε Aur, VV Cep, CH Cyg, β Lyr, υ Sgr, HD 192909; Koch: RX Cas, RW Cas, SX Cas, V382 Cyg, V1073 Cyg, TT Hya, AR Pav, RW Per, V Sgr, υ Sgr, V453 Sco; Kondo: UW CMa, AO Cas, U Cep, 31 Cyg, 32 Cyg, β Lyr, θ Mus, η Ori, δ Pic, μ Sgr, υ Sgr, V861 Sco, γ Vel, HD 153919; Plavec: RX Cas, W Cru, V367 Cyg, 17 Lep, χ Oph, AG Peg, RW Per, 3 Pup, μ Sgr, V356 Sgr, RZ Sct, W Ser, HD 161114, 163296, 206773. Additional interacting binaries, such as SS Cyg, were also observed with the IUE in other guest-observer programs but are not included here. X-ray emitting binaries observed with the IUE in a collaborative program between USA and Europe are given as follows (list prepared by R. Davis): good phase-coverage for the systems HZ Her, V1357 Cyg, V818 Sco, HD 77581, HD 153919, and varying degrees of phase-coverage for SS Cyg, AM Her, X Per, Sk 160, HD 152667, LMC X-4.

Highlights of X-ray binary observations obtained during the initial high-priority period of the IUE are presented in an article by Dupree *et al.* in a special issue of *Nature* (1978 October 5). Among preliminary reports is Hack's detection of the hot companion in ε Aur. Plavec reports detection of strong emission lines of hot plasma in W Ser, V367 Cyg, and W Cru. Most prominent lines in the region 1200 - 2000 Å are due to N V, Si IV, C IV, Si III, Al III, and He II. On the whole the spectra are similar to that of β Lyr. Koch and Sobieski detected similar spectra in RX Cas and SX Cas. Thus, it is established that a whole class of interacting binaries shows this type of spectrum. Plavec speculates on several possible explanations, including the presence of black holes. Kondo and McCluskey observed spectral profiles indicating mass flow in a number of cases, including υ Sgr, μ Sgr, V861 Sco, β Lyr, UW CMa, and U Cep. Those interested in learning more about these satellites' observations may contact Dr. R. Polidan (Copernicus), Code 410, and Dr. R. Perry (IUE), Code 685, both at Goddard Space Flt. Ctr., Greenbelt, MD 20771, USA.

11. Report of the Working Group on RS CVn Binaries
(D.S. Hall)

This Working Group was conceived at the end of the Grenoble Meeting to stimulate interest in RS CVn systems, make it known which astronomers are involved, define some of the important problems, minimize unnecessary duplication, and help with cooperative observing programs. To date 11 Circulars have been distributed to about 35 members in the Western Hemisphere and (with the help of M. Rodonò) to about 15 in the Eastern Hemisphere. Members are encouraged to announce research-in-progress and look for published results in the *Bibliography and Program Notes on Close Binaries*. We still need one light curve per year outside eclipse for the about 20 most important systems in order to follow the wave migration continuously; with less frequent observations, the phasing can become ambiguous. The "First RS CVn Workshop" was held 1978 April 6 and 7 at New Mexico Institute of Mining and Technology, Socorro, NM, with D.S. Hall scientific chairman, D.M. Gibson local chairman, and 27 participants. A summary of the discussions was distributed as Circular No. 9. There we arranged to have a special issue of the *Astron. J.* (December 1978) devoted to the large 1978 February 20 radio outburst of V711 Tau (HR 1099) and arranged to have a "Second RS CVn Workshop" at the Montreal Meeting, with P. Feldman local chairman.

G. LARSSON-LEANDER
President of the Commission

45. SPECTRAL CLASSIFICATIONS AND MULTI-BAND COLOUR INDICES (COMMISSIONS DES CLASSIFICATIONS SPECTRALES ET INDICES DE COULEUR A PLUSIEURS BANDES)

PRESIDENT: B. Hauck
VICE-PRESIDENT: A. Slettebak
ORGANIZING COMMITTEE: A.L. Ardeberg, R.A. Bartaya, A.P. Cowley, C. Jaschek, P.C. Keenan, E. Mendoza, V.L. Straizys

1. Introduction

During the period covered, two important meetings dealing specifically with spectral classification have been held: (i) IAU Symp. 80 on the HR diagram was organized in Washington in November 1977 on the occasion of the 100th anniversary of H.N. Russell. The proceedings have been edited by A.G.D. Philip and D.S. Hayes. (ii) IAU Coll. 47 "Spectral Classification in the Future" brought participants to Vatican City in July 1978 to commemorate the 100th anniversary of the death of A. Secchi S.J. The proceedings will be edited by M. McCarthy, A.G.D. Philip and G.V. Coyne. The proceedings of the following meetings have been published: IAU Symp. 70 (Be and Shell Stars), IAU Symp. 72 (Abundance Effects in Classification), IAU Coll. 32 (The Physics of the Ap Stars). All included important papers on spectral classification.

2.1 Work Involving Slit Spectrograms (A. Slettebak)

Perhaps the most significant work since the last General Assembly in the use of slit spectra for spectral classification is the preparation and completion of spectral atlases which bring the MK system up to date: the Morgan-Abt-Tapscott atlas of the early-type stars and the Keenan-McNeil atlas of the later types. These and other important atlases of slit spectra are listed in Section 4 of this report.

O- and B-type Stars. Walborn (17.114.025) has discussed the OBN and OBC stars, and given a review paper on progress and problems in the upper left of the HR diagram (IAU Symp. 80). Heap (19.135.029) has discussed the spectral classification of O-type planetary nuclei. Garrison et al. (20.114.044) have classified 1113 OB stars in the southern hemisphere. Lesh (IAU Coll. 47) has made a comparison of luminosity calibrations for MK classifications of OB stars. Bisiacchi et al. (18.114.165) have carried out TV-Vidicon multichannel spectral classifications of O and B stars. Sinnerstad (IAU Coll. 47) has made a spectroscopic and photometric study of B2-B5 main-sequence stars. Roman (*Astron. J.* 83, p. 172, 1978) has provided spectral types for early-type stars observed by Skylab. Spectral classification of Of and Be stars was discussed by Andrillat (IAU Coll. 47). Hill and Lynas-Gray (20.113.005) provided MK spectral types for 193 northern early-type stars at intermediate galactic latitudes. Jaschek (20.114.032) discussed the present status of spectral classification in the conventional wavelength range, with emphasis on the early-type stars, while Bidelman (20.114.041) considered spectral classification from Copernicus data and Kondo (20.114.042) discussed the 2800 Å Mg II lines and spectral classification. Cucchiaro et al. (18.114.053, 20.114.021) have developed an ultraviolet spectral classification system for the B-type stars using TD1A, S2/68 observations.

A- and F-type Stars. Hill et al. (18.155.020) carried out spectral classification of A- and F-type stars at the north galactic pole. A.P. Cowley (17.114.036) estimated spectral types of the bright F-type stars. Floquet (14.114.001) has provided spectral classification for some Ap and Am stars, while C.R. Cowley (IAU Coll. 47) introduced a classification scheme for the chemically peculiar late B- and A-type stars. Cucchiaro et al. (*Astron. Astrophys. Suppl.* 33, p. 15, 1978) have developed an ultraviolet spectral classification system for early A-type stars using TD1A, S2/68 observations.

Late-type Stars. Keenan is continuing his work to improve and extend the classification in the revised MK system of G, K and M stars. He is presently giving attention to the calibration of luminosity classes II, III and IV for stars of approximately solar composition. A practical classification for stars in the S-SC-C sequence has been devised by Keenan and Boeshaar (IAU Coll. 47), taking account of the work of Ake (IAU Symp. 80). Campbell, working with Garrison, has studied the classification of yellow giant stars in the near infrared using a reticon device on the Toronto classification spectrograph (18.158.221 and in press). Irwin, working with Bolton and Garrison, has synthesized low-dispersion spectra of yellow giant stars from stellar atmospheres theory and from high-dispersion spectra in order to study the effects of metallicity and microturbulence variations on the normal classification criteria (18.114.111 and in press).

Binary Stars. Hill et al. (14.114.047) have provided MK classifications for 186 close binary systems. Lutz and Lutz (19.118.009) have carried out spectral classification and *UBV* photometry of 91 visual binary systems.

Open Clusters and Associations. Abt (20.153.024), Abt and Levato (14.153.038, 17.152.010, 19.114.041, 20.153.025) and Levato and Malaroda (14.153.032, IAU Coll. 47) are continuing their work in spectral classification of stars in galactic clusters and associations. Abt (IAU Coll. 47) has also discussed the occurrence of abnormal stars in open clusters. Harris (17.153.023) has classified evolved stars in 25 open clusters. Hartoog (17.153.016) obtained MK types for 263 B- and A-type stars in southern open clusters to study the frequency of Ap and Am stars. Guetter (18.152.007, 18.152.019) did spectral classification of stars in the associations Ori OB1 and Lac OB1. Garrison and Kormendy (19.153.006) classified stars in the cluster Trumpler 37, of which μ Cep (M2Ia) appears to be a member. Garrison has also classified several stars on the main sequence of NGC 6231, the nucleus of I Sco, down to $V = 14.4$ (approximately A0).

Magellanic Clouds. Feast (IAU Coll. 47) has discussed the future of MK classification work in the Magellanic Clouds. Walborn (19.159.015) has provided spectral classification of O and B0 supergiants in the Magellanic Clouds, and also pointed out a very early O-star in the Small Magellanic Cloud (*Astrophys. J. Letters* 224, L133, 1978). Dubois, Jaschek and Jaschek (*Astron. Astrophys.* 60, p. 205, 1977) discuss MK classification in the Small Magellanic Cloud. Fehrenbach (IAU Coll. 47) has considered the supergiant stars in the Large Magellanic Cloud, and the significance of luminosity class I.

General. Important review papers on various aspects of slit spectral classification were presented at IAU Colloquium 47 by Code, Garrison, Loden, Spinrad, and Walborn.

Garrison is classifying the 1300 brightest stars in both hemispheres in order to provide homogeneous classifications on the system of the new MK atlases as well as secondary classifications all over the sky. He is also observing a large number of stars in the ultraviolet at low resolution (6Å) with the International Ultraviolet Explorer and will devise a classification scheme for them which will be independent of the MK system, but which will be compared with the MK system.

MacConnell and Alvarez have obtained spectrograms of 83 high-radial-velocity stars from an unpublished catalogue of Stock. The stars, with radial velocities between 130 and 450 km/sec, have been classified by MacConnell, with types between B8V and M6III, including many weak-metal stars.

Ardeberg (IAU Coll. 47) has discussed the practical application of the MK and related spectral classification systems to spectrograms of various resolutions.

Slettebak and Kuzma (IAU Coll. 47) have considered the effects of stellar rotation on spectral classification, from the point of view of (1) the effects of plate resolution, and (2) physical changes in the rotating stellar atmospheres.

2.2 Objective Prism Spectra (A. Cowley)

S.W. McCuskey is continuing spectral classification of stars in LF12 (~ 8^h, -25°) and LF14 (~ 12^h, -61°). In LF14 spectral types have been determined for 8717 stars (V > 12.5) and luminosity classes as well for those stars with V > 11 and later than F8. A catalogue including spectral types, V magnitudes and identification charts is being prepared. In LF12 only B8-A2 and G8 III-K3 III stars are so far completed.

C.B. Stephenson and N. Sanduleak report on surveys for Hα emission-line stars which might have been confused with OB stars in the earlier Cleveland-Hamburg surveys for luminous stars. Two recent publications now complete this work (19.114.052 and *Publ. Warner and Swasey Obs.* 2, no. 4, 1977). They are about to start a new Hα survey of the entire northern sky above and below 10° galactic latitude and north of declination -25°. This will be done with the optically refurbished Burrell Schmidt with a 4° prism covering 5000-6800 Å to look for both Hα emission stars and carbon stars.

W. Iwanowska and S. Krawcyk are using 250 Å/mm objective prism spectra of about 40 non-variable carbon stars to establish classification criteria and look for population effects in these stars.

Th. Schmidt-Kaler, G. Diaz-Santanilla, R. Rudolph and H. Unger have presented a summary of their work "Quantitative Classification of Objective Prism Spectra" at IAU Symp. 80. Further low dispersion objective prism classification work in Southern Milky Way fields and in the Magellanic Clouds is under way.

K.G. Henize reports on a Michigan-Mt. Wilson southern Hα survey (1929 emission-line stars $\delta < -25^\circ$ observed between 1949-1951) which has been published (17.114.042). The plates, now stored at the University of Texas, are available for historical reference concerning variable emission-line objects. Henize also gives a summary (19.032.557) of some of the findings of the Skylab S-019 UV Objective-Prism Survey which covers 9% of the sky. Widened spectra centered near 1550 Å give useful data for moderately reddened B0 stars to about V = 6.5, while unwidened spectra at 2000 Å can reach hot white dwarfs and subdwarfs as faint as V = 12.0.

Clariá and Stock (19.154.009) have used objective prism radial velocities to establish membership of late-type metal-poor stars and A stars in the halo of the globular cluster NGC 6397. Ibanez (19.112.012) had determined objective prism radial velocities of field Mira variables near the same cluster. Stock, working in the intermediate latitude region (RA 12.30-14.30, Dec. -30° to -35°) has determined positions, classifications, magnitudes and radial velocities for about 6000 stars. About 5% have velocities > 100 km s^{-1} relative to the LSR. Alvarez has determined velocities and types for 350 late-type giants in a field at 19.20, -32°. Stock and Vrba are working on classifications of all stars > 13.5 in fields surrounding the nearby dark clouds of CrA and Oph. Stock, Osborn and Clariá have given classifications, positions and magnitudes for 890 A and F supergiants in the LMC (17.159.005). Alvarez is studying the same stars in the ultraviolet for new type and abundance criteria. MacConnell and Bidelman have published classifications for 149 southern luminous stars (17.114.029). MacConnell is preparing a series of catalogues of new Hα emission stars, carbon stars and S stars found from his red survey of the southern galactic plane. There are about 1000 new emission-line objects, 40 carbon stars and about 170 MS and S stars.

N. Houk is continuing systematic reclassification of the southern HD stars on the MK system, using IIaO plates obtained with the 10° objective prism (108 Å/mm at Hγ) on the Curtis Schmidt telescope at CTIO. Volume 2 of the Michigan Spectral Catalogue (Dec. -53° to -40°; 30400 stars) is now being distributed. Volume 3, to be finished in 1980, will cover Dec. -40° to -26°, Some results from Volumes 1 and 2 are published in the Proceedings of IAU Symp. 80 and IAU Coll. 47.

Bidelman is preparing a comprehensive compilation of identifications and spectral classifications for the stars contained in the Caltech Two-Micron Sky Survey. In addition to data from the literature, existing Warner and Swasey objective prism plates are being examined and new plates taken for this work. Sang Gak Lee, working with Bidelman, has determined spectral types for almost 900 objects contained in the Lowell Observatory Northern Hemisphere Proper-Motion Survey. Most of her classifications, done with Warner and Swasey Observatory objective prism material, relate to previously unclassified stars.

Red and near-photographic infrared spectra have been obtained by Pesch with the 4° objective prism on the Burrell Schmidt for over 100 red stars with recently published trigonometric parallaxes (primarily from the U.S. Naval Observatory). One unusual object has been noted: G165-47, whose spectrum shows no TiO bands in spite of its large *V-I* colour. This star is similar to the well-known sdMp star G95-59.

N. Sanduleak and A.G.D. Philip have published a list of 474 carbon stars and 609 suspected late-type supergiants in the LMC (*Publ. Warner and Swasey Obs.* 2, no. 5). These authors are now working on a survey of OB stars in both Clouds down to $V \sim 16$ using the Curtis Schmidt telescope at CTIO. A catalogue of probable dwarf stars of type M3 and later in the direction of the south galactic pole is now in press (*Astron. J.* 1978, Sept. by Pesch and Sanduleak). It includes $V + I$ photometry.

Robertson has completed a Ph.D. thesis "The Kinematics and Space Density of M Dwarf Stars". In an objective prism survey of 875 square degrees he found 200 M dwarfs brighter than $V \sim 13$. The luminosity function derived from these data agrees with Luyten's.

Bartaya reports that on the basis of the Abastumani Observatory 70 cm meniscus telescope with 8° prism (dispersion 166 Å/mm near Hγ) the spectral classification of 87 bright member stars of 12 open clusters (NGC 129, 457, 1907-1912, 2168, 6866, 6882, 6885, 7092, 7209, 7790, χ Per, h Per) is completed. Spectral classification of about 11000 stars is performed in the regions of O associations Cyg OB4, Cepheus-Lac OB1 and Cas OB9. In 14 McCormick Fields located in the vicinity of NGP the spectra of 2600 stars are classified, in 3 fields the quantitative two-dimensional spectral classification of F, G-type 200 stars is completed. The third parameter has been determined for 120 stars. Approximately 10 stars turned out to be metal deficient (D.M. Depadel'). On the basis of Abastumani Astrophysical Observatory data and in collaboration with the latter, following the methods developed by V.D. Malyuto, in Tartu, the quantitative spectral classification with definition of the metal abundance parameter of about 100 stars in the region of the NGP is completed (V.D. Malyuto, *Bull. Abast. Astrophys. Obs.*, in press). In Garny Laboratory of Space Astronomy (GLSA) under the supervision of G.A. Gurzadyan a new method of spectral classification is developed. The character of the continuous spectrum in the UV and especially the length of the spectrum itself significantly depend on the spectral type of the stars. The spectral classification of about 2000 stars has been completed. In so doing the stars mostly fainter than $10^{m}.00$ and up to $13^{m}.5$ were covered. Results can be found in 18.114.031, *Comm. Byurakan Obs.* 48, 5 and 14).

2.3 Automatic Spectral Classification (R.A. Bartaya)

Methodical work on automatic spectral classification is under way at Bochum Observatory (Schmidt-Kaler et al., 1976, *Forschungsber. Lunds-Nordrhein-Westfalen*, N2595).

A review paper by Schmidt-Kaler (IAU Coll. 47) considers the problem of automatic spectral classification. Special consideration is given to the classification of low and intermediate dispersion spectra and to application with the aid of photoelectric scanning detectors.

West (18.031.408) has considered the situation in automatic classification, in particular, of objective prism spectra. The experience in application of two-coordinate scanning means a step towards realization of automatic spectral survey. The experimental programme is initiated at ESO.

Mendoza (IAU Coll. 47) reports that in Mexico automatic MK spectral classification is achieved. It is based upon standard spectrograms obtained for visual classification. Computer results are given for all the MK standard stars earlier than B1. Slight modifications have to be done for later spectral types. There is a good correlation with the MK system. However, the results might be improved by using a rapid-scanning Michelson Fourier Transform Spectrophotometer to obtain the stellar spectra. The analysis of the interferograms should also be made by a computer.

Appenzeller and Zekl (IAU Coll. 47) at Heidelberg Observatory have developed a method to derive spectral types and absolute luminosities of stars automatically from classification spectrograms by means of a computer.

Kinman (IAU Coll. 47) in his report has discussed the problem of the possibility of spectral classification by means of digital method. An appropriate method is developed at Kitt Peak National Observatory.

The COSMOS automatic plate-measuring machine at the Royal Observatory, Edinburgh is intended for a fast surveying of the modern large Schmidt telescope negatives (both with direct images and low-dispersion spectra). Among the capabilities of COSMOS is a technique to separate automatically the images of galaxies from those of stars (MacGillivray et al., 18.031.402; Pratt, 19.031.428).

A scanning microphotometer with an on-line data reduction for large field Schmidt plates (Fritze et al. 19.034.085) was developed and put into operation at Tautenburg Observatory.

Radoslavova (19.031.427) has prepared a programme for treating absorption spectra on a computer with the aim of accurate and fast translation of transmissions into intensities and for computing equivalent line widths.

Barry, Cromwell and Schoolman (19.031.219) have developed a fast, automated data reduction procedure for quantitatively measuring stellar spectra to determine the fundamental stellar parameters of temperature, luminosity and metallicity.

In the ESO programme of discovering and studying the magnetic stars of the Southern Sky (PDS-1000) a digital microdensitometer is used (19.031.254).

3. Classification Employing Multi-Colour Work (B. Hauck)

An important book on stellar photometry has been published (in Russian) since the last report by Straizys: Multicolour Stellar Photometry (20.003.150). It is an excellent and useful compilation of all information about the various photometric systems.

Many theoretical colours are calculated, in particular by Bell and Gustafsson for giant stars (*IAU Symp. 80* and *Astron. Astrophys. Suppl.* in press) in the *UBVR*, Geneva, *uvby*, DDO, *gnkmf*, Uppsala and Johnson et al. systems. Other theoretical colours are obtained for the Walraven system by Lub and Pel (19.113.012). Work is being carried out at the Institute of Basle with a wide variety of spectral scans combined with objective prism spectra and of model atmospheres for stars representing all major stellar types and population. Buser (*Astron. Astrophys.* 62, pp. 411 and 425) has obtained results for the *UBV* and *RGU* systems and, in collaboration with Kurucz, has given a physical calibration of the *UBV* system for the early-type stars. The calculations of broad-band colours are now being extended to late-type giants, white dwarfs and subdwarfs (Buser). A study is being made of relations between metal content and colour indices, while an investigation of the possible multi-colour systems for three-dimensional classification of stars later than K5 has been commenced. Stellar energy fluxes were also used by Crawford and Mandwevala (19.113.011) to obtain an interstellar reddening relation in the *UBV*, *uvby* and Geneva systems. Another problem concerning all systems is the obtaining of blanketing coefficients and many have been published: for the G, K and M giants (17.114.009), for the M0-M5 stars (18.114.169, 170), for the F5 stars (18.114.005) and for the UV spectrum of the B6 to A0 stars (19.114.008).

A. WIDE-BAND SYSTEMS

Many observations have been made in the *UBV* and its extension to the red and infrared. All measurements are collected at the Stellar Data Centre at Strasbourg. Some categories of stars were studied in this system: T Tauri in (17.113.002) and by Rydgren (17.113.035), who uses the J-H/H-R diagram to separate the effects of envelope emission and interstellar reddening; S stars by Feast et al. (17.113.007) who publish a mean bolometric absolute magnitude of -5.4 for these stars; C stars (17.113.008); barium stars by Catchpole et al.(20.114.027), who obtain spectra and *UBVI* colours for 122 Ba stars and derive a mean absolute visual magnitude of -1; M dwarf stars by Mould and Hyland (18.113.003) who use the J-H/H-R diagram to segregate the populations and derive temperature and absolute magnitude; more recently Am stars by Mendoza (*Astron. J.* 83, p. 606) and Mermilliod, who is studying the blanketing effect on *U-B* and *B-V* (to be published).

Graphical correlations with the MK types have been obtained by Davis (19.114.030), while Deutschmann et al. (17.113.032) have published intrinsic colours for stars from O to M and luminosity classes from V to Ia. Kuriliene and Straizys (*Bull. Vilnius* 44, p. 3) have published also new intrinsic colour indices of supergiants. Alexander (19.113.057) has studied the accuracy of *R-I* as a temperature indicator and Hartwick (19.113.039) uses $\delta(B-R)$ as a measure of the metal deficiency of the subdwarfs. Buser has redetermined the response function for the *UBV* and *RGU* systems and the interstellar reddening parameters (*Astron. Astrophys.*62, p. 411) and obtained a transformation formula between the two systems (*Astron. Astrophys.* 62, p. 425).

The *WBVR* system proposed by the Vilnius astronomers (*Trans. IAU XVIA-2*, p. 100) was tested observationally (14.113.055). In 1976 the Sternberg Astronomical Institute (Khaliulin et al.) and the Main Astronomical Observatory of the Ukraine (Pugach et al.) started observations of standard stars. Up to 1978 more than 500 stars were observed.

The properties of the Walraven system are reviewed by Lub and Pel (19.113.012). Theoretical colours and applications to the three-dimensional classification of stars from A to G are presented. The subdwarfs are well segregated in a $(B-L)$ vs $(V-B)$ diagram and $\delta(V-B)$ is a good blanketing parameter. Th. and J.H. Walraven (20.159.007) have published many data, together with a comparison with the *UBV* system and also standard mean sequence relations.

Concerning the Geneva system, Goy has used various diagrams to obtain a photometric separation of the components of spectroscopic binaries (19.119.017), *Astron. Astrophys.* 64, p. 445) while Cramer and Maeder (to be published) have determined absolute magnitudes and temperatures, with a great internal accuracy, of the B stars. The photometric parameter of the peculiarity for the Ap stars was determined for 234 stars and a photometric determination of the stars classified Ap or Am in the Bertaud-Floquet catalogue was made (Hauck, *Astron. Astrophys.* 69, p. 685). A calibration of Δm_2 in terms of [Fe/H] was published by Hauck (*Astron. Astrophys.* 63, p. 273). A study of the metallicity and rotation of the Am stars is now being made. Grenon (*Publ. Obs. Geneva*, Série B, no 5) has obtained calibrations in absolute magnitude, effective temperature and metal content for the G, K and M stars. Golay et al. (20.114.005) have applied the concept of "stellar boxes" and found that stars in the same box have nearly the same spectral type and nearly the same values as the parameters used in β, Copenhagen, UV and IR photometries. Golay (*Astron. Astrophys.* 62, p. 189) has shown that stars in the same box have the same absolute visual magnitude within a range less than ~0.1 mag. M. Jaschek et al. (IAU Coll. 47) have observed 380 stars at 80 Å/mm and concluded that stars belonging to the same box are spectroscopically identical at this dispersion. Fracassini and Pasinetti are at present studying the correlation between the blanketing parameter and absolute radius for 664 B5-F5V stars. For spectral types later than A5V, the single stars show average radii larger than those of multiple stars of the same spectral type.

Canterna (17.113.031) has used a new broad-band system (3910, 5085, 6330, 7885 Å) to obtain temperatures, metal abundances and a CN index for G and K giants. The metal abundances derived from the blanketing parameter are uncertain by a factor two.

B. MEDIUM-BAND SYSTEMS

Using the Vilnius system, Straizys (IAU Symp. 80) has proposed a method for classification based on five independent reddening-free parameters *Q*. The method is calibrated in T_e, log *g*, spectral type and M_v. Many observations of various kinds of stars were made: Am and Ap (*Bull. Vilnius* 45, p. 13), Straizys and Zitkevicius, *Soviet Astron. J.* 54, p. 987), population II (Bartkevicius and Sperauskas), white dwarfs (18.126.053). Other measurements were made in various parts of the sky (*Bull. Vilnius* 52, 18.131.231). Kuriliene (*Bull. Vilnius* 44, p. 15) and Straizys and Kuriliene (*Bull. Vilnius* 47, p. 3) published intrinsic colour indices of the Vilnius system for stars of different spectral types and luminosities. Bartkevicius is preparing a catalogue of F-M type stars with metal deficiency. Plans to join the Vilnius and Geneva systems are being discussed and theoretical investigations made by Straizys, North and Hauck.

The magnitudes and colours of the DDO system given in (17.113.026) are the new standard of this system. A test of accuracy of low-dispersion objective-prism spectral classification of late-type stars is given in (18.114.044) while in (20.114.025) a method to derive MK types is presented by Yoss.

Philip and Perry have published (Vistas in Astronomy) a bibliography concerning the *uvby*β system for the period 1950-1976. Many data were published, in particular in 17.113.057, 17.118.013, 18.113.010, 19.113.010, 19.113.001, 19.113.007, 20.113.027 and 20.113.028. Grönbech and Olsen (18.113.039) published a list of standard stars ($\delta < 10^o$). A calibration for the B stars is given by Crawford (*Astron. J.* 83, p. 48) while a luminosity calibration for the late-group (19.115.004) and F stars (*Astron. Astrophys.* 66, p. 335) is presented by Heck. For the OB stars luminosity calibrations are given in 18.115.002 by Klare and Neckel and also by Egret (*Astron. Astrophys.* 66, p. 275). Correlations with temperature are given in 19.113.031, 19.114.032, 20.114.055 and effects of stellar rotation are studied by W.H. Warren (17.113.001). A calibration of δm_1 in terms of [Fe/H] is given by Crawford in 18.113.023. A catalogue of 5083 dereddened stars was published (19. 113.056) giving log g, θe for most of the stars. Based on the new catalogue of Hauck and M. Mermilliod a preliminary updated version was prepared in June 1978 by Philip. Philip and Egret are applying Golay's method of stellar boxes to this new version.

The v, b and y filters were combined by Maitzen with two medium-band filters g_1 (5020 Å) and g_2 (5240 Å) to study the Ap stars (18.113.008). A new medium-band system with four colours (3530, 3980, 4930, 6550) is proposed (18.113.025). Data and study of solar-type stars in the 13-colour system are presented by Schuster (18.113.032).

C. NARROW-BAND SYSTEMS AND SCANS

In two papers concerning the BCD system, Chalonge and Divan (19.114.014, 015) have shown the possibility to segregate stars with different chemical composition, to calibrate this system in terms of [Fe/H] and to obtain for stars from O to F8 a good indicator of temperature.

Gerbaldi and Morguleff have performed an extensive analysis of the narrow-band spectrophotometric system started by Barbier and Morguleff for the BO to F5 stars of all luminosity classes. Correlations with the MK types were obtained (IAU Coll. 47). In the near future, Gerbaldi and Morguleff will perform a multivariate analysis for the 800 stars measured in this system. This system was also used for a study of the Ap stars (Gerbaldi).

Massa and Lillie (*Astrophys. J.* 221, p. 833) have applied the vector space method to the O stars and interstellar reddening and have obtained intrinsic Hβ values for the O and B stars.

K-line indices were obtained by Pedersen and Rudkjøbing (*Astron. Astrophys. Suppl.* 34, no. 3) for B, A and F stars. The Am stars were found to belong to an extended group of stars with relatively narrow K lines. Henry et al. (IAU Symp. 80) have also used K-line photometry to segregate the Am stars and to study the variations in metal abundance. Using the eight narrow bands system (0.7 to 1.1 μm), White and Wing (*Astrophys. J.* 222, p. 209) have classified the M supergiants.

Mendoza has employed a system of interference filters centered respectively on Hα and at λ7774 Å (neutral oxygen triplet). This system seems very suitable for the segregation of the Am stars (18.113.048 , *Rev. Mexicana Astron. Astrofis.* 1, p. 363), the Ap stars (*Rev. Mexicana Astron. Astrofis.* 2, p. 259) and young objects (18.113.048, *Rev. Mexicana Astron. Astrofis.* 1, p. 363; 2, p. 29; 2, p. 33; 2, p. 231; 2, p. 259).

Rao and Mallik (*Mon. Not. R. Astron. Soc.* 183, p. 211) have derived a luminosity calibration for F5-G2 supergiants using the product of equivalent width measures of OI 7774 Å and Fe I 7748 Å.

Many studies are based on scans. Kuan and Kuhi (17.114.037) presented the observations of O and Of stars, both kinds presenting evidence for effects of atmospheric extension. Effective temperatures of some Ap, Am and A stars were obtained by Babu (18.114.007), while Adelman (20.114.057), using Breger's catalogue (18.113.052), examined the energy distribution of Ap stars and pointed out the stars for which the features at λλ4200, 5200 and 6200 Å are present. At IAU Symp. 80 Greenstein presented results for multi-channel spectrophotometry of G, K and M stars of the main sequence and for fainter degenerate stars. The dMe are slightly above the main sequence and in the "cosmic scatter". At the same symposium, Gutierrez-Moreno and Moreno showed an analysis of photoelectric spectrophotometric observations in the Orion aggregate and the Upper Scorpion association. Concerning the cool stars, Gow (18.114.135, 120.114.053) observed barium, CH, and R-type carbon stars and Piccirillo the S stars (18.114.161). Maehara and Yamashita (17.122. 038; *Ann. Tokyo Astron. Obs.*, in press) have carried out scanner observations of Mira variables and related stars and found that the colours between 3700 and 5500 Å are a good indicator of the intensities of TiO bands. Yamashita et al. (20.114.065) have obtained new spectral types in the C-system for 22 carbon stars. At the Uttar Pradesh Observatory, Rautela and Joshi have used spectral scans of M-type stars to define two indices (M_{6250}-M_{5950}, M_{6500}-M_{5400}). These indices can be used respectively to separate the M stars into spectral subclasses and luminosity classes. Rao (*Mon. Not. R. Astron. Soc.* 185) has measured with an automated scanner strenghts of the G, CH and NH bands in weak G-based stars. New spectrophotometric calibrations of the absolute enerty distribution of standard stars can be found in (18.114.135).

Spectrophotometric observations have been made in the UV at the Byurakan Observatory and the most important results can be found in 19.114.018, 019, 020, 021, 022, 023, 024, 025, 026.

4. Catalogues and Atlases (C. Jaschek)

Since the activity in the field of Commission 45 is increasing regularly, the number of catalogues is similarly increasing. Attention is called to two publications *Astronomical Catalogues 1951-1975* by M. Collins (20.002.025) and the reports of the Working Group of Commission 45 for the collection of photometric and spectroscopic data, which are published regularly (bi-annually) in the *Bulletin d'Information du Centre de Données Stellaires* at Strasbourg. In the latter the newest catalogues are listed, together with their availability on magnetic tape.

In what follows, we shall divide the catalogues into two classes (a) long observing lists and (b) lists of objects drawn from different sources. The note (CDS) means that the catalogue is only available at the Centre de Données, on magnetic tape.

CATALOGUES TYPE (A), SPECTROSCOPIC

L.O. Lodén et al.: A spectral survey of the Southern Milky Way III. O-B9 and M stars, l = 280° to 306° (17.114.008). Part IV is in preparation.

A. Cucchiaro et al.: Spectral classification from the ultraviolet line features of S2/68 spectra (18.114.053, 20.114.021, *Astron. Astrophys. Suppl.* 33, p. 15 and 35, no. 1)

V.I. Voroshilov et al.: Catalogue of *B, V* magnitudes and spectral classes of 1800 stars (18.113.001)

J. Breger and A.M. Fringant: A search for faint blue stars in high galactic latitude (19.002.019).

J. Andersen: A uniform edition of the Stockholm Southern Milky Way Survey (20.002.002).

R.F. Garrison et al.: MK spectral classifications for southern OB stars (20.114.044).

V.I. Voroshilov et al.: Catalogue of *B, V* magnitudes and spectral classes of 720 stars (20.002.049).

N. Houk: University of Michigan Catalogue of two-dimensional spectral types for the HD stars. Vol. II. 30400 stars.

R.A. Bartaya: Catalogue of MK classifications of 10396 stars (*Bull. Abastumani Astron. Obs.* 50, 1978).

E.H. Geyer: Southern Milky Way spectral survey for stars earlier than A5 ($227 < l < 310$; $|b| < 9^{\circ}5$). 30000 stars. Bonn.

N.B. Kalandadze and M.D. Metrevili: Spectra of 600 O, B and A stars (in preparation)

R.F. Garrison: A catalog of MK types of the 1300 brightest stars. Toronto. (in preparation)

P.C. Keenan: List of stars of types G, K and M classified on the revised MK system (about 500 stars). Perkins Observatory (in preparation).

P.C. Keenan and P. Boeshaar: List of a representative selection of type S and SC stars. Perkins Observatory (in preparation).

W.P. Bidelman: G.P. Kuiper's classifications of stars of large proper motion north of -50°. Cleveland (in preparation).

CATALOGUES TYPE (B), SPECTROSCOPIC

M. Breger: Catalogue of spectrophotometric scans (18.113.053)

G. Goy: A general O-type star catalogue. Third edition (18.114.054)

M. Morel et al.: A catalogue of [Fe/H] determinations (18.114.130)

C.B. Stephenson: A general catalog of S stars (18.114.141)

W. Buscombe: Third general catalogue of MK stellar classifications (20.002.004)

M. Jaschek: Catalogue of selected spectral types in the MK system. (CDS)

P.M. Kennedy: MK classification extension. 1977 edition. (CDS)

G.P. McCook and E.M. Sion: A catalogue of spectroscopically identified white dwarfs (20.002.048)

A. Uesugi: Revised catalogue of stellar rotation velocities. Preliminary edition. Univ. of Tokyo (1978).

I. Lundström and B. Stenholm: A planned new catalogue of the galactic WR stars. Lund Observatory (in preparation).

A. Strobel and J. Strobel: Catalogue of abundance determinations. Torun (in preparation).

R. Glebocki: Catalogue of stars having CaII H and K emissions. Gdansk (in preparation).

CATALOGUES TYPE (A), PHOTOMETRIC

B. Grønbech and E.H. Olsen: Four colour *uvby* photometry for bright O to GO type stars south of declination +10° (18.113.010)

L. Topaktas: *RGU* photometry of a field in the direction of galactic longitude 330° (18.113.014)

F. Rufener: Second catalogue of stars measured in the Geneva Observatory photometric system (18.113.056)

S. Wramdemark: *B* and *V* magnitudes and spectrophotometric quantities of stars in a field centered at l = 36°, b = +8°3 (18.113.077)

B. Grønbech and E.H. Olsen: Photoelectric Hβ photometriy for bright O to GO type stars south of δ = +10° (19.113.010)

M. Golay and N. Mandwewala: Catalogue of photometric star boxes in the Geneva system (20.002.038)

G. Klare and T. Neckel: *UBV*, Hβ and polarization measurements of 1660 southern OB stars (19.113.008)

W.H. Warren, Jr. and J.E. Hesser: Catalog of individual *UBV* and *uvby*β photoelectric observations of stars in the Orion OBI association (19.152.010)

J.K. Knude: Photoelectric *uvby* and Hβ photometric of 750 A and F stars in 63 selected areas within $|b|$ < 30° (20.113.028)

A.M. Spaenhauer: *RGU* three-colour photometry of a field in Aquila near NGC 6755 (20.155.023)

H. Steppe: The Basle catalogue of blue objects in higher galactic latitudes in SA 51, 54, 82 and in the Halo field around M5

Th. Walraven and J.H. Walraven: Five-channel photometric observations of stars in the Magellanic Clouds and in the Milky Way (20.159.007)

R.F. Wing: Classification and photometry of giants and supergiants K4-M8 in the "Bright Star Catalogue". Perkins Observatory.

CATALOGUES TYPE (B), PHOTOMETRIC

J.C. Mermilliod: Catalogue of *UBV* photometry and MK spectral types in open clusters (18.153.023)

J.C. Mermilliod and B. Nicolet: A general catalogue of *UBV* photoelectric photometry (20.002.003)

B. Nicolet: Catalogue of homogeneous data in the *UBV* photoelectric photometric system (*Astron. Astrophys. Suppl.* 34, p. 1, 1978)

M. Morel and P. Magnenat: Catalogue of *UBVRIJKLMNH* measurements (*Astron. Astrophys. Suppl.* 34, p. 477, 1978)

B. Hauck and M. Mermilliod: *uvby*β photometric catalogue. Institut d'Astronomie, Lausanne (new edition in preparation)

ATLASES

H. Johnson: The atlas of stellar spectra (19.002.044)

W. Seitter: Atlas for objective prism spectra. Bonner spectral atlas II (original dispersion 645, 1280 A/mm) (19.002.068)

P. Keenan and R.C. McNeil: An atlas of spectra of the cooler stars: types G, K, M, S and C (18.114.132)

J. Landi-Dessy, M. Jaschek and C. Jaschek: An atlas of grating spectra at intermediate dispersion (original dispersion 42 A/mm). Cordoba University, Argentina

Y. Yamashita et al.: An atlas of representative stellar spectra (original dispersion 73 A/mm). University of Tokyo Press (1977)

A. Cucchiaro, M. Jaschek and C. Jaschek: An atlas of ultraviolet stellar spectra (λλ1350-2740). Liège-Strasbourg Observatories.

A.M. Hubert-Delplace and H. Hubert: Un atlas d'étoiles Be. Observatoire de Paris-Meudon and Observatoire de Haute-Provence (in preparation).

W.W. Morgan, H.A. Abt and J.W. Tapscott: Revised MK atlas for stars earlier than the Sun

MISCELLANEOUS

D.S. Mathewson et al.: Polarization catalogue. (CDS)

M. Fracassini and L.E. Pasinetti: Catalogue of stellar diameters. Milano-Merate Observatories (in preparation).

5. Conclusion

Many important works have been published during the period covered and most of them are listed above. It is interesting to note the large increase of studies on spectral classification in the ultraviolet, the ever-increasing use of scans for classification and related problems of coolest stars and the great amount of data published during this period. Finally I think it necessary to mention that the spectral type and colours of the Sun are always a very controversial problem which has been discussed in some papers (e.g. Barry (21.114.004), Barry et al. (*Astrophys. J.* 222, p. 1032), Hardorp (21.114.006), Schuster (18.113.032) and Garrison (IAU Coll. 47)).

B. HAUCK
President of the Commission